AF361157

Cell and Model
Membrane Interactions

Cell and Model Membrane Interactions

Edited by
Shinpei Ohki

State University of New York at Buffalo
Buffalo, New York

PLENUM PRESS • NEW YORK AND LONDON

Library of Congress Cataloging-in-Publication Data

Cell and model membrane interactions / edited by Shinpei Ohki.
       p.   cm.
    "Proceedings of a symposium on Cell and Model Membrane
  Interactions, held April 22-27, 1990, in Boston, Massachusetts"-
  -T.p. verso.
    Includes bibliographical references and index.
    ISBN 0-306-44097-0
    1. Membrane fusion--Congresses.   I. Ohki, Shinpei.
  QH601.C33   1992
  574.87'5--dc20   91-43343
                                                          CIP

Proceedings of a symposium on Cell and Model Membrane Interactions,
held April 22–27, 1990, in Boston, Massachusetts

ISBN 0-306-44097-0

© 1991 Plenum Press, New York
A Division of Plenum Publishing Corporation
233 Spring Street, New York, N.Y. 10013

Preface

Membrane interaction is a large research area involving various disciplines. A symposium entitled "Cell and Model Membrane Interactions" which took place in Boston, MA during the 155th American Chemical Society Meeting, April 25, 1990, focused on membrane adhesion and fusion. The topics were explored in studies involving lipids, virus envelopes and cell membranes. Especially discussed, were the roles of polymers, lipids, and proteins on these membrane interactions.

Fusion of membrane is an important molecular event which plays a pivotal role in many dynamic cellular processes, such as exocytosis, endocytosis, membrane genesis, viral infection processes, etc. The process includes adhesion of the membranes, fusion, and finally reorganization of the components of the two membranes.

The basic notion shared during the symposium was that membrane hydrophobicity, especially local membrane hydrophobicity is one of the important factors contributing to membrane fusion. Most of the papers are collected here and they are arranged approximately in the same order as they were presented at the symposium. These papers are the most up-to-date and representative work at the forefront in each membrane interaction field.

I sincerely hope the reader will gain further understanding on membrane interactions especially, membrane and vesicle fusion phenomena through this symposium proceedings volume.

Shinpei Ohki

June 1991

Contents

Physical Basis Underlying Membrane Adhesion and Fusion

DETERMINATION OF LIPID ASYMMETRY AND EXCHANGE IN MODEL MEMBRANE SYSTEMS

C. Tilcock[1], S. Eastman[2] and D. Fisher[3]

[1] Department of Radiology, University of British Columbia
Vancouver, BC, Canada V6T 2B5
[2] Department of Biochemistry, University of British Columbia
Vancouver, BC, Canada V6T 1W5
[3] Department of Biochemistry and Molecular Cell Pathology Laboratory
Royal Free Hospital School of Medicine, University of London
London NW3 2PF, UK

INTRODUCTION

Several biological membrane systems, including the erythrocyte and
platelet plasma membranes[1-3], inner mitochondial membrane[4] and endoplasmic
reticulum[5] are known to exhibit asymmetry with respect to lipid
distribution across the lipid bilayer. For example, it is well established
that in the erythrocyte plasma membrane, the majority of the phosphatidylcholine and sphingomyelin are present in the outer leaflet of the
bilayer while the aminolipids phosphatidylethanolamine and phosphatidylserine are located primarily on the inner leaflet. Despite studies which
implicate a role for specific lipid-transport proteins[6-8] in the
transbilayer movement of the aminolipids phosphatidylserine (PS) and
phosphatidylethanolamine (PE), the mechanisms responsible for the
generation and maintenance of lipid asymmetry in biological membranes
remain ill defined. The functional significance of such lipid asymmetry is
also not clear, although there is evidence that loss of phosphatidylserine asymmetry is correlated with increased macrophage uptake of
erythrocytes[9] and also platelet activation[10].

Studies in model membrane systems indicate that intrabilayer lipid
transport rates in response to putative modulators of lipid asymmetry such
as intrinsic and extrinsic proteins[11,12], phase separation phenomena [13] and
oxidative processes[14] are often less than that observed for biological
systems[15,16]. Given the observation that fatty acids act as proton
ionophores[17], it is of interest that fatty acids have been shown to
redistribute rapidly across a model membrane system in response to an
applied pH gradient[18]. This observation has been recently extended to the
negatively-charged diacylphospholipids phosphatidylglycerol (PG)[19] and
phosphatidic acid (PA)[20], which redistribute across a lipid bilayer in
response to an applied pH gradient to accumulate on the high pH side of
the bilayer.

In this article a novel general procedure for the quantitative
determination of lipid asymmetry is described, based upon partition in
aqueous two-phase polymer systems. Partitioning in dextran-polyethylene
glycol two-phase systems is an established method for the fractionation of
cells and organelles on the basis of subtle differences in the interaction

of surfaces with the polymer systems[21]. Addition of anions such as alkali phosphates or sulfates to aqueous two-phase polymer systems results in the establishment of a liquid-junction (Donnan) potential between the two-phases (top phase positive) due to differential ion adsorption to each phase[22]. This electrostatic potential difference may be utilized to measure the charge on a particulate such as a lipid vesicle[23-25]. The utility of the phase partition method is demonstrated using examples of pH-gradient induced asymmetry in lipid vesicles containing stearylamine (SA), PA, PG and cardiolipin (CL). The effects of temperature and lipid composition upon both the rate of formation and extent of asymmetry in model systems is described. Lastly, an example is given of one model system in which the interaction between two membrane surfaces is modulated by the lipid asymmetry within the individual membranes.

MATERIALS AND METHODS

<u>Preparation of Lipid Vesicles</u>

Lipids were combined in the appropriate molar ratios from stock solutions in chloroform, solvent removed initially under nitrogen then by storage under reduced pressure (< 0.1 mm Hg) for at least one hour. For partition measurements, the lipids were spiked with 1-2 µCi of tritiated dipalmitoyl phosphatidylcholine 3H-DPPC prior to removal of solvent. Multilamellar vesicles were prepared by dispersing 25 µmoles of lipid in 2 ml of either 10 mM sodium citrate pH 4, 10 mM sodium phosphate pH 7 or 10 mM sodium phosphate pH 8.5 by extensive vortexing at room temperature then freeze-thawed five times from liquid nitrogen. Lipid vesicles of 100 nm average diameter were prepared by extrusion through polycarbonate filters under nitrogen pressure as previously described[26]. A pH gradient was established across the vesicle bilayers by column chromatography on Sephadex G50M using the appropriate external buffer as eluant.

<u>Partition Measurements</u>

A phase system of 5% (w/w) dextran T-500 (Pharmacia) and 5% (w/w) poly(ethylene glycol), PEG 8000 (BDH), was prepared in 10 mM sodium phosphate at both pH 7 and pH 8.5. The phases were mixed and allowed to equilibrate at 25°C overnight. The PEG-rich upper phase and dextran-rich lower phase were separated and stored at -20°C. For partition measurements, typically 30 µl of vesicle preparation (0.3-0.4 µmoles lipid) was added to 1.5 ml of PEG top phase and 1.5 ml of dextran bottom phase in a 10 x 75 mm tube at 25°C. The contents were mixed by repeated inversion for 1 min. Triplicate 50 µl samples were removed for total counts. The tubes were allowed to stand at 25°C for 25 min then triplicate 50 µl samples of top phase and triplicate 20 µl samples of bottom phase were removed for counting. The difference between counts in top and bottom phases represents counts at the interface; corrections due to adsorption to the tube walls or at the air-water interface are negligible.

RESULTS AND DISCUSSION

<u>Effect of Vesicle Size and Composition</u>

An essential difference between the partitioning of small molecular weight solutes and particulates is that whereas solutes partition between the two phases, particulates partition between one of the phases and the microinterfaces between streams and droplets of the polymer phases as they demix. This very fact that particles adsorb to the interfaces, means that partitioning of particulates exhibits kinetics. As phase separation proceeds progressively more particles are delivered to the bulk interface until, at equilibrium, all particulates end up at the bulk interface between the two phase-separated polymers. This clearly has important operational significance with regard to choice of the appropriate time point at which to sample the two phases.

In general terms a particle associates with an interface when the change in surface free energy upon association with the interface is of the same order or greater than the average thermal energy. If Brownian and shear forces dominate, then the particle will be swept off the interface. The change in surface free energy that occurs depends both on the interfacial tension between the two polymer phases and also the surface area of the absorbed particle, thus partitioning is sensitive to the size of the particulate over a certain regime[27].

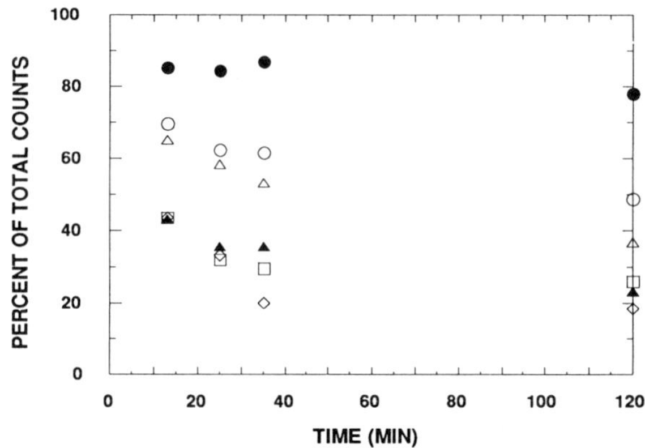

Figure 1. The kinetics of the partitioning of EPC/EPG (6:4) vesicles in a dextran T-500/PEG 6000 (5%/5% w/w) system containing 0.11M sodium phosphate pH 6.8. The filter pore size (nm) used for sizing of the vesicles was (●) 50, (O) 100, (△) 200, (▲) 400 and (□) 600; (◇) corresponds to unsized multilamellar vesicles. (Figure redrawn from ref.29)

The effect of vesicle size upon the kinetics of partitioning of vesicles composed of egg phosphatidylcholine (EPC) and egg phosphatdyl-glycerol (EPG) (6:4 mole ratio) in a phase system of high interfacial potential is shown in Figure 1. It is clear that there is no appreciable difference in the rate at which vesicles (radiolabelled) are cleared from the top phase for vesicles sized through 400 or 600 nm filters, or for multilamellar vesicles (>1 micron diameter). For vesicles sized through 200 and 400 nm filters clearance to the bulk interface was slower and for 50nm vesicles, the majority of the vesicle (>80%) remained in the top phase at 2 hr post mixing. Theoretical considerations indicate that for phase systems with an interfacial tension of 5 x 10^{-3} erg/cm^2, particles of diameter less than or equal to approximately 30nm diamter will not adsorb to the interface and so remain suspended in one phase[28]; consistent with the results of Figure 1.

Albertsson's generalized application of the Brønsted equation[21] indicates that the partition of a particulate should be exponentially dependent upon the surface area of the particulate and this is indeed borne out by experimentation (Figure 2). The practical consequence of these findings is that in order not to confuse size and charge effects it is necessary to perform partition with model membranes of defined size[29].

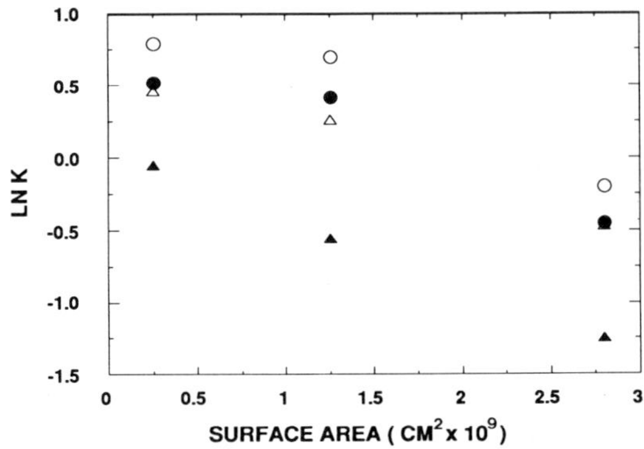

Figure 2. Plot of the natural logarithm of the partition coefficient K versus vesicle surface area for EPC/EPG (6:4) vesicles sized through 100, 200 and 400 nm pore size filters at (O)13, (●)25, (△)45 and(▲)120 min after mixing. The partition coefficient K is defined as (percent counts in top phase)/ (total - percent counts in top phase).

Calibration curves for the partition of 100nm diameter unilamellar vesicles containing dioleoyl phosphatidylcholine (DOPC) and various molar ratios of dioleoyl phosphatidic acid (DOPA) and dioleoyl phosphatidylglycerol (DOPG) are shown in Figure 3. Consistent with Brønsted theory, there is an approximately linear relation between the logarithm of the partition coefficient and the mole percent of charged lipid species in the membrane, indicating the exponential sensitivity to charge. The region of greatest sensitivity was for vesicles containing between approximately 2 and 6 mole percent of the charged species. In order to increase the sensitivity of the phase systems to vesicles containing smaller amounts of charged lipids, phase systems closer to the critical point (for two-phase formation) can be used. Of course, for systems closer to the critical point, other non charge-related factors such as the interfacial tension may play a greater role in the partitioning. It is noteworthy that the results of Figure 3 indicate that whereas the absolute value of the partition coefficient differs for vesicles containing DOPA or DOPG, the slope of the standard curves is very similar. One interpretation of this is that the phase systems are equally sensitive to changes in surface charge whether derived from DOPG or DOPA, but that the lipids are differentially wetted by the lower dextran-rich phase. This increased wetting as a function of surface charge should be manifested as a decrease in contact angle; measurements upon systems containing egg phosphatidylglycerol indicate that this is indeed the case (Figure 4).

Phase partition is sensitive to variation in surface charge over a wide range as illustrated in Figure 5 for 100nm vesicles containing up to

40 mole percent of DOPA, DOPG or dioleoyl phosphatidylserine (DOPS). In this phase system the top phase partition of vesicles containing DOPS or DOPA exhibited a maximum between 10 and 20 mole percent of the charged species. It is reasonable to suggest that the partition of charged vesicles in these phase systems represents a balance between electrostatic forces favoring partition of the negatively charged vesicles into the (relatively positive) PEG-rich upper phase and increased wetting of the membrane surface which favors partition to the lower dextran-rich phase.

Advantages and Disadvantages of Phase Partitioning for Determination of Lipid Asymmetry

Aqueous two-phase polymer partition is well suited to the determination of lipid asymmetry in model membranes for several reasons. As indicated in the previous section, partition is exponentially sensitive to changes in surface charge over variable composition regimes, dependent upon the composition of the phase system. The method is quantitative and also non-specific in the sense that any charged species may be determined. Controls have established that phase systems do not cause permeability changes in model membrane systems or in themselves induce changes in lipid asymmetry. Phase partition is non-perturbatory in that no probe molecule need be added to the system. It will be shown in the next section that phase systems are sensitive to charge-related factors present only on the external face of the membrane and hence there is no potential confusion due to sampling both sides of the membrane. Additionally the methodology is straightforward.

The principal disadvantage is the fact that the method requires a time-dependent partition before sampling the two phases to determine the distribution, thus partitioning cannot be used to follow real-time changes in lipid asymmetry. Partitioning at low temperature (conditions under which intrabilayer flip-flop is slow) minimizes the extent to which changes in asymmetry occur during the partitioning process itself. An operational consideration is that partition is sensitive both to the molecular weight and polydispersity of the polymers used, therefore a certain amount of batch-to-batch variation is to be expected. This consideration is obviated by preparing a sufficiently large batch of phase so that all experiments within a given series are performed using exactly the same phase system.

It should perhaps be stressed that relatively little is understood about the mechanisms by which phase systems detect changes in surface charge. As previously discussed, partitioning of particulates in two-phase systems is principally between one of the phases and the interface such that the kinetics of partitioning depend upon the size of the particulate[27,28]. Large aggregates of particles are readily adsorbed to the interface and cleared as the two phases demix. This is of particular relevance to phase partitioning because it is known that polyethylene glycol causes the aggregation of neutral phosphatidylcholine vesicles[30] and it would be expected that the addition of charge may affect the extent of vesicle aggregation and hence clearance to the bulk interface. Recent evidence suggests a depletion or partial exclusion of polyethylene glycol from the surface of membrane systems[31], thus it is not obvious whether both phases sample the membrane surface to the same extent or even what aspect of the surface the phase systems detect. It is not clear whether the phase systems detect charged species on the membrane surface or a property of the surface that is related to the change in surface charge. It may be that the phase systems are not sensitive to changes in charge per se but rather changes in the degree of hydration or polarization of water at the membrane surface.

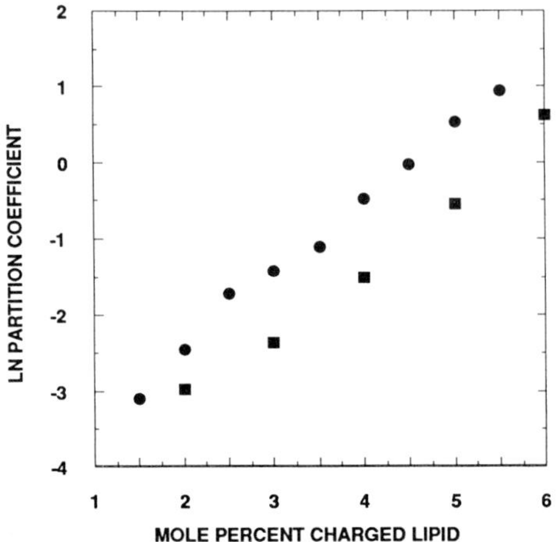

Figure 3. Standard curves for the partition of (●) DOPC/DOPA and (■) DOPC/DOPG 100 nm diameter unilamellar vesicles in 5%/5% (w/w) PEG 8000/dextran T-500, 0.01M sodium phosphate pH 7. The partition coefficient is defined as (percent counts in top phase)/(total – percent counts in top phase) sampled at 25 min after initial mixing.

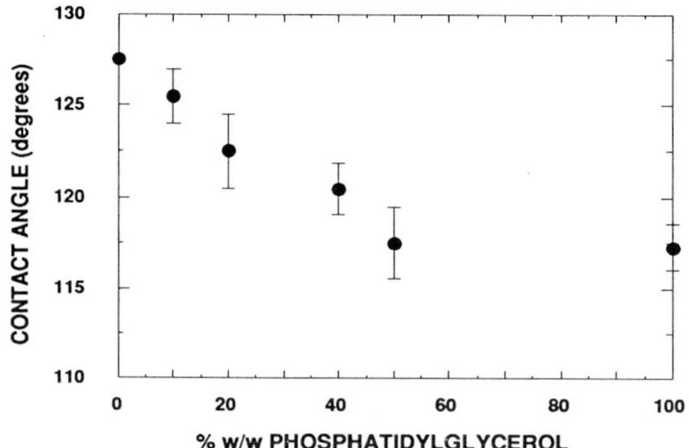

Figure 4. Changes in contact angle with egg phosphatidylglycerol (EPG) content for multilamellar vesicles composed of EPG and egg phosphatidylcholine at 15°C. The phase system is PEG 6000/dextran T-500 (5%/5% w/w) containing 0.01M sodium phosphate, 0.15M sodium chloride pH 6.8. For methodology see ref.29.

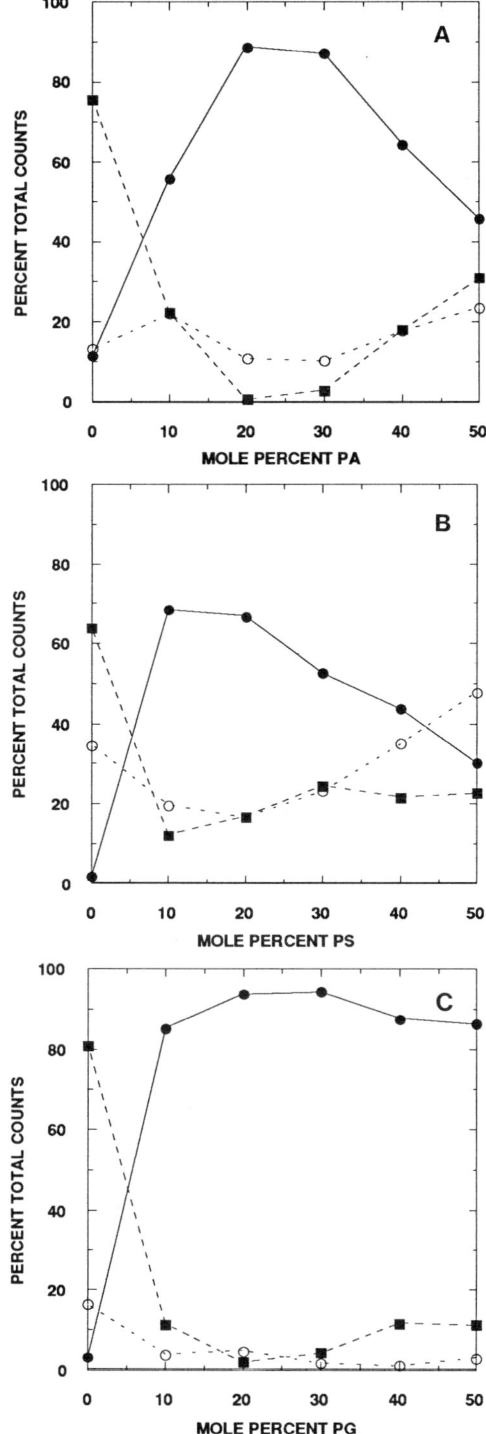

Figure 5. Effect of lipid composition on the partition of (A) PC/PA, (B) PC/PS and (C) PC/PG 100 nm unilamellar vesicles in PEG 8000/ dextran T-500 (5%/5% w/w) containing 0.01M sodium phosphate pH 7. Vesicle associated counts in the (●) top and (O) bottom phases and also at the (■) interface are shown versus the mole percent of the charged species.

<u>Generation of Lipid Asymmetry</u>

Water-soluble weak bases such as methylamine can permeate across a
lipid bilayer in the neutral form to achieve transmembrane concentrations
that satisfy the the Henderson-Hasselbalch equation and so can be used to
measure the pH gradient across a membrane[32]. Similarly, the neutral form
of long chain primary amines such as stearylamine can more readily pass
across a lipid bilayer than the protonated form. Thus in the presence of a
pH gradient across a vesicle bilayer with the vesicle interior acidic
relative to the exterior, stearylamine will redistribute across a bilayer
such that the protonated form is located primarily on the inner surface of
the bilayer[18]. In the presence of a 3 pH unit difference between internal
and external compartments there would be approximately 1000 times as much
stearylamine on the inner than the outer leaflet of the lipid bilayer.
The converse is true of a monacylcarboxylic acid or negatively-charged
mono- or diacylphospholipid, where the negatively charged deprotonated
species would be expected to accumulate on the high pH side of the bilayer
(Figure 6).

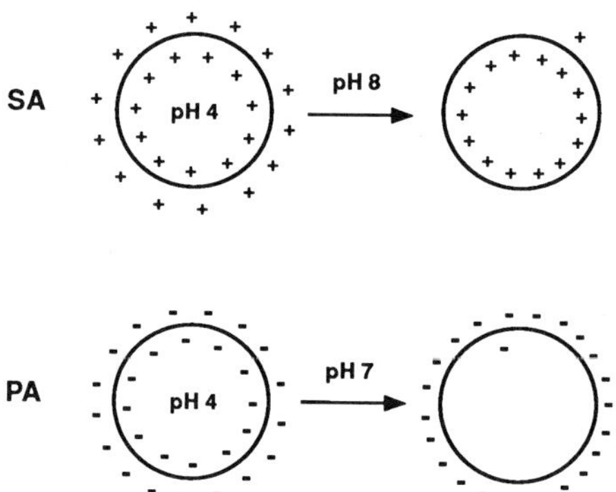

Figure 6. The effect of an applied pH gradient upon the redistribution of
stearylamine (SA) and phosphatidic acid (PA) in unilamellar
lipid vesicles

The partition of 100 nm neutral PC vesicles is illustrated in
Figure 7A which shows that these vesicles partition approximately 70
percent to the interface with approximately 25-30 percent of the vesicles
in the bottom phase and with only a trace amount of lipid in the top
phase. Addition of stearylamine, a positively charged component, causes
transfer of the vesicles to the bottom phase in response to the
electrostatic potential[24] (Figure 7B) such that vesicles with 5 mole
percent stearylamine partition approximately 70 percent to the bottom
phase and 30 percent to the interface. However when vesicles containing
5 mole percent stearylamine are prepared in the presence of a pH gradient
(pH 4 inside/pH 8.5 outside), the vesicles now partition in a similar
manner to pure PC vesicles (Figure 7C), consistent with a redistribution
of the positively charged aminolipid from the outer to the inner monolayer
of the lipid bilayer.

The negatively charged acidic phospholipids phosphatidic acid (PA)
and phosphatidylglycerol (PG) may be induced to redistribute across a
lipid bilayer in the presence of a suitable pH gradient (Figure 8). In
this instance, 100nm PC/PA vesicles (containing 2.5 mole percent PA) were
prepared (pH 4 inside/pH 7 outside) such that the PA would redistribute
from the inner to the outer monolayer whereas 100nm PC/PG vesicles

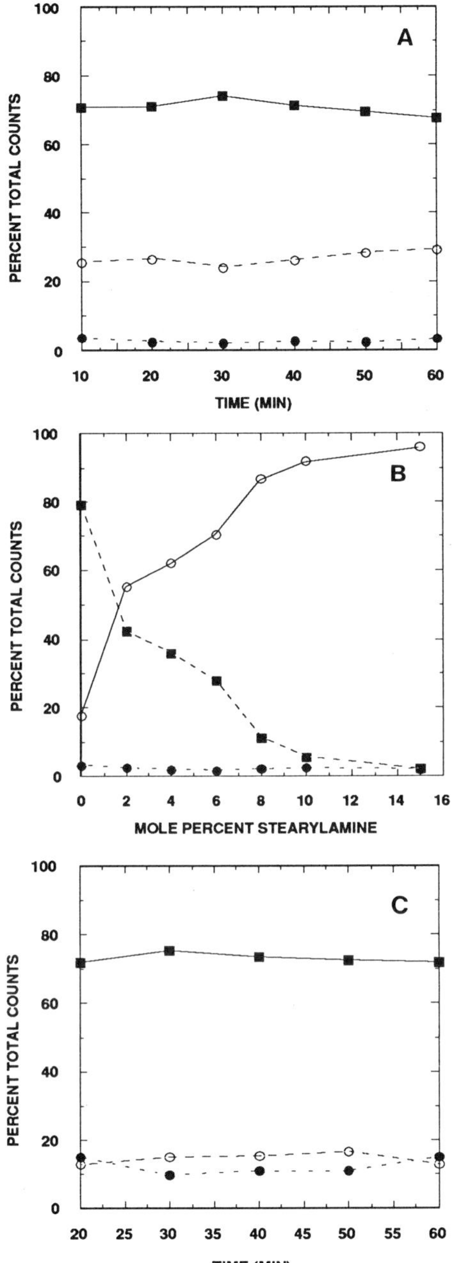

Figure 7. (A) Time-course for the partition of egg phosphatidylcholine (EPC) 100 nm unilamellar vesicles in PEG 8000/dextran T-500 (5%/5% w/w) containing 0.01M sodium phosphate, 0.15M sodium chloride pH 8.5. Counts in the (●) top, (O) bottom and at the (■) interface are shown at various times after mixing. (B) Standard curve for the partition of EPC/stearylamine 100nm unilamellar vesicles without an applied pH gradient sampled at 25 min after initial mixing. (C) Time-course for the partition of EPC/stearylamine (95:5 mole ratio) 100nm unilamellar vesicles prepared with an applied pH gradient (pH 4 internal/pH 8.5 external). Phase system and symbols as in Figure 7A.

(containing 7 mole percent PG) were prepared (pH 7 internal/pH 4 external)
so that the PG would accumulate on the inner monolayer of the vesicle.
The vesicles with applied pH gradient were incubated at either 25°C or
45°C prior to partitioning. For both PA and PG-containing systems it is
clear that after 60 min incubation at 45°C the generated asymmetry is
better than 80 percent with respect to the charged component and that
there is a marked temperature dependence to the redistribution. Recent
studies[19,20] have shown that the pH-gradient induced redistribution of both
PG and PA across a lipid bilayer can be approximated by first-order
kinetics and that the activation energy of redistribution is approximately
28-30 kcal.mol^{-1} consistent with previous studies[33]. This relatively high
activation energy suggests that it is the penetration of the hydrophilic
headgroup into the hydrophobic core of the membrane bilayer (rather than
diffusion across) that is rate-limiting. This interpretation is also
suggested by studies indicating that the rate of transport of monoacyl PG
is very similar to that for a diacyl PG[19].

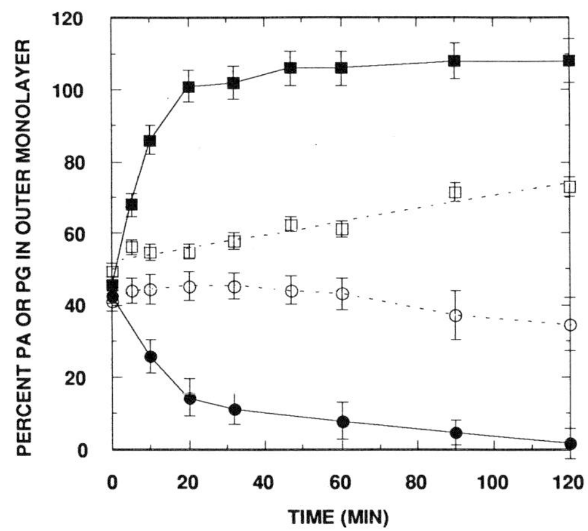

Figure 8. Time-course for the partition of PC/PA 100nm vesicles containing
2.5 mole percent PA prepared in the presence of a pH gradient
(pH 4 internal/pH 7 external) and PC/PG 100nm vesicles
containing 7 mole percent PG prepared in the presence of a pH
gradient (pH 8 internal/pH 4 external) in PEG 8000/dextran
T-500 (5%/5% w/w) containing 0.01M sodium phosphate pH 7 after
incubation at 25°C (□) PC/PA, (O) PC/PG and at
45°C (■) PC/PA, (●) PC/PG.

Preliminary studies indicate that cardiolipin (CL) may also
redistribute in response to a pH gradient albeit under extreme conditions.
Figure 9 shows a partition standard curve for 100 nm PC/CL vesicles as a
function of CL content. Increasing negative charge favors partition to the
top phase in response to the electrochemical potential between the phases
as expected. For compositions between 2 and 4 mole percent CL, the top
phase partition is greater than observed for either PG or PA
(Figure 3), consistent with the fact that each CL molecules has two
negative charges per headgroup compared to only one for either PA or PG.
After incubation at 60°C in the presence of an applied pH gradient (pH 4
internal/pH 7 external), 100nm PC/CL vesicles (containing 2.5 mole percent
CL) exhibit partition values consistent with the majority of the CL being
on the outer monolayer of the vesicle bilayer. It is intriguing to
speculate that there may be a relation between pH gradient induced CL
asymmetry in such model membrane systems and asymmetry in the CL-rich
inner mitochondrial membrane.

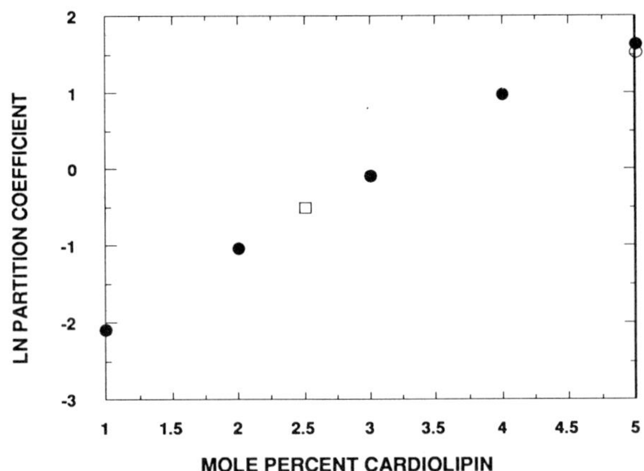

Figure 9. Partition standard curve for PC/CL 100 nm vesicles in PEG
8000/dextran T-500 (5%/5% w/w) containing 0.01M sodium
phosphate pH 7 sampled at 25 min after initial mixing. Also
indicated is the partition of a PC/CL (97.5/2.5 mole ratio)
system (□) prepared in the presence of a pH gradient
(pH 4 internal/pH 7 external) and incubated at 60°C for
1 hr prior to partitioning (○).

Studies of the effects of lipid composition on PG transport[19]
indicate (i) that the pH-dependence of transport is consistent with
movement of the neutral protonated species, (ii) the rate of transport
increases with increasing negative surface charge on the membrane (which
may be rationalized in terms of a decrease interfacial pH which in turn
affects the pKa of dissociable groups on the membrane surface[34]), (iii)
lipids appears to move across the bilayer as monomers with no evidence for
co-transport of lipid in the opposite direction and (iv) the rate of
formation but not the extent of asymmetry is decreased by increased lipid
saturation or addition of cholesterol.

The effect of lipid asymmetry upon lipid exchange in model systems

When PC/PS vesicles (containing 20 mole percent PS) are mixed with
PC/SA vesicles (containing 5 mole percent SA) the vesicles undergo a
spontaneous aggregation due to electrostatic attraction between the
vesicles which results in a rapid increase in absorbance. Over a period of
60 seconds, the vesicles disaggregate and the absorbance decreases.
Dual label experiments have shown these changes are accompanied by a
unidirectional transport of SA into the PS-containing vesicles without
vesicle fusion[35]. If asymmetric PC/SA vesicles are prepared in the
presence of a pH gradient such that the SA is accumulated on the inner
monolayer of the PC/SA vesicles, then there is no aggregation observed
when these vesicles are added to PC/PS vesicles (Figure 10). When the pH
gradient is dissipated by addition of nigericin and valinomycin, the SA
asymmetry decays, SA appears on the outer surface of the lipid vesicles
and the two populations of vesicles now aggregate and disaggregate as
previously observed.

Transmembrane pH gradients exist across several intracellular compartments
including the inner mitochondrial membrane, lysosomes and sercretory
vesicles and, although speculative, it is clearly possible that the pH
gradients may play a role in either the generation or maintenance of

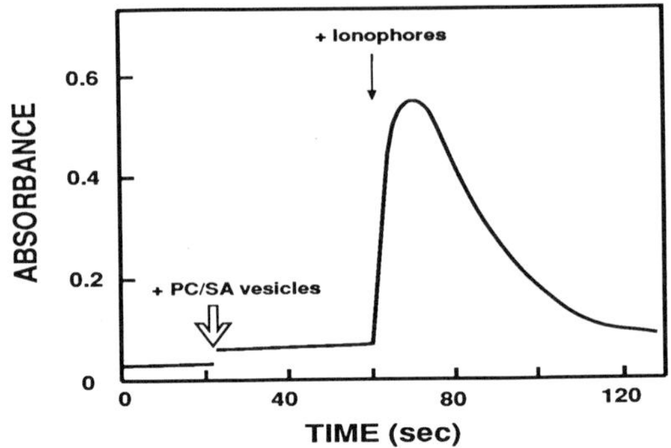

Figure 10. Turbidity measurements at 550 nm of 100nm unilamellar dioleoyl phosphatidylcholine (DOPC)/ bovine brain phosphatidylserine (8:2) vesicles mixed with DOPC/steary lamine (95:5) vesicles. The PC/streaylamine vesicles were prepared in the presence of a pH gradient (pH 4 internal/pH 7 external) so that the streaylamine was kinetically trapped on the inner leaflet of the lipid bilayer and added at the timepoint indicated by the open arrow. After addition of ionophores (valinomycin and nigericin) at the time indicated by the solid arrow, the pH gradient across the PC/stearylamine vesicles is collapsed, the streaylamine redistributes between the inner and outer leaflets and vesicle aggregation and exchange occur as before.

asymmetry of certain lipid species or directing lipid flow between compartments. For example previous studies[18] suggest that any free fatty acids in an intracellular organelle with an acidic lumenal pH will be rapidly transferred from the lumenal to the cytoplasmic side of the membrane where it would be available for intracellular transport or else may potentiate membrane fusion[36].

ACKNOWLEDGMENTS

This work was supported by grants from the B.C. Health Care Research Fund, The Medical Research Council of Canada and by a travel grant from the British Council. CT is an MRC Scholar.

REFERENCES

1. H.J.Chap, R.F.A.Zwaal and L.L.M.van Deenen, Action of highly purified phospholipases on blood platelets. Evidence for an asymmetric distribution of phospholipids in the surface membrane. Biochim.Biophys.Acta.,467:146 (1977)
2. J.A.F.Op den Kamp, Lipid asymmetry in membranes. Ann.Rev.Biochem., 48:47 (1979)
3. R.F.A.Zwaal, Membrane and lipid involvement in blood coagulation. Biochim.Biophys.Acta., 515:163 (1978)
4. J.J.R.Krebs, H.Hauser and E.Carafoli, Asymmetric distribution of phospholipids in the inner membrane of beef heart mitochondria. J.Biol.Chem., 254:5308 (1979)
5. R.G.Sleight and R.E.Pagano, Transport of a fluorescent phosphatidylcholine analog from the plasma membrane to the Golgi apparatus J.Cell.Biol., 99:742 (1984)

6.   M.Seigneuret and P.F.Devaux, ATP-dependent asymmetric distribution of spin-labeled phospholipids in the erythrocyte membrane: Relation to shape changes. Proc.Natl.Acad.Sci., 81:3751 (1984)

7.   A.Zachowski, E.Favre, S.Cribier, P.Herve, and P.F.Devaux, Outside-inside translocation of aminophospholipids in the human erythrocyte is mediated by a specific enzyme., Biochemistry, 25:2585 (1986)

8.   J-Y.Calvez., A.Zachowski., A.Hermann., G.Morrot and P.F.Devaux., Asymmetric distribution of phospholipids in sepctrin-poor erythrocyte vesicles. Biochemistry 27:5666 (1988)

9.   L.McEvoy., P.Williamson and R.A.Schlegel., Membrane phospholipid asymmetry as a determinant of erythrocyte recognition by macrophages. Proc.Natl.Acad.Sci. USA, 83:3311 (1986)

10. E.M.Bevers.,P.Comfurius.,J.L.M.L.Van Rijn., H.C.Hemker and R.F.A.Zwaal. Generation of prothrombin-converting activity and the exposure of phosphatidylserine at the outer surface of platelets., Eur.J.Biochem., 122:429 (1982)

11. B.de Kruijff and P.Baken, Rapid transbilayer movement of phospholipids induced by an asymmetrical perturbation of the bilayer., Biochim.Biophys.Acta., 507:38 (1978)

12. W.J.Gerritsen, P.A.J.Hendricks, B.de Kruijff and L.L.M.van Deenen, The transbilayer movement of phosphatidylcholine in vesicles reconstituted with intrinsic proteins from the human erythrocyte membrane. Biochim.Biophys.Acta., 600:607 (1980)

13. B.de Kruijff and K.W.A.Wirtz, Induction of a relaively fast transbilayer movement of phosphatidylcholine in vesicles. A 13C NMR study., Biochim.Biophys.Acta., 468:318 (1977)

14. J.M.Shaw and T.E.Thompsom, Effect of phospholipid oxidation products on transbilayer movement of phospholipids in single lamellar vesicles., Biochemistry, 21:920 (1982)

15. J.E.Rothman and E.P.Kennedy, Rapid transmembrane movement of newly synthesized phospholipids during membrane assembly. Proc.Natl.Acad.Sci.USA, 74:1821 (1977)

16. D.B.Zilversmit and M.E.Hughes, Extensive exchange of rat liver microsomal phospholipids., Biochim.Biophys.Acta., 469:99 (1977)

17. J.Gutnecht, Proton/hydroxide conductance through phospholipid bilayer membranes:effects of phytanic acid.Biochim.Biophys.Acta.,898:97(1987)

18. M.J.Hope. and P.R.Cullis, Lipid asymmetry induced by transmembrane pH gradients in large unilamellar vesicles.,J.Biol.Chem.,262:4360 (1987)

19. T.E. Redelmeier, M.J.Hope and P.R.Cullis, On the mechanism of transbilayer transport of phosphatidylglycerol in response to transmembrane pH gradients., Biochemistry 29:3046 (1990)

20. S.Eastman, M.J.Hope, and P.R.Cullis, The transbilayer transport of phosphatidic acid in response to a transmembrane pH gradient. Biochemisty, in press, (1990)

21. P-Å. Albertsson. Partition of Cell Particles and Macromolecules, 3rd. Edn. Wiley-Interscience, New York

22. R.Reitherman, S.D.Flanagan. and S.H.Barondes, Electromotive phenomena in partition of erythrocytes in aqueous polymer two-phase systems, Biochim.Biophys.Acta.,297:193 (1973)

23   E,Ericksson. and P.Å. Albertsson, The effect of lipid composition on the partition of liposomes in aqueous two-phase systems., Biochim.Biophys.Acta.,507:425 (1978)

24. P.T.Sharpe, The partition of charged liposomes in aqueous two-phase systems, Mol.Cell.Biochem.,68:151(1985)

25. C.Tilcock, R.Chin, J.Veiro, P.Cullis. and D.Fisher, Detection of surface charge-related properties in model membrane systems by aqueous two-phase partition., Biochim.Biophys.Acta 986:167 (1989)

26. M.J.Hope, M.B.Bally, G.Webb and P.Cullis, Production of large unilamellar vesicles by a rapid extrusion procedure. Characterization of size distribution, trapped volume and ability to maintain a membrane potential., Biochim.Biophys.Acta.,812:55 (1985)

27. H.Walter, D.Fisher and C.Tilcock., On the relation between surface area and partitioning of particulates in two-polymer aqueous phase systems., FEBS Lett., 270:1 (1990)

28. D.E.Brooks, K.Sharp and D.Fisher, Theoretical aspects of partitioning in "Partitioning in Aqueous Two- Phase Systems: Theory, Methods, Uses and Applications", (H.Walter, D.Brooks and D.Fisher,eds.) pp 11-84 Academic Press, New York (1985)

29. C.Tilcock, P.Cullis, T.Dempsey, B.Youens and D.Fisher, Aqueous two-phase polymer partitioning of lipid vesicles of defined size and composition., Biochim.Biophys.Acta.,979:208 (1989)

30. C.Tilcock and D.Fisher., The interaction of phospholipid membranes with poly(ethylene glycol). Vesicle aggregation and lipid exchange., Biochim.Biophys.Acta. 688:645 (1982)

31. K.Arnold, O.Zschoernig, D.Barthel and W.Herold, Exclusion of poly(ethylene glycol) from liposomal surfaces. Biochim.Biophys.Acta.,1022:303 (1990)

32. Deamer,D.W. in "Intracellular pH: Its Measurement, Regulation and Utilization in Cellular Functions", pp. 173-187, Alan R.Liss, New York (1982)

33. R.Homan and H.J.Pownall, Transbilayer diffusion of phospholipids: dependence on headgroup structure and acyl chain length., Biochim.Biophys.Acta 938:155 (1988)

34. F.C.Tsui.F.C., D.M.Ojicius and W.L.Hubbell, The intrinsic pKa values for phosphatidylserine and phosphatidylethanolamine in phosphati dylcholine host membranes., Biophys.J.,49:459 (1986)

35. S.J.Eastman, J.Wilschut, P.Cullis and M.J.Hope, Intervesicular exchange of lipids with weak acid and weak base characteristics: influence of transmembrane pH gradients. Biochim.Biophys.Acta.,981:178 (1989)

36. W.J.Zaks and C.E.Creutz in: "Molecular mechanisms of membrane fusion", (Ohki,S., Doyle,D., Flanagan,T.D., Hui,S.W. and Mayhew,E., eds.), 1988, pp 325-340. Plenum Press, NY

PARTITIONING OF GRAMICIDIN A' BETWEEN COEXISTING PHASES

WITHIN PHOSPHOLIPID BILAYERS

Andrew R.G. Dibble, Mark D. Yeager, and Gerald W. Feigenson

Section of Biochemistry, Molecular and Cell Biology
Cornell University
Ithaca, NY

ABSTRACT

We investigated the partitioning behavior of gramicidin A' in various binary phospholipid mixtures in which gel and liquid crystalline phase coexistence had been induced. The quenching of the gramicidin A' tryptophanyl fluorescence by a spin-labeled phosphatidylcholine was used to determine the equilibrium ratio of gramicidin A' concentration in the liquid crystalline phase to that in the gel phase (i.e. the partition coefficient, K_p). Three multilamellar vesicular systems, differing in the order and rigidity of the gel phase, were compared. The phospholipid content of the gel phase was rich in dipalmitoylphosphatidylcholine or distearoylphosphatidylcholine, or was entirely $Ca(dioleoylphosphatidylserine)_2$. In all cases the gel phase was depleted of gramicidin A', with $K_p = 10 \pm 2$, 25 ± 5, and ≥ 30, respectively. These data suggest that the more ordered the gel phase, the less it can accommodate a transmembrane polypeptide.

INTRODUCTION

The fact that gel and liquid crystalline phases can coexist in biological membranes has been known since the scanning calorimetry (Steim *et al.*, 1969) and x-ray diffraction (Engelman, 1970) experiments which proved that the lipids of *A. laidlawii* are in a bilayer. Electron microscopy (Kleeman and McConnell, 1976), transport protein activity (Thilo *et al.*, 1977), and fluorescence quenching (Florine and Feigenson, 1987b) experiments have shown that proteins can be partially cleared from the gel phase when gel and liquid crystalline phases coexist. It is possible that such protein clearing from a Ca^{2+}-induced gel phase is involved at a stage of a membrane fusion process (Portis *et al.*, 1979). However, it has not yet been firmly established that biological membranes undergo phase transitions under physiological conditions.

We are interested in a quantitative description of the partitioning behavior of proteins when gel and liquid crystalline phases coexist. Such information makes it possible to compare not only the partitioning behavior of different proteins, but also how a given protein behaves in the presence of different gel phases. Our method is a fluorescence quenching technique developed several years ago in this laboratory (London and Feigenson, 1981b). Early experiments showed that gramicidin A' strongly favors the liquid crystalline phase when both a liquid crystalline and a gel phase coexist (London and Feigenson, 1981b). Attempts to quantitate this partitioning behavior were hindered because an accurate phase diagram for the binary phospholipid mixture used was not available. Subsequent experiments showed that the membrane-bound coat protein from M13 bacteriophage partitions strongly out of a Ca^{2+}-induced phosphatidylserine gel phase (Florine and Feigenson, 1987b). Because the phase behavior of the lipid system used was known, it was possible to determine the equilibrium ratio of protein concentration in the liquid crystalline phase to that in the gel phase ($K_P \sim 25$).

This study examines the partitioning behavior of gramicidin A' in three binary phospholipid mixtures whose phase behavior is known. When two-phase coexistence was induced, the gel phase was composed predominantly of DPPC[†] or DSPC, or entirely of $Ca(DOPS)_2$. Electron paramagnetic resonance spectroscopic measurements have shown that these gel phases differ in their degree of order and rigidity, with $Ca(DOPS)_2$ being more ordered than DSPC, which in turn is more ordered than DPPC (Florine and Feigenson, 1987a). A major goal of this study was to determine the relationship, if any, between the relative order of a gel phase and its ability to accommodate gramicidin A'. We found that, for the gel phases analyzed here, the relative solubility of gramicidin A' increased as the order of the gel phase decreased.

[†]Abbreviations: DPPC, 1,2-dipalmitoyl-*sn*-glycero-3-phosphocholine; DSPC, 1,2-distearoyl-*sn*-glycero-3-phosphocholine; DOPS, 1,2-dioleoyl-*sn*-glycero-3-phosphoserine; (7,6)PC, 1-acyl-2-[2-(6-carboxyhexyl)-2-octyl-4,4-dimethyl-oxazolidinyl-3-oxy]glycero-3-phosphocholine; Pipes, piperazine-1,4-bis(2-ethanesulfonic acid); [(7,6)PC], the overall mole fraction of (7,6)PC; [(7,6)PC]$_S$, the mole fraction of (7,6)PC at the solidus boundary of the phase diagram; [(7,6)PC]$_L$, the mole fraction of (7,6)PC at the liquidus boundary of the phase diagram; F, observed or calculated fluorescence intensity; F_0, observed fluorescence intensity in the absence of (7,6)PC and in a gel phase; F_S, observed fluorescence intensity at the solidus boundary of the phase diagram; F_L, observed fluorescence intensity at the liquidus boundary of the phase diagram; [G], the mole fraction of total lipids in the gel phase; K_p, the partition coefficient, i.e. the ratio of gramicidin A' concentration in the liquid crystalline phase to that in the gel phase at equilibrium.

EXPERIMENTAL PROCEDURES

Materials

DPPC, DSPC, and DOPS were purchased from Avanti Polar Lipids, Inc. (Birmingham, AL). The spin-labeled phosphatidylcholine (7,6)PC, was prepared as previously described (London & Feigenson, 1981a). These lipids were judged to be >98% pure by thin-layer chromatography of 0.1 mg of each lipid on Absorbosil Plus P plates (Applied Science, State College, PA) using chloroform/methanol/concentrated ammonium hydroxide (65/25/5, v/v/v). The natural mixture of gramicidins A, B, and C from *Bacillus brevis* was purchased from Sigma Chemical Co. (St. Louis, MO), and used without further purification. This mixture of peptides is referred to as gramicidin A'. The water was purified with a Milli-Q system (Millipore Corp., Bedford, MA). Chloroform (HPLC grade), methanol (HPLC grade), and benzene (analytical reagent grade) were from Mallinckrodt Inc. (Paris, KY). The buffer, Pipes, was Biochemika Microselect grade and was obtained from Fluka Chemical Corp. (Ronkonkoma, NY). All other chemicals were reagent grade.

Preparation of Multilayer Dispersions of Binary Phospholipid Mixtures

Aliquots of stock solutions of lipid (6-10 mM) in chloroform were delivered to borosilicate culture tubes to produce binary lipid mixtures of (7,6)PC and DPPC, DSPC, or DOPS, at mole fractions of (7,6)PC from 0 to 1. An aliquot of a stock solution of gramicidin A' (50-100 μM) in methanol was added to some of these tubes so that the peptide/lipid molar ratio was 1/200. Blanks containing lipid without peptide were also prepared in order to determine the background fluorescence. Samples were dried to a thin film under a stream of nitrogen gas, then lyophilized from benzene/methanol (19/1, v/v). Each freeze-dried sample was hydrated for 1 h under argon gas with buffer (20 mM Pipes, 100 mM KCl, pH 7.0). Hydration was at 48 $^{\circ}$C and 1 mM lipid for DPPC/(7,6)PC dispersions, 61 $^{\circ}$C and 1 mM lipid for DSPC/(7,6)PC dispersions, or 20 $^{\circ}$C and 2 mM lipid for DOPS/(7,6)PC dispersions. Midway through the hydration period, samples containing DPPC or DSPC were briefly vortexed above the transition temperature. At the end of the hydration period, samples containing DPPC or DSPC were cooled at a rate not exceeding 0.5 $^{\circ}$C/min to the temperature at which the fluorescence measurements were to be made. At the end of the hydration period, samples containing DOPS were briefly vortexed and brought to 1 mM lipid by a two-fold dilution with buffer. The diluting buffer contained 40 mM $CaCl_2$ if the sample was to be treated with Ca^{2+}. The DOPS/(7,6)PC multilayer dispersions were equilibrated with Ca^{2+} by fifteen cycles of freezing at -10 $^{\circ}$C and thawing in a water bath at ambient temperature, as previously described (Florine and Feigenson, 1987a; Florine and Feigenson, 1987b).

Fluorescence Spectroscopy

Fluorescence measurements were made with a home-built spectrofluorometer (Caffrey and Feigenson, 1981) utilizing conventional 90° optics and equipped with double monochromators in both excitation and emission optics. The excitation and emission wavelengths were 285 and 345 nm, respectively. Nominal excitation and emission bandwidths were 2 and 8 nm, respectively.

Immediately before the measurement of fluorescence intensity, each sample was diluted ten-fold with buffer to a final lipid concentration of 0.1 mM. At this lipid concentration light scattering did not contribute to the fluorescence signal. The diluting buffer was identical to the buffer of the sample in composition and temperature. Background fluorescence intensity from the phospholipids was determined and used to correct the observed fluorescence intensity of the samples containing gramicidin A'. This background ranged from 10-50% of the total signal, within the two-phase region of the phase diagram. Samples containing DPPC or DSPC were placed into a quartz microcuvette in a temperature-controlled cuvette holder. Each sample containing DOPS was thawed (completing its last freeze-thaw cycle) immediately before dilution and determination of fluorescence intensity. This was done to avoid complications arising from the aggregation of DOPS vesicles treated with Ca^{2+}. This aggregation becomes significant after 0.5 to 1 h at room temperature (Florine and Feigenson, 1987a). Also, DOPS vesicles treated with Ca^{2+} stick to glass, resulting in variable transfer efficiencies of the DOPS/(7,6)PC dispersions (Florine and Feigenson, 1987a). Thus, after each fluorescence measurement (using a disposable acrylic cuvette; Sarstedt, Inc., Princeton, NJ) a 0.1 ml aliquot was removed from the cuvette for subsequent determination of phospholipid concentration by phosphorus analysis as in Kingsley and Feigenson (1979). This determination of transfer efficiencies was used to correct the fluorescence measurements.

Determination of the Partition Coefficient, K_p

The quenching of the tryptophanyl fluorescence of gramicidin A' by (7,6)PC occurs only when the spin-labeled phospholipid is in contact with the peptide (London and Feigenson, 1981a). For this reason, the local lipid environment of gramicidin A' can be determined by measuring fluorescence intensity at various mole fractions of (7,6)PC in a binary lipid mixture whose phase behavior is known. The phase diagrams for the binary lipid mixtures studied here have been determined by differential scanning calorimetry for DPPC/(7,6)PC and DSPC/(7,6)PC dispersions (Huang *et al.* , 1988), and by electron paramagnetic resonance spectroscopy and x-ray diffraction for DOPS/(7,6)PC dispersions equilibrated with 20 mM Ca^{2+} (Florine and Feigenson, 1987a). The solidus boundaries of these phase diagrams are such that, at the temperatures at which these binary mixtures were analyzed, the gel phase is predominantly DPPC or DSPC, or is entirely $Ca(DOPS)_2$. Thus, the gel phase is either greatly or completely depleted of (7,6)PC. The coexisting liquid crystalline phase is enriched with the spin-labeled lipid. The partition coefficient, K_p, was determined directly from the fluorescence quenching data within the two-phase region of the phase diagram, by using the relation

$$F = F_L + \frac{[G]\,(F_S - F_L)}{K_p\,(1 - [G]) + [G]} \tag{1}$$

where F is the observed (or calculated, see below) fluorescence, $F_{S(L)}$ is the observed fluorescence at the solidus (liquidus) boundary of the phase diagram, and [G] is the mole fraction of total lipids in the gel phase (London and Feigenson, 1981b). The value of [G] is calculated as $\{[(7,6)PC]_L - [(7,6)PC]\}/\{[(7,6)PC]_L - [(7,6)PC]_S\}$ (the lever rule) where $[(7,6)PC]_{S(L)}$ is the mole fraction of (7,6)PC at the solidus (liquidus) boundary of the phase diagram. The partition coefficient, the only unknown, was determined by fitting the experimental data to theoretical curves of F *vs.* [(7,6)PC] within the two-phase region of the phase diagram.

RESULTS

In order to avoid self-quenching of gramicidin A' tryptophanyl fluorescence during the quenching experiments, the dependence of fluorescence intensity on gramicidin A' concentration was investigated at room temperature with phospholipid multilamellar vesicles composed entirely of DPPC, DOPS, or Ca(DOPS)$_2$. The fluorescence was found to depend linearly on gramicidin A' concentration over the entire range of peptide/lipid molar ratios studied, 1/300 to 1/50, for all three lipid systems (data not shown). Therefore, throughout these peptide/lipid molar ratios, self-quenching of tryptophanyl fluorescence did not occur significantly in multilamellar vesicles which were in a fluid phase of DOPS, in a thermally-induced gel phase of DPPC, or in a Ca^{2+}-induced gel phase of Ca(DOPS)$_2$. For the fluorescence quenching experiments, a gramicidin A'/lipid molar ratio of 1/200 was chosen in order to obtain a sufficient fluorescence signal without significant self-quenching at any mole fraction of (7,6)PC. Note that if $K_P \gg 1$, the protein/lipid molar ratio in the liquid crystalline phase can be quite high, especially near the solidus boundary of the phase diagram. Thus, it is possible for self-quenching, or other phenomena dependent upon protein concentration, to occur at some mole fractions of (7,6)PC while not at others.

Gramicidin A' did not have a significant effect on the phase behavior of the binary lipid mixtures used in this experiment, at a peptide/lipid molar ratio of 1/200. The effect of gramicidin A' on the solidus boundaries of the DPPC/(7,6)PC and DSPC/(7,6)PC phase diagrams can be estimated by calculating the freezing point depression of the DPPC gel phase or the DSPC gel phase when the peptide is present. With the assumptions that the peptide mixes ideally in the liquid crystalline phase but is insoluble in the gel phase, the freezing point depression was calculated to be on the order of only 0.1 $^{\circ}$C.

The fluorescence quenching data for each of the binary lipid mixtures analyzed, as well as the best fits of equation 1, are shown in Figures 1-3. Gramicidin A' was found to partition strongly into the liquid crystalline phase with all three binary phospholipid mixtures; with $K_P = 10 \pm 2$ for gramicidin A' in DPPC/(7,6)PC at 34 $^{\circ}$C (Figure 1), $K_P = 30 \pm 5$, 22 ± 2, and 26 ± 2 for the peptide in DSPC/(7,6)PC at 35, 40, and 46 $^{\circ}$C, respectively (Figure 2), and $K_P \gtrsim 30$ for the peptide in DOPS/(7,6)PC equilibrated with 20 mM Ca^{2+} at 20 $^{\circ}$C (Figure 3). The estimated error indicates the range of K_P values which reasonably fit the data. In the final case, the partition coefficient may be much greater than 30 (see Discussion).

DISCUSSION

The tendency of membrane-bound proteins to partition from gel to fluid phospholipid is generally recognized. This is based largely on the clearing of intramembrane particles from gel phase lipid domains, as observed by freeze-fracture electron microscopy (Kleeman and McConnell). By making a quantitative measurement of such partitioning behavior it becomes possible to compare the degree of clearing of different polypeptides from a common gel phase, or to compare the relative solubility of a given polypeptide in various gel phases. This study emphasizes the latter type of analysis.

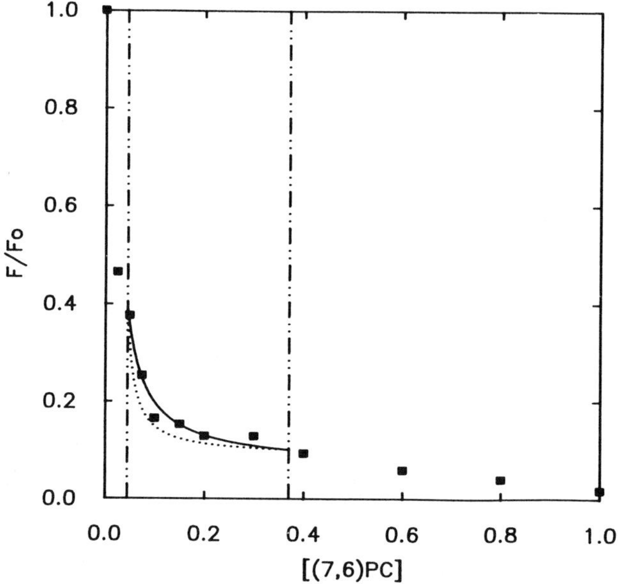

Fig. 1. Fluorescence quenching of gramicidin A' in multilamellar vesicles composed of DPPC/(7,6)PC at 34 $^{\circ}$C. The abscissa is the mole fraction of (7,6)PC in the vesicles, and the ordinate is the ratio of observed or calculated fluorescence (F) to observed fluorescence in the pure DPPC gel phase (F_0). All data points are averages of duplicate samples. The theoretical curves were calculated by use of equation 1 (text) setting [(7,6)PC]$_S$ = 0.045, [(7,6)PC]$_L$ = 0.37, F_S/F_0 = 0.388, F_L/F_0 = 0.102, and K_P = 10 (——) or 25 (······). The latter curve is included to allow comparison of these data with those of Figures 2 and 3. The solidus and liquidus phase boundaries are indicated by the vertical broken lines.

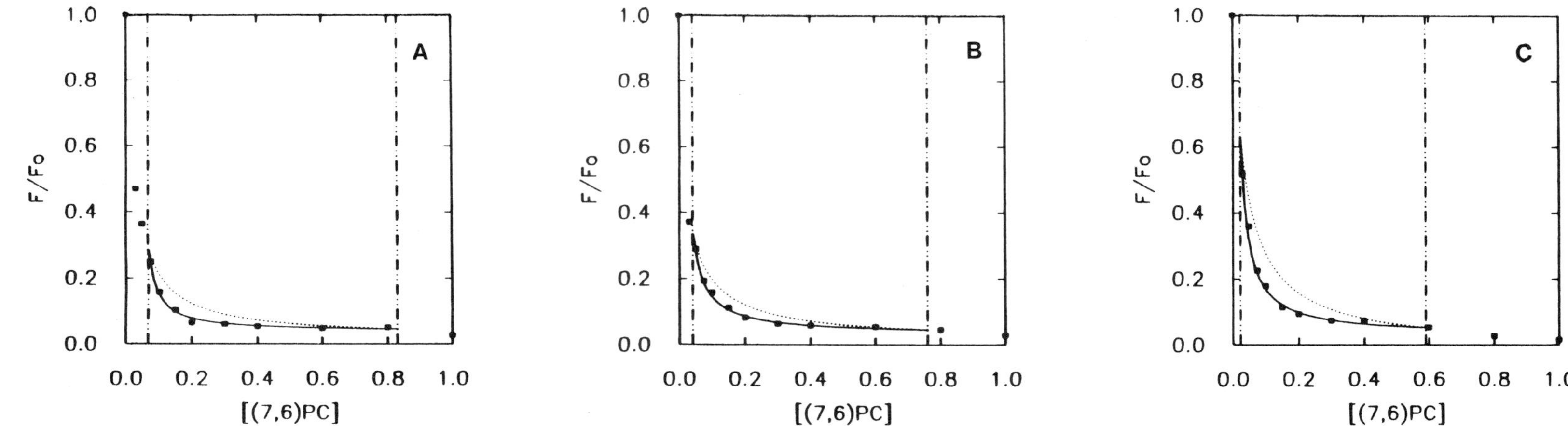

Fig. 2. Fluorescence quenching of gramicidin A' in multilamellar vesicles composed of DSPC/(7,6)PC at 35 °C (A), 40 °C (B), and 46 °C (C). The abscissa is the mole fraction of (7,6)PC in the vesicles, and the ordinate is the ratio of observed or calculated fluorescence (F) to observed fluorescence in the pure DSPC gel phase (F_0). All data points are averages of duplicate samples. The theoretical curves were calculated by use of equation 1 (text) setting (A) $[(7,6)PC]_S$ = 0.067, $[(7,6)PC]_L$ = 0.83, F_S/F_0 = 0.285, F_L/F_0 = 0.046, and K_P = 30 (——) or 10 (······); (B) $[(7,6)PC]_S$ = 0.040, $[(7,6)PC]_L$ = 0.76, F_S/F_0 = 0.334, F_L/F_0 = 0.047, and K_P = 22 (——) or 10 (······); and (C) $[(7,6)PC]_S$ = 0.023, $[(7,6)PC]_L$ = 0.59, F_S/F_0 = 0.630, F_L/F_0 = 0.053, and K_P = 26 (——) or 10 (······). The curves with K_P = 10 are included to allow comparison of these data with those of Figure 1. The solidus and liquidus phase boundaries are indicated by the vertical broken lines.

We found gramicidin A' to be efficiently excluded from the gel phase with all three binary phospholipid mixtures studied. The best fits of equation 1 are when $K_P = 10 \pm 2$ for gramicidin A' in DPPC/(7,6)PC, 25 ± 5 for the peptide in DSPC/(7,6)PC, and at least 30 (see below) for the peptide in DOPS/(7,6)PC with 20 mM Ca^{2+}. We infer that the packing of gramicidin A' into a gel phase is more unfavorable the more orderly the gel, with the peptide fitting best into the DPPC-rich gel, significantly less well into the DSPC-rich gel, and less well yet into $Ca(DOPS)_2$. This conclusion relies on the assumption that gramicidin A' has similar mixing behavior in the liquid crystalline phase with all three binary phospholipid mixtures. This is a reasonable assumption, based on the large fraction of (7,6)PC in each of the liquid crystalline phases.

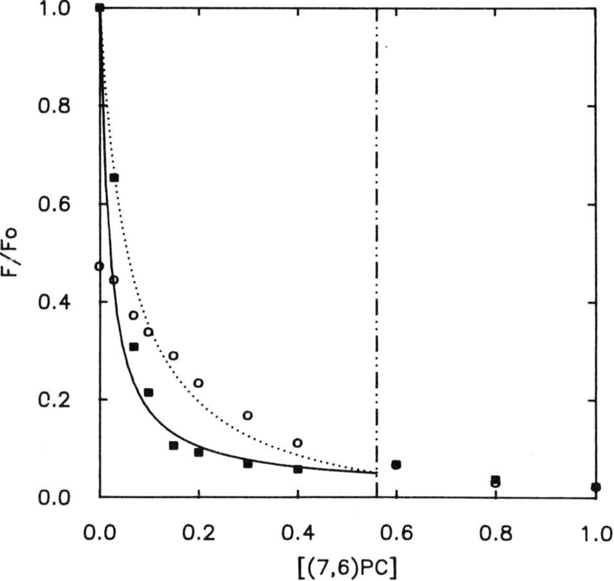

Fig. 3. Fluorescence quenching of gramicidin A' in multilamellar vesicles composed of DOPS/(7,6)PC at 20 $^{\circ}$C equilibrated without Ca^{2+} (O) or with 20 mM Ca^{2+} (■). The abscissa is the mole fraction of (7,6)PC in the vesicles, and the ordinate is the ratio of observed or calculated fluorescence (F) to observed fluorescence in the pure $Ca(DOPS)_2$ phase (F_0). All data points are averages of duplicate samples, except the [(7,6)PC] = 0 point, which is an average of quadruplicate samples. The theoretical curves were · calculated by use of equation 1 (text) setting $[(7,6)PC]_S = 0.00$, $[(7,6)PC]_L = 0.56$, $F_S/F_0 = 1.00$, $F_L/F_0 = 0.051$, and $K_P = 30$ (——) or 10 (······). The latter curve is included to allow comparison of these data with those of Figure 1. The solidus and liquidus phase boundaries are indicated by the vertical broken lines.

For the DPPC/(7,6)PC and DSPC/(7,6)PC dispersions, useful fluorescence quenching measurements could be made only throughout a limited temperature range. This was because accurate estimates of K_P cannot be made unless the fluorescence at the solidus phase boundary is significantly greater than that at the liquidus phase boundary. We found that (7,6)PC was extremely efficient at quenching gramicidin A' tryptophanyl fluorescence in gel phases. As a consequence, at room temperature most of the fluorescence was quenched at mole fractions of (7,6)PC below $[(7,6)PC]_S$. We attribute this phenomenon to a nonrandom distribution of lipids about the peptide in the gel phase, with gramicidin A' preferentially in contact with (7,6)PC. In order to obtain higher fluorescence at the solidus boundary of the phase diagram, $[(7,6)PC]_S$ was lowered by raising the temperature at which the fluorescence quenching data were collected. The best determinations of K_P could be made at approximately 34 ^{o}C for DPPC/(7,6)PC, and between about 40 and 46 ^{o}C for DSPC/(7,6)PC. Data were also collected with the latter system at 35 ^{o}C in an attempt to detect any temperature dependence of K_P. However, throughout this temperature range, no strong temperature dependence was observed, with $K_P = 30 \pm 5$ (35 ^{o}C), 22 ± 2 (40 ^{o}C), and 26 ± 2 (46 ^{o}C). This lack of a strong dependence of K_P on temperature implies that it is reasonable to compare, without adjustment, all of the partition coefficients determined in this study.

The fluorescence quenching curve obtained with gramicidin A' in DOPS/(7,6)PC equilibrated with 20 mM Ca^{2+} could not be well described by a single value of K_P. At mole fractions of (7,6)PC where less than about 25% of the total lipid is in the liquid crystalline phase, the data points lie above the best fit of equation 1. All other data points lie below this curve. This type of nonuniform fluorescence quenching curve was observed only with the DOPS/(7,6)PC dispersions with Ca^{2+}, and was reproducible. Three possible explanations for this effect are that at low mole fractions of (7,6)PC, (i) some gramicidin A' is kinetically trapped in the $Ca(DOPS)_2$ phase, even after fifteen freeze-thaw cycles; (ii) the small amount of liquid crystalline phase has a high ratio of perimeter to area, resulting in reduced quenching of the peptide because a significant fraction of its nearest neighbors are in the $Ca(DOPS)_2$ phase; and (iii) gramicidin A' is concentrated in the small amount of liquid crystalline phase to such an extent that the peptide aggregates, with the net effect being reduced quenching due to decreased contact of (7,6)PC with the peptide. Additional experiments are planned to test these possibilities. If any of these explanations are found to be correct, then the K_P of gramicidin A' is greater than 30 in this system.

In an earlier study, we determined the partitioning behavior of M13 bacteriophage coat protein in DOPS/(7,6)PC equilibrated with 20 mM Ca^{2+} (Florine & Feigenson, 1987b). That fluorescence quenching curve is remarkably like the analogous one found here for gramicidin A', with the best fit of equation 1 being when $K_P \gtrsim 25$. The similarity of the partitioning behavior of the coat protein and gramicidin A' suggests that neither the chemical nature of the amino acid residues nor their secondary structure determines the partitioning behavior. The coat protein spans the bilayer by means of a β-strand (Datema *et al.*, 1988), while membrane-bound gramicidin A' is a $\pi_{(L,D)}$ helix (Urry, 1971; Urry *et al.*, 1971). One similarity is that both molecules have about fifteen amino acid residues which interact with the bilayer. Thus, the decisive factor which determines the degree of polypeptide clearing from a given gel phase may be the amount of lipid contact per polypeptide.

ACKNOWLEDGEMENTS

This work was supported by a grant from the National Institutes of Health, U.S. Public Health Service (HL-18255). A.D. and M.Y. were supported in part by National Institutes of Health Research Service Award 5T32GM07273. Present address of M.Y.: Department of Biochemistry, Sciences II, University of Geneva, Ch-1211 Geneva 4, Switzerland. We thank Lisa Miller for typing the manuscript.

REFERENCES

Caffrey, M., and Feigenson, G.W. (1981) *Biochemistry* **20**, 1949-1961.

Datema, K.P., Spruijt, R.B., Wolfs, C.J.A.M., and Hemminga, M.A. (1988) *Biochim. Biophys. Acta* **944**, 507-515.

Engelman, D.M. (1970) *J. Mol. Biol.* **47**, 115-117.

Florine, K.I., and Feigenson, G.W. (1987a) *Biochemistry* **26**, 1757-1768.

Florine, K.I., and Feigenson, G.W. (1987b) *Biochemistry* **26**, 2978-2983.

Huang, N., Florine-Casteel, K., Feigenson, G.W., and Spink, C. (1988) *Biochim. Biophys. Acta* **939**, 124-130.

Kingsley, P.B., and Feigenson, G.W. (1979) *Chem. Phys. Lipids* **24**, 135-147.

Kleeman, W., and McConnell, H.M. (1976) *Biochim. Biophys. Acta* **419**, 206-222.

London, E., and Feigenson, G.W. (1981a) *Biochemistry* **20**, 1932-1938.

London, E., and Feigenson, G.W. (1981b) *Biochim. Biophys. Acta* **649**, 89-97.

Portis, A., Newton, C., Pangborn, W., and Papahadjopoulos, D. (1979) *Biochemistry* **18**, 780-790.

Steim, J.M., Tourtellotte, M.E., Reinert, J.C., McElhaney, R.N., and Rader, R.L. (1969) *Proc. Natl. Acad. Sci. U.S.A.* **63**, 104-109.

Thilo, L., Trauble, H., and Overath, P. (1977) *Biochemistry* **26**, 1283-1290.

Urry, D.W. (1971) *Proc. Natl. Acad. Sci. U.S.A.* **68**, 672-676.

Urry, D.W., Goodall, M.C., Glickson, J.D., and Mayers, D.F. (1971) *Proc. Natl. Acad. Sci. U.S.A.* **68**, 1907-1911.

MEMBRANE CONTACT INDUCED BETWEEN ERYTHROCYTES

BY POLYCATIONS, LECTINS AND DEXTRAN

W. Terence Coakley, Homa Darmani and Alec J. Baker

School of Pure and Applied Biology
University of Wales College of Cardiff
Cardiff CF1 3TL, Wales, U.K.

1. INTRODUCTION

Normal eukaryotic cell function requires that the cell's extensive
arrays of intracellular membrane remain separate and stable as they come,
through thermal movements, into close proximity to each other. In other
situations, as when vesicles approach the Golgi body, during exocytosis or
during the acrosome reaction in sperm (Russell, 1979) local membrane
contact is a prerequisite for membrane fusion. Cell plasma membranes make
local contacts during cell locomotion and during differentiation (Andre and
Bongrand, 1990). Modification of membrane interactions is required _in vitro_
for fusion of cells or protoplasts when they are subjected to non-
physiological solutions such as high concentrations of polyethylene glycol
(Pontecorvo, 1975) or of dextran (Kameya et al., 1981). There is thus a
general interest in understanding how membranes avoid coming into close
contact with each other, in understanding how repulsive interactions can be
overcome so that membranes may come into close contact, and in establishing
whether contact, when it occurs, is laterally continuous or is localised.

In the present paper the modifications of membrane interactions by
contact-inducing polymers such as dextran or polycations will be examined.
The characteristics of such adhesion will be compared with those of
adhesions achieved with the more specific lectin molecules. Previous work
on polymer-induced cell adhesion will be reviewed. Particular attention
will be paid to the following questions which have been addressed in this
laboratory: (i) Whether the plasma membranes involved in the seam of
contact lie in a parallel configuration and in a stable energy minimum or
whether regions of close membrane approach are separated by regions of
larger separation so that the seam of contact is characterised by spatially
periodic contact regions: (ii) Whether, following initial cell-cell contact
at a point, contact(s) continue(s) to form without pause or whether there
is a characteristic (polymer dependent) delay before a sudden movement to
mutual cell-cell engulfment. Progress to date in explaining the observed
contact patterns is described. Areas where further information is required
to develop an internally consistent explanation of the adhesion patterns
for the different molecules are identified.

2. ERYTHROCYTE SURFACE

2(i) <u>Normal Cell Surface</u>

The human erythrocyte has been used as a model system in the present
work both because there is already an extensive literature on erythrocyte
adhesion and because it may reasonably be expected that a set of basic
biophysical responses to changes in membrane interactions will be easier to
detect in the relatively simple erythrocyte than in differentiated
eukaryotic cells where biochemical responses, which can further complicate
interpretation, may be triggered.

The erythrocyte glycocalyx, a molecular sieve of thickness in the range
5.5 - 10 nm at the outer surface of the cell (Donath and Pastushenko, 1979;
Kahane et al., 1978) is largely made up of glycophorin A, B and C and of
band 3 transmembrane protein. Band 3 has a molecular weight of 90,000. A 43
kDa N-terminal fragment is exposed on the cytoplasmic side of the membrane
while the C-terminal also appears to be intracellular (Jay and Cantley,
1986). These authors proposed that band 3 has 8 helical membrane-spanning
regions and a hairpin loop into the membrane. Approximately 8% of the
molecular weight of band 3 is carbohydrate and almost all of that is in a
single extracellular-phase oligosaccharide chain. Since most of the
polypeptide is not exposed on the outer membrane face the greatest
contribution of band 3 to the glycocalyx is probably due to this
oligosaccharide.

The glycophorins carry 90% of the sialic acid and thus most of the
outer cell surface negative electric charge (Bretscher and Raff, 1975). A
small amount of charge is carried by sialoglycolipids (Ballas et al.,
1986). Glycophorin A is the most abundant (400,000 copies per cell)
erythrocyte sialoglycoprotein. It traverses the membrane once, probably as
a dimer, and has a short cytoplasmic domain (Bennett, 1985). 60% of its 31
kDa molecular weight is carbohydrate (Carter, 1985). About half of its
amino acids are exposed in the extracellular phase. These are glycosylated
by 16 short oligosacchaide chains generally terminating in sialic acid
groups (Marchesi, 1979) which account for 75% of the total cell surface
sialic acid.

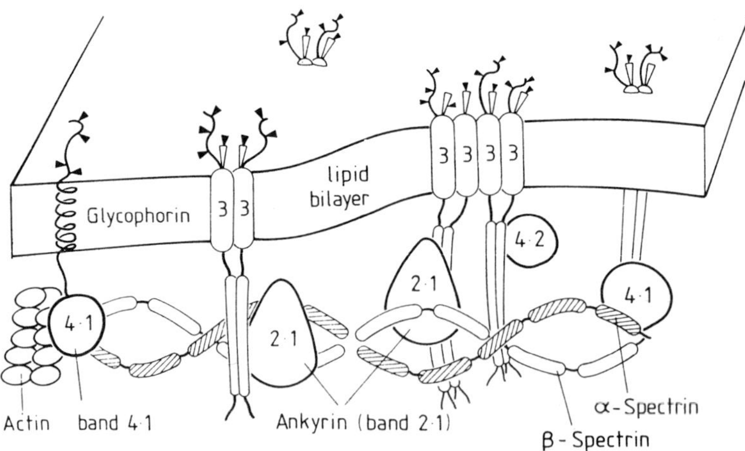

Fig. 1. Schematic diagram of the erythrocyte membrane (Clague et al.,
1989). See text for details.

The erythrocyte membrane skeleton is an elastic network on the
cytoplasmic face of the cell membrane. It plays a role in stabilising cell
shape (Coakley et al., 1980, Lange et al., 1982). The principal membrane
skeleton protein, spectrin is composed of two subunits of molecular weight
260,000 and 225,000 (Bennett, 1989). The spectrin heterodimer has a contour
length of 100nm. Head to head association of the heterodimers forms
spectrin tetramers. Negatively-stained spread membrane-skeleton
preparations show five to seven long (ca. 180nm) thin filaments,
corresponding to purified spectrin tetramers, attached by their ends to
short acin filament junctions, thus generating a series of, mostly five and
six sided, polygons (Fowler, 1986; Bennett, 1989). The membrane skeleton is
linked to the membrane by interactions of the protein ankyrin with spectrin
and with the cytoplasmic face of band 3. Since band 3 proteins (1.2 million
per cell, accommodating as many as 300,000 tetramers) outnumber the 100,000
specrin tetramers much of the band 3 protein is not attached to the
cytoskeleton (Bennett, 1985; Carraway and Carothers Carraway, 1989). There
is also an association between spectrin and the membrane by linkage of band
4.1 to the actin-spectrin junction and to a membrane protein which may be
the minor glycoprotein glycophorin C (Bennett, 1989). A schematic model of
the structure of the erythrocyte membrane is shown in Fig. 1. There are
some differences of emphasis in current models of the erythrocyte membrane
e.g. Bennett (1989) links ankyrin to glycophorin C while recognising that
band 3 may also be a binding site, Carraway and Carothers Carraway (1989)
link ankyrin to glycophorin A which they show as having an association with
band 3. Clague et al. (1989) associate the ankyrin with 'glycophorin' and
show the glycophorin as being free of band 3 interaction.

There are 30,000 actin oligomers distributed over the cell surface area
of $134\mu m^2$ (Bennett, 1989) giving a separation of the order of 70nm between
actin attachment sites. Fowler (1986) pointed out that fully extended
spectrin molecules (180nm) are 3 times longer than necessary to span the
distance between the adjacent actin filaments in the unperturbed membrane.
The 100,000 spectrin tetramer binding sites, the number in excess of
300,000 band 3 tetramers and dimers and the 1,000,000 glycophorins will
have average separation distances less than the 70nm separation for the
30,000 actin oligomer binding sites above. These separation distances are
all significantly less than the separation of the order of a micron
detected between macromolecule-induced cell-cell contacts (Sec. 5 and Table
5).

2(ii) <u>Enzyme Modified cell surface</u>

The erythrocyte outer surface molecules can be modified by exposure to
proteinases (which fragment the proteins) or to neuraminidase (which
removes charge bearing sialic acid). Gokhale and Medha (1987) examined the
effect of a range of proteinases on band 3 and on the glycophorins. The
work was part of a study of the agglutination of erythrocytes by the lectin
concanavalin A (con A) which binds to the band 3 oligosaccharide. They
found that the degradation of band 3 by pronase, papain, trypsin,
neuraminidase and chymotrypsin did not correlate with their enhancement of
cell agglutinability. On the other hand, the cleavage of glycophorins,
especially glycophorin A, and the release of sialic acid (in the peptide-
bound form) was well correlated with agglutination enhancement by
proteinases. The release of free sialic acid from the glycophorins by
neuraminidase also correlated with agglutination. Because of the number of
times band 3 traverses the membrane (Sec. 2(i)) Gokhale and Mehta (1987)
argued that when pronase cleaves band 3 the fragments remain embedded in
the membrane causing no loss of the con A receptor. This conclusion was
supported by finding little difference in the number of con A binding sites
on normal and on pronase or trypsin treated erythrocytes. They concluded
that the con A agglutination requirement for enzymic pretreatment of cell

surfaces required modification of glycophorin A rather than removal of the lectin receptor.

There is a close correlation between electrophoretic mobility change and loss of sialic acid for a variety of cell glycocalyx modifying enzymes (Schnebli et al., 1976). The electrophoretic mobility of pronase treated cells is higher than that of neuraminidase treated cells (ranges 40%-60% and 5%-20% of control values respectively, Seaman, 1975; Schnebli et al., 1976). Since, despite the smaller loss of charge, pronase is more efficient than neuraminidase in facilitating erythrocyte agglutination by a variety of agents pronase may, in addition to reducing electrostatic repulsion, also act by reducing steric hindrance (Sec. 4) to close cell-cell approach (Schnebli et al., 1976). It has been shown, by phase partition, that neuraminidase treatment of erythrocytes leads to an increase in hydrophobicity (Gascoigne and Fisher, 1984; Shelton et al., 1985) while contact angle measurements have shown that pronase increases the hydrophobicity of erythrocytes to a greater extent than does neuramindase (van Oss et al., 1975). Increased hydrophobicity enhances non-specific adhesion of cells to surfaces (Rosenberg and Kjelleberg, 1987). The increased hydrophobicity of pronase pretreated cells may augment the effect of reduced steric and electrostatic repulsion in modifying cell-cell repulsion.

3. MACROMOLECULAR ADHESION OF ERYTHROCYTES; CELLULAR ASPECTS

Early studies of polymer and of lectin induced adhesion concentrated on establishing the effective concentration and polymer molecular weight range over which adhesion occurred. The influence of ionic strength and of enzyme modification of the cell surface was examined. The form of the agglutinate where the cells were stacked as a concave or convex rouleau or where cell doublets had a spherical form also received attention (Skalak et al., 1981; Tilley et al., 1987).

3(i) <u>Polycations</u>

The agglutination of human erythrocytes by polycations has been examined in detail. There is a critical (cell-concentration and polymer-size dependent) polymer concentration threshold for agglutination. Polycation agglutinated cells often (Katchalsky et al., 1959, Coakley et al., 1985) present a convex outer surface in contrast to the rouleaux form of serum agglutinated cells. Cells exposed to polylysine (which interacts with the cell surface sialic acids (Danon et al., 1965)) agglutinate when the cell electrophoretic mobility shows a detectable increase (less negative) from control values (Katchalsky et al., 1959; Hewison, 1988). Agglutination occurs readily under conditions where the electrophoretic mobility is zero or is at positive values greater in modulus than the negative electrophoretic mobility of control cells so that cross-linking of cells rather than neutralisation of surface charge by the polycation is the important interaction for agglutination. Noting that very little agglutination and only a small decrease in electrophoretic mobility occurs when cells are exposed to 1mg/ml of 4 kDa polylysine while marked aggregation and a significant change in electrophoretic mobility occurred on exposure to 20µg/ml of 14 kDa polylysine Coakley et al. (1985) concluded that protrusion of polylysine from the glycocalyx was a requirement for agglutination. Prolonged pretreatment of cells with neuraminidase almost eliminates agglutination by polylysine while milder enzymic pretreatment increases agglutination compared with control cells (Marikowsky et al., 1965). The small amount of polylysine-induced agglutination which follows erythrocyte pretreatment with pronase takes the form of overlapping pairs of cells or cells in edge-edge contact rather than the convex-ended stacks of agglutinated control cells (Hewison, 1988).

3(ii) <u>Dextran Agglutinated Cells</u>

Jan (1979) reviewed early work on the adhesion of erythrocytes by
dextran, a polymer of glucose. Cell-cell adhesion occurs in solutions of
dextran with a molecular weight of 40kDa or higher. Adhesion is first
observed at a dextran concentration of about 1%w/v, adhesion increases to a
maximum at 4%w/v and then declines so that a suspension remains
monodisperse in 10% w/v detran. Treatment of cells with neuraminidase
reduces the polymer molecular weight and polymer concentration thresholds
for adhesion to 20 kDa and 0.4%w/v respectively. Neuraminidase treatment
also allows the maximum aggutination level seen in normal cells at 4%w/v to
be maintained without sign of decline to 12%w/v dextran (Jan and Chien,
1973a). A reduction of ionic strength causes reduction of adhesion of
normal cells but does not influence neuraminidase treated cells (Jan and
Chien, 1973b). Darmani and Coakley (1990) showed that the seam of
agglutination for cells in 2%w/v 450 kDa dextran consisted of parallel
membranes while pronase pretreated cells showed spatially periodic contacts
(Sec. 5).

Chien et al. (1977) showed, that the amount of detran interacting with
the cell surfaces in the region of the contact seam was less than the sum
of the amounts of dextran interacting with membrane surfaces exposed to the
suspending bulk phase and interpreted the information as evidence for a
crosslinking interaction of individual molecules with two surfaces. Evans
and Needham (1988) provided good experimental evidence that the space
between interacting liposome membranes is devoid of dextran (i.e. dextran
is a non adsorbing polymer on lipid bilayers) and successfully applied
theoretical treatments of polymer depletion theory to the measured
interactions between liposomes. Donath et al. (1989) analysed the
electrophoretic mobility of erythrocytes in dextran and concluded that
there was some depletion of dextran close to the glycocalyx. van Oss and
Coakley (1988) argued that monopolar repulsion between detran molecules
free in suspension and dextran molecules attached to the glycocalyx would
result in a pressure which would push cells together.

3(iii) <u>Lectin agglutination of erythyrocytes</u>

Lectins (carbohydrate-binding proteins of non-immune origin that
agglutinate cells or precipitate polysaccharides or glycoconjugates (Dixon,
1981)) have a small number of binding sites which interact with specific
saccharide receptors on the cell surface. The threshold concentration for
lectin agglutination of erythrocytes is low, being of the order of 1µg/ml
for wheat germ agglutinin (WGA) at a cell concentration of 3×10^7/ml
(Darmani et al., 1990). Glycophorin A is the major WGA receptor. Binding is
eliminated when the sialic acid groups are removed by neuraminidase.
Erythrocyte agglutination by concanavalin A (which binds to band 3) is
enhanced when cells are pretreated with neuraminidase or with pronase. The
enhancement is attributed to reduction of agglutination inhibition by
glycophorin A rather than to a direct effect on band 3 (Gokhale and Mehta,
1987), Sec. 2(ii) above). The binding of WGA and of concanavalin A to their
receptors in intact cells (and when the receptors are reconstituted in
model membranes) has a half life of many minutes (Grant and Peters, 1984).
These authors concluded that the slow binding results from short-distance
receptor rearrangements rather than from long-distance translocation of
receptors. Darmani et al. (1990) monitored the time dependence of the shape
of WGA induced erythrocyte agglutinates. They found that the edge-edge cell
contact was dominant in cells sampled after 2min incubatation. Cell
sampling at intervals during the following 60min showed that contact area
was increasing slowly with time. The spreading of contact area was
accompanied by formation of spatially discrete contacts (Sec. 5).

Table 1. Threshold concentrations, c, for the agglutination of
         erythrocytes by polylysine (Coakley et al., 1985), dextran (Jan
         and Chien, 1973a) concanavalin A (Darmani and Coakley, 1991),
         WGA (Darmani et al., 1990).

Polymer	M (kDa)	Cells/ml	c (µg/ml)
Polylysine	4	7×10^6	1,000
"	14	"	8.0
"	60	"	0.25 - 0.05
"	210	"	0.25 - 0.05
Dextran	40	10^8	30,000
"	80	"	5,000
Con A	101	1.5×10^7	<110
WGA	35	"	1.0

Table 1 shows that the threshold concentrations for agglutination by the
polycations and the lectins are a number of orders of magnitude lower than
those for the uncharged nonspecific polymer dextran.

3(iv) Video microscopy of cell interactions in suspension

Tilley et al. (1987) described a technique which enabled cells to be
observed in suspension for times of the order of minutes. A suspension of
cells in dextran or polylysine solution was observed in a rectangular
cross-section (300µm pathlength) glass microslide. A microscope was turned
through 90° on its back so that the plane of the microscope stage was
vertical. The microslide was placed on the vertical stage. A 1MHz
ultrasonic standing wave field was set up along the microslide to hold the
cells in suspension by radiation force. Concave-ended cell doublets and
linear rouleaux developed in 0.5 - 1.5% w/v 450 kDa dextran by a gradual
(2.5-17s) increase in the area of cell contact. In 2-7% dextran and in
20µg/ml of 14 kDa polylysine mutual adhesion was a two stage process
(Tilley et al., 1987; Darmani and Coakley, 1990). Cells first formed a
strong local contact which persisted (without apparently growing in area)
for a number of seconds following which the cell surfaces moved suddenly to

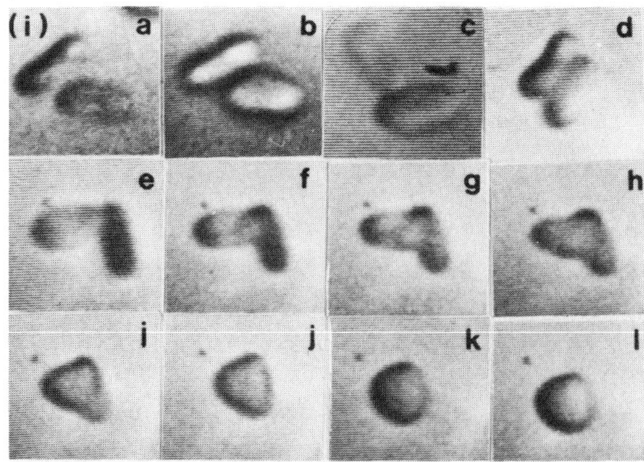

Fig. 2. Cells in 5%w/v 450 kDa dextran. (a) Time zero, cells in suspension
0.04s before initial contact (b,c,d) stable doublet at 4, 16 and 32s
respectively; (e) 40s, the frame before the final adhesion stage begins;
(f) 40.12s; (g) 4.2s; (h) 40.28s; (i) 40.4s; (j) 40.52s; (k) 41.02s; (l)
41.52s; the final spherical doublet is formed (Tilley et al., 1987).

form a spherical doublet. The average initial contact and engulfment time
for cells in 7% dextran were 18s and 2.7s respectively. The corresponding
values for cells in 20µg/ml polylysine were 2.7 and 0.3s (Tilley et al.,
1987). Pronase pretreated cells exposed to 2%w/v 450 kDa dextran showed a
shorter (18s) initial contact time than did non-pretreated control cells
(35s; Darmani and Coakley, 1990). Unlike the situation for cells in
dextran or polylysine it was not possible to detect any sudden (of the
order of 1s) extension of cell contact area when cells were suspended in
WGA (Darmani et al., 1990). The maintenance of a stable interaction for
many seconds before beginning a rapid mutual-engulfment movement in dextran
or polylysine implies that the cells were in a secondary interaction energy
minimum where a repulsive potential barrier precluded movement to a deeper
primary minimum.

The measured delay times from which the above averages were calculated
varied significantly for a given macromolecular solution suggesting that
overcoming the potential barrier is a stochastic process. The delay may be
required for change in the macromolecular concentration or cross-bridging
to occur in the cell-cell gap or may be a requirement for delay until a
thermal fluctuation of sufficient magnitude occurred to allow the membranes
to overcome the repulsive barrier in the manner suggested by Leiken et al.,
(1987) for movement against a repulsive barrier at bilayer surfaces.

4. MEMBRANE INTERACTIONS

4(i) Equilibrium Processes

Physical chemical interpretations of erythrocyte adhesion experiments
were often guided by the classical Derjaguin, Landau, Verwey and Overbeek
(DLVO) theory of colloid stability (Verwey and Overbeek, 1949) in which the
equilibrium normal separation of two parallel surfaces approaching each
other would coincide either with a secondary minimum in the interaction
energy profiles (where the long range van de Walls attractive force was
balanced by the electrostatic repulsion between two like-charged surfaces)
or in a primary interaction energy minimum at smaller separation.

The electrostatic normal pressure between two parallel surfaces in an
electrolyte is given [at separation h significantly larger than L_e], to a
good approximation (Hiemenz, 1986), by

$$P_e = P_{e0}\exp(-h/L_e)\ldots\ldots\qquad (1)$$

where Pe_0 is a function of the ionic strength of the medium and of the
surface potential. The characteristic length, L_e, is proportional to the
square root of the ionic strength and is 0.8nm for the 145mM buffered
saline solution in which control erythrocytes are normally suspended (Table
2). The classical colloid science expressions (Hiemenz, 1986) have been
modified for biological membrane systems to account, in the case of the van
der Waals attraction, for the small thickness of a membrane bilayer (Nir,
1976) and, in the case of electrostatic repulsion, for the distribution of
cell surface charge within a glycocolyx rather than on a plane surface
(Donath and Voigt, 1983).

In an experimental investigation of bilayer-bilayer interactions LeNeveu
et al. (1976) discovered a very strong short-range hydration repulsion
force which also decayed exponentially with bilayer separation distance.

The importance of the cell glycocalyx in cell-cell interactions was
emphasised by Bongrand and Bell (1984) and by Bell et al. (1984). These
authors considered that the repulsive barrier between cells arises mostly
from a combination of electrostatic repulsion and of a glycocalyx steric

stabilisation effect which arises as the macromolecular layers of
approaching cells overlap and water of hydration is squeezed out of the
intercellular gap. The repulsive force results from a combination of the
osmotic tendency of solvent to return into the intercellular layer and from
the steric compression of the macromolecules. The repulsive potential was
given as

$$E(h) = -\gamma/h. \ \exp(-h/L_G) \ldots \quad (2)$$

where γ is a compressibility coefficient of the glycocalyx and L_G is a
measure of the combined thickness of the glycocalyx layers (Table 2).

Macromolecular cross-linking of cells was considered to play an
important role in dextran induced cell adhesion (Jan, 1979). Gallez and
Coakley (1986) described the attractive macromolecular crosslinking force
of 14 kDa polylysine in the form

$$P_m = P_{mO}\exp(-h/L_m) \ldots \quad (3)$$

where P_{mO} was taken to be 6.10^6 N/m^2 and the decay length L_m was taken as
4nm.

Some insights into the modification of cell interactions by polymers
might be expected from the results of force balance (Israelachvili and
McGuiggan, 1988; Patel and Tirrell, 1989) measurements of the influence of
polymers on the interaction between mica surfaces. Fig 3(b) shows how
dextran (0.03%w/v, 2.0MDa), a polymer which does not adsorb to mica
surfaces, introduces a repulsive pressure at larger separations and a
polymer-depletion attractive force at small separations (Perez and Proust,
1985). Since dextran does not adsorb to mica no attractive macromolecular-
crosslinking interaction would be expected between the surfaces. Force
balance measurements (Fig 3(a)) of the effect of 90kDa polylysine in 0.1M
KNO$_3$ show a long range repulsion (h = 120nm) which increases monotonically
with decreasing h (Luckham and Klein, 1984). There is no sign here of
attractive macromolecular cross-linking of surfaces. Force balance studies
with adsorbed electrically neutral polyethyleneoxide also show a growing
repulsion (essentially a steric repulsion effect) as h is decreased
(Luckham and Ansarifar, 1990). An attractive component was seen only when
efforts were made to reduce polyethyleneoxide surface coverage (thus
offering an opportunity for a free end of a polymer adsorbed to one surface
to engage with the second surface). There was qualitative agreement
between the shape of the experimental attractive interaction profile and a
theoretical treatment of crosslinking in under-covered surfaces (Luckham
and Ansarifar, 1990). In summary, force balance studies do not readily
show the long range attractive force necessary to stabilise initial cell-
cell contact in polylysine or the attractive force which achieves the
collapse to mutual cell engulfment (Sec 3(iv)). Special steps were
necessary to demonstrate an attractive cross-linking force with adsorbing
polyethyleneoxide. Fig 3(b) shows a repulsive potential for detran which
could provide the barrier necessary to prevent immediate progression to
cell-cell engulfment.

Table 2 shows how cell-cell interactions may attenuate as h increases.
The electrostatic repulsion and glycocalyx repulsion effects are of similar
magnitude at h = 10nm but the contribution calculated for stereorepulsion
is totally dominant at h = 30nm because of the large value of L taken (Bell
et al., 1984) for the latter interaction. This apparent dominance of
glycocalyx stereorepulsion at larger values of h needs to be reconciled
with the experimental observation that a reduction of ionic strength from
0.15 of 0.05 inhibits agglutination of erythrocytes by 80kDa dextran (Jan
and Chien, 1973b). The polycation repulsion and macromolecular

Table 2. Estimates of the characteristic length L and the normal pressure P (at a range of separations h (in nm)) for a number of different interactions. The thickness of the glycocalyx is taken as 5nm so that h = 10 nm represents the plane of glycocalyx contact.

INTERACTION:	L (nm)	$P_{h=10}$ (N/m^2)	$P_{h=30}$ (N/m^2)	$P_{h=100}$ (N/m^2)
Electrostatic Repulsion[a]	0.8	1.5×10^5	2×10^{-6}	–
Glycocalyx Steric Rep.[b]	10	2×10^5	2×10^4	13
Polycation Repulsion[c]	2–30	2×10^6	6×10^5	
Macromolec crosslinking[d]	25	4×10^4	2×10^4	10^3

[a](Eqn.1, Hiemenz, 1986) Ionic strength = 0.15; potential at glycocalyx edge taken as 8mV (Donath and Voigt, 1983).
[b]L, is twice the glycocalyx thickness, (P_h is calculated from Eqn.2 (Bell et al., 1984).
[c]Repulsion of polylysine (90 kDa) coated mica surfaces; L is large for the initial compression and small for subsequent compression/decompression cycles. (Luckham and Klein, 1984).
[d]Polyethylene oxide (1.1 MDa) on mica for limited surface coating; L estimated from measurements of Klein and Luckham (1984).

crosslinking data of Table 2 are based on the force balance experiments discussed above. Some relevant properties of the macromolecules of interest here are given in Table 3.

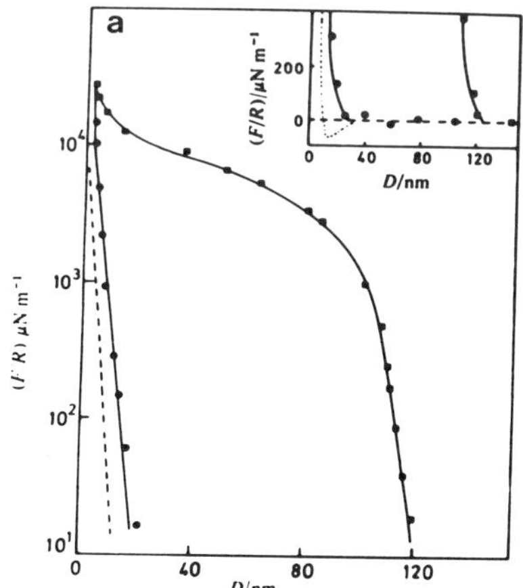
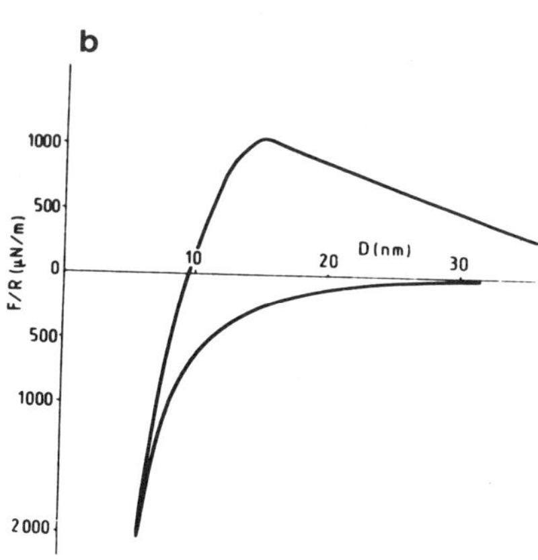

Fig. 3. Force/Radius (= (Interaction energy between plane surfaces)/2π) against separation for mica surfaces in: (a) 10μg/ml solution of polylysine in 0.1M KNO$_3$ (pH 3.5), on first approach (■) and first withdrawal (●). The broken line is interaction in the absence of polymer. The inset shows the profiles at small values of F (Luckham and Klein, 1984): (b) Differences between Force/Radius curves for 0.1M NaCl and for salt containing 0.3 gm/1 (upper curve) an 0.1 gm/1 (lower curve) of 2MDa dextran (Perez and Proust, 1984).

Table 3. Scale of influence of macromolecules of interest.

Macromolecule:	Mw (kDa)	: Estimates of characteristic scales of influence		
Dextran[a]	: 20	: Radius of gyration	=	4nm
"	: 450	: Radius of gyration	=	16nm
Polylysine[b]	: 90	: Range of repulsion on mica surface	=	60nm
WGA[c]	: 36	: Distance between active sites	=	5nm
con A[d]	: 102	: Distance between active sites	=	5nm

[a]Snabre et al. (1985). [b]Luckham and Klein (1984). [c]Wright (1977).
[d]Edelman et al. (1972).

5. GEOMETRY OF SEAM OF ERYTHROCYTE AGGLUTINATION

The seam of contact in polymer induced adhesion of erythrocytes has been
examined in recent studies in this laboratory. Essentially two types of
seam geometry have been identified. In one type the membranes of adhered
cells are parallel with each other along the seam of contact. In the second
case contact occurs at spatially periodic local points which are separated
by regions where the membranes are not in close contact. The lateral
separation distance of contact regions along the membrane depends on the
contact-inducing molecule and, for a given molecule, on the properties
(control or enzyme modified) of the cell surface.

The light micrographs of dextran-induced and of polylysine-induced
clumps of erythrocytes (Fig. 4) show spatially periodic contacts. The cells
which were agglutinated by dextran had been pretreated with pronase to
modify the properties of the cell surface. Erythrocyte clumps formed in
suspension by exposure to polylysine and then placed on a glass microscope
slide were often drawn down onto the glass by polylysine and distorted so
the periodic contacts were sometimes resolved only with difficulty (e.g.
contrast the unfixed cells of Fig. 4 row(ii) and Fig. 5c). Cell clumps
which were fixed in suspension by glutaraldehyde were not distorted when
placed on the glass (Fig. 5a,b). Fig.5a shows a regular pattern of contacts

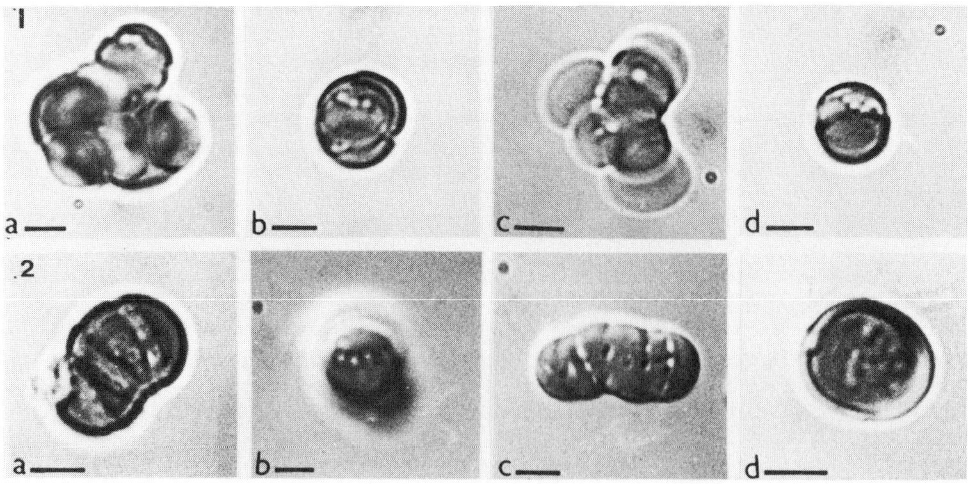

Fig. 4. Light micrographs of: (Row i) cells pretreated with pronase and
then exposed to 2%w/v 450 kDa dextran. (Row ii) Normal erythrocytes exposed
to 5µg/ml 275kDa polylysine (Darmani and Coakley, 1990).

at the intercellular seam. The example in Fig. 5b shows that the pattern
occurred only where the cell overlapped. Periodic contact spacing was seen
in freeze-fracture electron micrographs of cryo-fixed cells which had not
been exposed to chemical fixative (Fig.5e). Transmission electron
microscopy shows regular spacing of contact regions between cells in a
polylysine induced clump (Fig. 5d). The wavy profile is confined to areas
where cell surfaces oppose each other i.e. the erythrocyte membrane on the
outside of the clump shows no undulation. When the cell clumps were
sheared to expose contact surfaces scanning electron micrographs confirm
that the pattern is confined to discrete regions of the cell surface (Fig.
5f,g). The average separation between contact points, measured for cell
agglutinated in 20µg/ml 14kDa polylysine was 0.83µm (Coakley et al., 1985).
The spacing did not appear to change detectably for agglutinates produced
by higher molecular weight polycations. Recent more complete measurements
show that longer separations are consistently found for erythrocyte
agglutinates formed very close to the threshold concentration for
agglutination by 20kDa polylysine (N.E. Thomas, personal communication).

The seam of contact of polycation agglutinated erythrocytes was first
examined by Katchalsky et al. (1959). The electron micrographs of these
authors showed interruption of contact but they but made no comment on
spatial periodicity between contact regions. Hewison (1988) observed
widespread spatial periodicity of contact regions when cells were fixed
solely in low concentrations (0.05-0.005%) of osmium tetroxide (the sole
fixative used for the electron micrographs of Katchalsky et al., 1959).
Hewison (1988) also showed that periodicity of cell contacts occurred but
was not widespread in cells fixed with 0.5% osmium tetroxide (the higher
fixative concentration of Katchalsky et al. (1959)). Since unfixed cells
such as those of Fig. 4 show periodic contacts Hewison (1988) concluded
that the lack of comment on spatial periodicity by Katchalsky et al. (1959)
was due to the higher concentration of fixative used in that work.

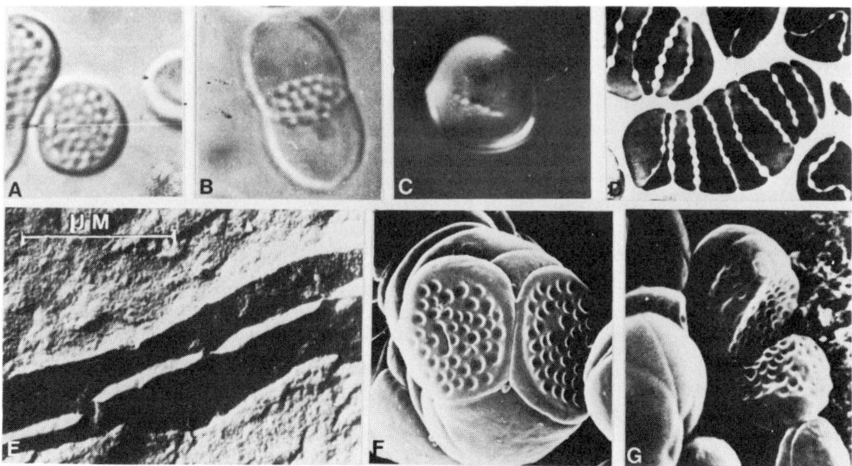

Fig. 5. (a) Light micrograph of erythrocytes glutaraldehyde-fixed following
exposure to 20µg/ml polylysine showing contact patterns (b) two cells (as
in (a)) with a small region of cell overlap, (c) Light micrograph of an
unfixed cell clump showing contact details between cells washed once after
exposure to 10µg/ml polylysine; (d) Transmission electron micrographs of
cells treated as in (a); (e) Freeze fracture electron micrograph of cryo-
fixed cells showing three contacts separated by distances of less than 1µm;
(f,g) Scanning electron micrographs of cell contact surfaces exposed by
shearing glutaraldehyde-fixed cell clumps ((a-d and f; Coakley et al.,
1985); (e; Hewison et al., 1988); (g; Hewison, 1988).

Hewison et al. (1988) examined the polylysine-induced seam of contact when surfaces with different mechanical properties e.g. an erythrocyte and a yeast cell surfaces were agglutinated. Electron micrographs revealed a spatially periodic undulatory contact pattern on the erythrocyte surface.

In contrast to the situation for polycations the contact seam for erythrocyte agglutinates formed on exposing normal human erythrocytes to dextran in solution is usually plane parallel. When the cell surface is treated with pronase before exposure to dextran spatial periodicity of contact regions is observed (Darmani and Coakley, 1990). Fig. 6 shows examples of agglutinates formed after different durations of cell pre-exposure to pronase. The characteristic distance between contact regions decreases for increasing duration of exposure to pronase (Figs. 6 and 7). The results shown in Figs. 6 and 7 represented the first clear indication of the dependence of the contact region separation distance (the wavelength) on experimental conditions. The influence of dextran molecular weight, dextran concentration and duration of pronase pretreatment of cells was examined. The results are shown in Table 4. No adhesion occurred, even for pronase pretreated cells, for a dextran molecular weight of 10,000. The Table shows that spatially periodic contacts are a feature of almost all cases where pronase treated cells were exposed to dextran. The experimental condition of most interest in terms of providing insight to mechanism are those where wavelength varies markedly with change in experimental variable. For the situations examined in Table 4 the dependence of

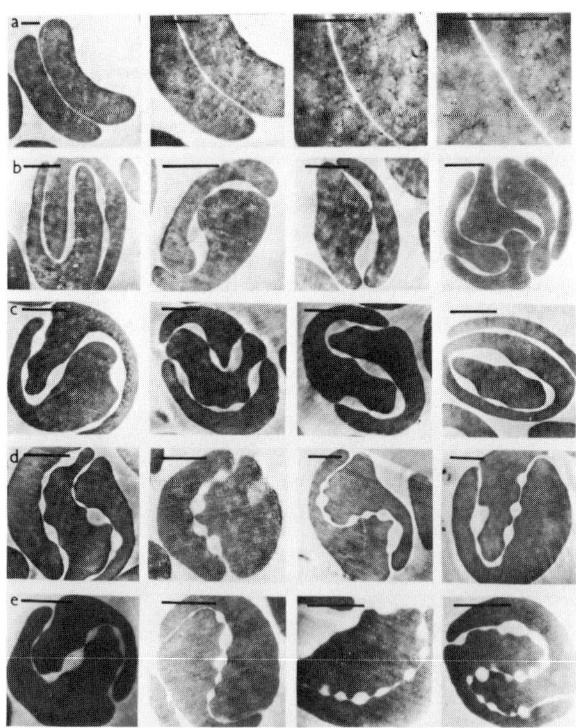

Fig. 6. Transmission electron micrographs of cells exposed to 2% w/v of 450 kDa dextran. Row (a) control cells showing, at different magnifications, the continuous parallel-membrane seam between cells in a clump. Bar represents 1µm. Rows (b-e) show examples of erythrocytes pretreated with 0.5 mg/ml pronase for 0.5-1.0 min (27% surface charge removal), 5 min (53% charge removal) and 2h (56% charge removal) respectively. Bar represents 2µm (Darmani and Coakley, 1990).

Table 4. The dependence of contact separation on dextran molecular
weight and concentration for cells pretreated with 0.5mg/ml
pronase for different times. The notation L, M and S refers to
wavelengths of the order of 2 - 3.0µm, 1 - 2µm and less than
1.0µm respectively. P refers to parallel membranes without
thickness variations in the membrane-membrane separation, MO
refers to monodisperse suspensions where no adhesion occurred.

Dextran Molecular Weight (kDa)	Pronase Pretreatment Time (min)	Dextran concentration (% weight/vol)			
		2	4	6	8
450,000	0	P	L,M	L,M	P
	5	L	S	S	S
	15	L,M	S	S	S
	30	M	S	S	S
	60	S	S	S	S
71,500	0	P	P	L	P
	5	M	S	S	S
	15	S	S	S	S
	30	S	S	S	S
	60	S	S	S	S
20,000	0	MO	MO	MO	MO
	5	MO	MO	MO	S
	15	MO	P	S	S
	30	MO	S	S	S
	60	MO	S	S	S

wavelength on experimental conditions was most marked for cells pretreated
to different extents with pronase and then exposed to 450kDa dextran.
Electrophoretic mobility measurements were used to provide an index (in
that the measurements presumably reflect the extent to which the
glycophorins are degraded) of the influence of pronase pretreatment on the
cell surface. It is unlikely that charge removal and the accompanying

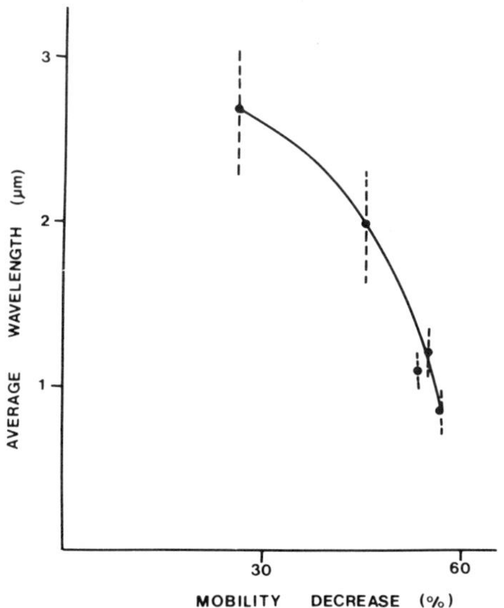

Fig. 7. Average spatial periodicity (with 95% conf. limits) against
decrease in electrophoretic mobility (Darmani and Coakley, 1990).

reduction of intercellular repulsion is the dominant change accompanying
pronase pretreatment as preliminary results show that pretreatment of cells
with neuraminidase (removing up to 80% of the surface charge) does not
result in as marked a wavelength change as does the pronase pretreatments
which remove 40% of surface charge.

Darmani et al. (1990) recently investigated the seam of contact formed
during agglutination of human erythrocytes by the lectin wheat germ
agglutinin (WGA). They found a slow spreading of contact with time (Sec. 3
iii)) and that the contact spreading proceeded by the formation of periodic
contacts separated by distances of the order of 0.8µm (Fig. 8a). The rate
of contact spreading could be increased 30-fold by continuously agitating
the cells during incubation. It was argued (Darmani et al., 1990) that the
intercellular water layer at the contact edge became narrow during
agitation and the narrow layer became unstable (Sec. 6) to give periodic
contacts. Neither WGA receptor labelling or freeze fracture electron
microscopy provided any evidence of precontact lectin-induced spatially
periodic concentration of receptors in the plane of the membrane (Darmani
et al., 1990).

Internalization of plasma membrane vesicles of diameter of the same
order as the inter-contact spacing occurred with prolonged incubation of
erythrocytes in WGA. This internalization may have the same origin as the
internalisation process exploited by coating rat embryo surfaces with WGA
in order to effect vesicular transport of an enzyme to the Golgi apparatus
(Tan and Morris-Kay, 1986). The fact that membrane internalisation by
vesiculation occurs supports the conclusion that discrete rather than
continuous cell contact is achieved by WGA treatment. It has also been
reported that WGA causes increased apposition of two-cell mouse
blastomeres. This is followed by formation of numerous small intercellular
spaces (Johnson, 1986) necessarily implying local contact formation.

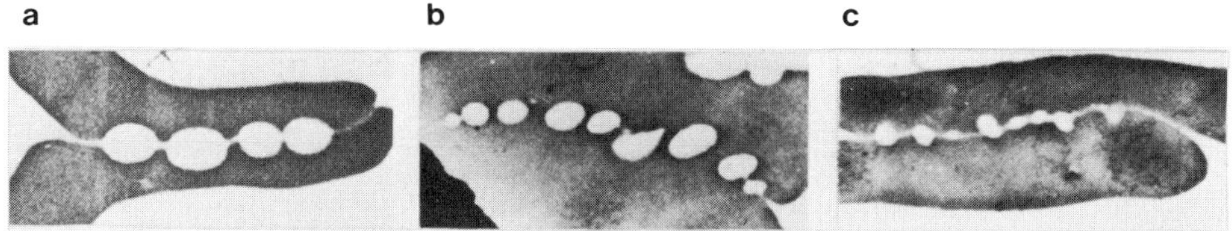

Fig. 8 Examples of spatially periodic contact in cells exposed to (a)
20µg/ml WGA (Darmani et al., 1990) (b) 110µg/ml con A following pronase
pretreatment of cells (c) 110µg/ml con A following neuraminidase
pretreatment (Darmani and Coakley, 1991). The bar is 2µm.

WGA forms contacts by interaction with the sialic acid residues of the
erythrocyte glycocalyx. In contrast neuraminidase or proteolytic enzyme
pretreatment of erythrocytes is required before concanavalin A induced cell
agglutination occurs (Gokhale and Mehta, 1987). While the principal con A
receptor is associated with band 3 protein protease-enhanced adhesion is
associated with the removal of glycophorin (Gokhale and Mehta, 1987; Sec.
2(ii)). The spreading of con A cell-cell contact takes place on the same
timescale as that of WGA. The contact is formed at discrete regions
separated for pronase or neuraminidase pretreated cells by a distance of
0.66 or 0.50µm respectively (Fig. 8b,c)

Sung (1984) noted that erythrocyte agglutinates induced by the lectin
Dolichos biflorus contained smooth parallel membranes with characteristic
intercellular spaces and that in addition 'ruffled intercellular spaces
were observed'. Sung's 'ruffled intercellular spaces' show spatially
periodic regions of close membrane apposition similar to those seen in this
laboratory for WGA and con A treated cells.

6. SPATIAL PERIODICITY OF ERYTHROCYTE CONTACTS AND INTERFACIAL INSTABILITY

While the understanding and measurement of the static forces involved in
membrane interactions continues to develop (Sec. 4) a number of
theoretical (Dimitrov, 1982; Prevost et al. 1983; Gallez et al., 1984;
Dimitrov and Jain, 1984, Gallez and Coakley, 1986) and experimental
(Coakley et al. 1985; Coakley and Gallez, 1989; Darmani and Coakley, 1990)
studies have been directed to understanding the influence of dynamic
processes, due to net attractive intermembrane forces, on the final form of
the membrane-membrane contact. Many studies in the area of drop and of
bubble approach and coalescence have shown that the behaviour of the
intervening liquid films strongly influence the outcome of the
interactions. Specifically, a film thinning under the influence of a net
attractive pressure may reach a 'transition thickness' at which the
thickness fluctuations of the thin film grow spontaneously leading either
to rupture of the thin film at spatially periodic points or to formation of
a thin film with smaller thickness (as in black film formation). Growth of
such fluctuations on the narrowing water layer between approaching
membranes might result in either (i) a small essentially uniform separation
which may be described by modern extensions (Sec. 4) of the classical DLVO
theory of colliod stability or (ii) rupture of the aqueous layer to give
spatially periodic contacts. By substituting published estimates of
membrane constants into thin film instability theory Dimitrov (1982) showed
that spontaneous growth of thickness fluctuations might reasonably be
expected on the thin aqueous layer between membranes which are being drawn
towards each other.

The condition for the spontaneous rupture of a non-thinning intermembrane aqueous layer is not very different from that of a thinning layer
while analysis of the former process provides physical insight to the
rupture process. The instability criterion for a non-thinning fluid between
two continuous phases is given (Gallez and Coakley, 1986) by

$$k^2\sigma_T + k^4 B - 2dP_t/dh < 0 \ \ldots\ldots\ldots \ (4)$$

where $k = 2\pi/\lambda$ (λ is the wavelength of the disturbance which will grow to
rupture the thin film); the total surface tension $\sigma_T = 2\sigma_S + \sigma_R + \sigma_A$
where σ_S is the pure surface tension of the film and σ_R and σ_A are
contributions to the surface tension from sources of repulsive and
attractive interactions respectively (Gallez and Coakley, 1986 and Sec. 4).
B is the bending elasticity modulus; dP_t/dh is the derivative with respect
to the film thickness 'h' of the net disjoining pressure (i.e. the sum of
the net attractive and repulsive normal interactions). Eqn (4) applies to
aqueous thin films containing surfactant (Prevost and Gallez, 1984).
Dimitrov (1982) has pointed out that films with surfactant share properties
of films bound by biological membranes in that (i) the surfactant interface
is essentially immobile tangentially and (ii) the tension in surfactant and
membrane boundaries depends on area change, unlike the case for a pure
liquid. It is therefore not unreasonable to use Eqn. 4 as a guide to the
stability of a membrane bound thin film.

A number of the interaction pressures can be expressed (Sec. 4) as

$$P(h) = P.\exp(-h/L) \ \ldots\ldots\ldots \ (5)$$

where L is a characteristic length and P is a constant. Gallez and Coakley (1986) have already examined the particular problem of accounting for a 0.8µm wavelength in agglutination of cells in a 20µg/ml solution of 14 kDa polylysine. They used glycocalyx compressibility estimates from Bell et al. (1984) to calculate a stereo-repulsion normal interaction between cells and took the characteristic length (L_G in Eqn. 2) to be 5nm. For values of h in the range 4 < h <130nm the stereo-repulsion contribution to dP_t/dh dominated the van der Waals attraction and would, in the absence of polycations, have stabilised a thin water layer between the cells. Instability was assumed to occur, in the presence of polycation, at h = 20nm. The polycation crosslinking normal pressure was taken to be equal to the stereo repulsion pressure at h = 20nm and the value of (L_M Eqn. 3) for crosslinking by 14kDa polylysine was taken to be 4nm. Substituting these values into the equation

$$w = (k^2 h^3/24\mu).(2dP_t/dh - k^2\sigma_T - k^4 B) \ldots\ldots\ldots (6)$$

gave a dominant disturbance growth rate w, of $3,000s^{-1}$ or $30s^{-1}$ when µ, the viscosity of the intercellular layer including the contribution from the glycocalyx, was taken as 1 cPoise or 1 Poise respectively. Corresponding growth times of 1 ms or 100 ms are consistent with the observed doublet formation time of 300ms (Tilley et al., 1987; Sec.3(iv)). The calculated dominant wavelength of 0.63µm (Gallez and Coakley, 1986) was close to the experimentally observed value of 0.83µm (Coakley et al., 1985).

7. DISCUSSION AND SUMMARY

    Attention has been drawn above to two experimental approaches which reveal aspects of erythrocyte agglutination by polymers which need to be taken into consideration in developing a full explanation of the biophysics of such interactions. The first experimental approach (microscopic observations of cells during agglutination in suspension) reveals that, at appropriate concentrations, dextran and polylysine initially make contact and hold such contact for measurable times before plunging to mutual engulfment (Sec. 3(iv)). Such behaviour does not occur during agglutination by lectins. The results suggest that for dextran and polylysine a long range attractive force holds the cells close together, that there is a potential barrier which needs to be penetrated and then, for both polymers, a primary potential well which completes the agglutination process.

    The factors influencing the final shape (spherical doublet or slightly convex rouleau) of the agglutinate in polymer (dextran or polylysine) induced agglutination are not linked in a one-to-one way with the average lateral separation of contacts e.g. spherical cell agglutinates in 5%w/v 450 kDa dextran and rouleau-like agglutinates in low concentrations of polylysine can have long contact separation distances while a short wavelength is seen in spherical agglutinates at intermediate concentrations of polylysine and in rouleau agglutinates of pronase pretreated cells in 6%w/v 20kDa dextran. Therefore agglutinate shape and contact separation distance reflect different properties of the cell-cell interaction profile.

    The seam of contact of agglutinated cells was examined in the second experimental approach. An understanding of the formation of local discrete contacts during membrane-membrane interactions is likely to be relevant to a number of _in vivo_ situations including membrane fusion as in the production of sheets of hybrid vesicles from sperm heads during the acrosome reaction (Russell et al., 1979) and the general area of cell locomotion and contact formation (Vasiliev, 1987; Andre and Bongrand, 1990; Gingell, 1990). Table 5 summarises the information on erythrocyte seams obtained here to date. It shows, for the polymers examined, that parallel membrane contact was the exception rather than the rule. The scale of the

40

Table 5. Summary of average lateral spacing of contacts (in microns) for cells exposed to different polymers and for cells pretreated with glycocalyx modifying enzymes (N.E.D. = not examined in detail).

Polymer	:	Polycation	:	Dextran	:	WGA	:	con A	:
Molecular Weight (kDa)	:	14	:	20 - 450	:	36	:	102	:
Normal cells	:	0.83	:	Parallel membranes: & long wavelength		0.79	:	N.E.D.	:
Pronase pretreated	:	N.E.D.	:	0.5 - 3.0	:	N.E.D.	:	0.66	:
Neuraminidase pretreated.	:	N.E.D.	:	$\approx$1.0 - 3.0	:	N.E.D.	:	0.50	:

lateral separation of close contact regions (Table 5) is much larger than typical repeat distances of erythrocyte membrane structures (Sec. 2(i)). The cases where Table 5 contains single values for the average spacing reflect the fact that a detailed search has not yet been completed for a dependence of the lateral spacing on macromolecular concentration, macromolecular size (where appropriate) or cell surface pretreatment. For example, the effects of mild enzyme pretreatments have not been examined for lectin or polylysine induced agglutination. It is, however, now known (N. E. Thomas, personal communication) that the lateral separation distance changes for polycation-agglutinated cells near the polymer concentration threshold for agglutination for the polymers studied. The average separation between contacts ranged from 3.0µm to 0.5µm (Table 5) suggesting that average lateral separation reflects differences in the membrane interaction profiles for the particular situations examined.

The three important components of the system involved in erythrocyte agglutination are; (i) the cell surface which may be regarded as a source of steric and electrostatic repulsions that can be modified (along with hydrophobic interactions) by enzymic pretreatment (Sec. 2(i), Sec. 4), (ii) the suspending phase, whose ionic strength can modify electrostatic interactions and (iii) the macromolecule, whose size, concentration and charge will influence lateral contact separation through modifying membrane-membrane interactions in the case of polycations and polysaccharides (it remains to be shown whether lectins significantly modify membrane-membrane interactions or operate through crosslinking spontaneous undulations on the thin intercellular water layer).

Analysis of the stability of a thin aqueous film requires knowledge of the attractive and repulsive normal interactions and their dependence on membrane separation (Sec. 6). High positive gradients in the attractive normal interaction profiles are destabilising. An interaction energy profile which reconciles the real-time microscopic observations of dextran and polylysine induced agglutination (Sec. 3(iv)) with a thin film instability dependent spatial periodicity of contacts would show a secondary attractive minimum as h decreases, a repulsive maximum and a high positive curvature as it moved towards an attractive minimum with decreasing h. Mica-surface force-balance studies of polymer-induced interactions provide an example of a repulsive barrier and a high curvature approach to an interaction potential well for dextran (Fig. 3(b)) for the case where dextran is a non-adsorbing molecule on mica. Force balance measurements also show (with care to allow crosslinking) an interaction minimum for adsorbed polyethylene oxide. In contrast to the well

established experimental observation that polycations agglutinate
erythrocytes and can act generally (in low concentrations) as flocculants
no attractive interaction is seen between polycation layers on two mica
surfaces (Fig. 3(a)). Studies with poly (2-vinylpyridine) also show no
evidence of attractive interaction until coated surfaces have been in
intimate contact with each other and are then pulled apart (Marra and Hair,
1988). This contrast with cell behaviour suggests that results with force
balances may not provide immediate explanations for all aspects of polymer
induced cell agglutination.

A comprehensive test of an interfacial instability approach to the
separation of discrete contact regions would, for example, require a model
of the effect of enzyme pretreatment which was consistent with their
influence on contact separation for both dextran (radius of gyration of
16nm for 450kDa polymers and for lectins (globular proteins with 5nm
separation of binding sites (Table 3)). Consolidation of the data
summarised in Table 5 concentrating, as has been the case for dextran
treated cells (Figs. 6 and 7; Table 4), on situations where changes in
experimental variables change the lateral contact separation, together with
improved characterisation of the erythrocyte surface properties such as
hydrophobicity will further advance the biophysical understanding of
erythrocyte adhesion by macromolecules.

8. ACKNOWLEDGEMENTS

HD held a SERC research studentship, AJB is supported by a PHLS
studentship.

9. REFERENCES

Andre, P. and Bongrand, P., 1990, Cell-cell contacts, in: "Biophysics of
    the Cell Surface", R. Glaser and D. Gingell, eds., Springer-Verlag,
    Berlin.
Ballas, S. K., Reilly, P. A. and Murphy, D. L., 1986, The blood group U
    antigen is not located on glycolipid B, Biochim. Biophys. Acta, 884:337.
Bell, G. I., Dembo, M. and Bongrand, P., 1984, Cell Adhesion: Competition
    between nonspecific repulsion and specific bonding, Biophys. J.,
    45:1051.
Bennett, V., 1985, The membrane skeleton of human erythrocytes and its
    implications for more complex cells, Ann. Rev. Biochem., 54:273.
Bennett, V., 1989, The spectrin-actin junction of erythrocyte membrane
    skeletons, Biochim. Biophys. Acta, 988:107.
Bongrand, P. and Bell, G. I., 1984, Cell-cell adhesion: parameters and
    possible mechanisms, in: "Cell surface dynamics: Concepts and models",
    C. DeLisi, A. Perelson and F. Wiegel, eds., Marcel Dekker, New York.
Bretscher, M. S. and Raff, M. C., 1975, Mammalian plasma membranes, Nature,
    258:43.
Carraway, K. L. and Carothers Carraway, C. A., 1989, Membrane-cytoskeleton
    interactions in animal cells, Biochim. Biophys. Acta, 988:147.
Carter, R. D., Lannom, H. K. and Dill, K., 1985, A ^{13}C-methylation study of
    of glycophorin A in intact erythrocytes by ^{13}C-NMR spectroscopy,
    Biochim. Biophys. Acta, 845:396.
Chien, S., Simchon, S., Abbott, R. E. and Jan K.-M., 1977, Surface
    adsorption of dextran on human red cell membranes. J. Colloid. Interface
    Sci., 62:461.
Clague, M. J., Harrison, J. P. and Cherry, R. J., 1989, Biochim. Biophys.
    Acta, 981:43.
Coakley, W. T. and Deeley, J. O. T., 1980, Effects of ionic strength, serum
    protein and surface charge on membrane movements and vesicle production
    in heated erythrocytes, Biochim. Biophys. Acta, 602:355.

Coakley, W. T. and Gallez, D., 1989, Membrane-membrane contact; involvement of interfacial instability in the generation of discrete contacts, _Bioscience Reports_, 9:675.

Coakley, W. T., Hewison, L. A. and Tilley, D., 1985, Interfacial instability and the agglutination of erythrocytes by polylysine, _Eur. Biophys. J._ 13:123.

Danon, D., Howe, C. and Lee, L. T., 1965, Interaction of polylysine with soluble components of human erythrocyte membranes, _Biochim. Biophys. Acta_, 101:201.

Darmani, H. and Coakley, W. T., 1990, Membrane-membrane interactions: parallel membranes or patterned discrete contacts, _Biochim. Biophys. Acta_, 1021:182.

Darmani, H. and Coakley, W. T., 1991, Contact patterns in concanavalin A agglutinated erythrocytes, _Cell Biophys._, in press.

Darmani, H., Coakley, W. T., Hann, A. C. and Brain, A., 1990, Spreading of wheat germ agglutinin induced erythrocyte contact area: Evidence for formation of local spatially discrete contacts, _Cell Biophys._, 16:105.

Dimitrov, D. S., 1982, Instability of thin liquid films between membranes, _Colloid & Polymer Sci._, 260:1137.

Dimitrov, D. S., and Jain, R.K., 1984, Membrane stability, _Biochim. Biophys. Acta_, 779:437.

Dixon, H. B. F., 1981, Defining a lectin, _Nature_, 292:192.

Donath, E. and Pastushenko, V., 1979, Electrophoretical study of cell surface properties. The influence of the surface coat on the electric potential distribution and on general electrokinetic properties of animal cells, _Bioelectrochem. Bioenerg._, 6:543.

Donath, E., Pratsch, L., Baumler, H. -J., Voigt, A. and Taeger, M., 1989, Macromolecule depletion at membranes, _Studia Biophysica_, 130:117.

Donath, E. and Voigt, A., 1983, Charge distribution within cell-surface coats of single and interacting surfaces - a minimum free electrostatic energy approach - conclusions for electrophoretic mobility measurements, _J. theor. Biol._, 101:569.

Edelman, G. M., Cunningham, B. A., Neeke, G. N., Becker, J. W., Waxdal, M. J. and Wang, J. L., 1972, The covalent and three dimensional structure of concanavalin A, _Proc. Nat. Acad. Sci._ (U.S.A.), 69:2580.

Evans, E. A. and Needham, D. 1988, Intrinsic colloidal attraction/repulsion between lipid bilayers and strong attraction induced by non-adsorbing polymers, _in_: "Molecular mechanisms of membrane fusion", S. Ohki, D. Doyle, T. D. Flanagan, S. W. Hui, and E. Mayhew, eds., Plenum Press, New York.

Fowler, V. M., 1986, Cytoskeleton: New views of the red cell network, _Nature_, 322:777.

Gallez, D. and Coakley, W. T., 1986, Interfacial instability at cell membranes, _Prog. Biophys. molec. Biol._, 48:155.

Gallez, D., Prevost, M. and Sanfeld, A., 1984, Repulsive hydration forces between charged lipidic bilayers. A linear stability analysis, _Colloids Surf._, 10:123.

Gascoigne, P. S. and Fisher, D., 1984, The dependence of cell partition in two-polymer aqueous phase systems on the electrostatic potential between the phases, _Biochem. Soc. Trans._, 12:1085.

Gingell, D., 1990, Cell contact with solid surfaces, _in_: "Biophysics of the cell surface" R. Glaser, and D. Gingell, eds., Springer-Verlag, Berlin.

Gokhale, S. M. and Mehta, M. G., 1987, Glycophorin A interferes in the agglutination of human erythrocytes by concanavalin A, _Biochem. J._, 241:505.

Grant, C. W. M. and Peters, M. W., 1984, Lectin-membrane interactions: Information from model systems, _Biochim. Biophys. Acta_, 779:403.

Hewison, L. A., 1988, "Spatial periodicity of cell-cell contact: An interfacial instability approach". Ph.D. Thesis, Univ of Wales.

Hewison, L. A., Coakley, W. T. and Meyer, H. W., 1988, Spatially periodic discrete contact regions in polylysine-induced erythrocyte-yeast adhesion, Cell Biophys., 13:151.

Hiemenz, P. C., 1986, "Principles of Colloid and Surface Chemistry", Marcel Dekker, New York.

Israelachvili, J. N. and McGuiggan, P. M., 1988, Forces between surfaces in liquids, Science, 241:795.

Jan, K.-M., 1979, Red cell interactions in macromolecular suspension, Biorheology, 16:137.

Jan, K.-M. and Chien, S., 1973a, Role of surface electric charge in red blood cell interactions, J. Gen. Physiol., 61:638.

Jan, K.-M. and Chien, S., 1973b, Influence of the ionic composition of fluid medium on red cell aggregation, J. Gen. Physiol., 61:655.

Jay, D. and Cantley, L., 1986, Structural aspects of the red cell anion exchange protein, Ann. Rev. Biochem., 55:571.

Johnson, L. V., 1986, Wheat-germ agglutinin induces compaction-like and cavitation-like events in 2-cell mouse embryo, Developmental Biol., 113:1.

Kahane, I., Polliak, A., Rachmilewitz, E. A. and Skutelsky, E., 1978, Distribution of sialic acids on the red blood cell membrane in ß thalassaemia, Nature, 271:674.

Kameya, T., Horn, M. E. and Widholm, J. M., 1981, Hybrid shoot formation from fused Daucus carota and Daucus capillifolius protoplasts, J. Pflanzenphysiol., 104:459.

Katchalsky, A., Danon, D., Nevo, A. and de Vries, A., 1959, Interactions of basic polyelectrolytes with the red blood cell 2. Agglutination of red blood cells by polymeric bases. Biochim. Biophys. Acta, 33:120.

Klein, J., and Luckham,P.F., 1984, Long-range attractive forces between two mica surfaces in an aqueous polymer solution, Nature, 308: 836

Lange, Y., Hadesman, R. A. and Steck, T. L., 1982, Role of the reticulum in the stability and shape of the isolated human-erythrocyte membrane, J. Cell Biol., 92:714.

Leiken, S. L., Kozlov, M. M., Chernomordik, V. S. and Chizmadzhev, Y. A., 1987, Membrane fusion: Overcoming of the hydration barrier and local restructuring, J. theor. Biol., 129:411.

LeNeveu, D. M., Rand, R. P. and Parsegian, V. A., 1976, Measurement of forces between lecithin bilayers, Nature, 259:601.

Luckham, P. F. and Ansarifar, M. A., 1990, Biomedical aspects of the direct measurement of the forces between adsorbed polymers and proteins, Br. Polymer J., 22:233.

Luckham, P. F. and Klein, J., 1984, Forces between mica surfaces bearing adsorbed polyelectrolyte, poly-l-lysine, in aqueous media, J. Chem. Soc. Faraday Trans. I, 80:865.

Marchesi, V. T., 1979, Functional proteins of the human red blood cell membrane, Semin. Hematol., 16:3.

Marikovsky, Y, Danon, D. and Katchalsky, A., 1966, Agglutination by polylysine of young and old red blood cells, Biochim. Biophys. Acta, 124:154.

Marra, J. and Hair, M. L., 1988, Forces between two poly(2-vinylpyridine)-covered surfaces as a function of ionic strength and polymer charge, J. Phys. Chem., 92:6044.

Nir, S., 1976, van der Waals interactions between surfaces of biological interest, Progr. Surf. Sci., 8:1.

Patel, S. S. and Tirrell, M., 1989, Measurement of forces between surfaces in polymer fluids, Annu. Rev. Phys. Chem., 40:597.

Perez, E. and Proust, J. E., 1985, Effects of a non-adsorbing polymer on colloid stability: force measurements between mica surfaces immersed in dextran solution, J. Physique Lett., 46:L-79.

Pontecorvo, G., 1975, Production of mammalian somatic-cell hybrids by means of polyethylene-glycol treatment, Somat. Cell Genet., 1:397.

Prevost, M. and Gallez, D., 1984, The role of repulsive hydration forces on
    the stability of queous black films: Application to vesicle fusion, <u>J.
    Chem. Soc. Faraday Trans. II</u>, 80:517
Prevost, M., Gallez, D. and Sanfeld, A., 1983, Electrodynamic stability of
    unsymmetrical aqueous films, <u>J. Chem. Soc. Faraday Trans II.</u>, 79:961.
Rosenberg, M., and Kjelleberg, S., 1987, Hydrophobic interaction: Role in
    bacterial adhesions, <u>Adv. Microbial Ecol.</u>, 9:353.
Russell, L., Peterson, R., Freud, M., 1979, Direct evidence for formation
    of hybrid vesicles by fusion of plasma and outer acrosomal membranes in
    boar spermatozoa, <u>J. Exp. Zool.</u>, 208:41.
Schnebli, H. P., Roeder, C. and Tarcsay, L. 1976, Reaction of lectins with
    human erythrocytes, <u>Exp. Cell Res.</u>, 98:273.
Seaman, C. V. F., 1975, <u>in</u>: "The Red Blood Cell", Vol. 2, D. M. Surgeoner,
    ed., <u>Academic Press, New York</u>.
Shelton, I., Gascoigne, P. S., Turner, D. M. and Fisher, D., 1985,
    Influence of sialic acid on the surface hydropohicity of membranes,
    <u>Biochem. Soc. Trans.</u>, 12:1085.
Skalak, R., Zarda, P. R., Jan K-M and Chien, S., 1981, Mechanics of rouleau
    formation, <u>Biophys. J.</u>, 35:771
Snabre, P., Grossmann, G.H. and Mills, P., 1985, Effects of dextran
    polydispersity of red blood cell agregation, <u>Colloid and Polymer Sci.</u>,
    263:478.
Sung, L. A., 1983, "The interaction of <u>Helix pomatio</u> and <u>Dolichos bifluorus</u>
    lectins with human group A blood cells", Ph.D. Thesis, Columbia Univ.,
    New York.
Tan, S. S. and Morris-Kay, G. M., 1986, Analysis of the cranial neural
    crest cell-migration and early fates in postimplantation rat chimeras,
    <u>J. Embryol. Exp. Morphol.</u>, 98:21.
Tilley, D., Coakley, W. T., Gould, R. K., Payne, S. E, and Hewison, L. A.,
    1987, Real time observations of polylysine, dextran and polyethylene
    glycol induced mutual adhesion of erythrocytes held in suspension in an
    ultrasonic standing wave field, <u>Eur. Biophys. J.</u>, 14:499.
van Oss, C. J. and Coakley, W. T., 1988, Mechanisms of successive modes of
    erythrocyte stability and instability in the presence of various
    polymers, <u>Cell Biophys.</u>, 13:141.
van Oss, C. J., Gillman, C. E. and Neumann, A. W., 1975, "Phagocytic
    Engulfment and Cell Adhesion", Dekker, New York.
Vasiliev, J. M., 1987, Actin cortex and microtubular system in morpho-
    genesis: cooperation and competition, <u>in</u>; "Cell Behaviour: Shape,
    Adhesion and Motility", J. E. M. Heaysman, C. A. Middleton and F. M.
    Watt eds., J Cell Sci; Supplement 8. Company of Biologists, Cambridge,
    U.K.
Verwey, E. J. W. and Overbeek, J. Th. G., 1949, "Theory of The Stability of
    Lyophobic Colloids", Elsevier, Amsterdam.
Wright, C. S., 1977, The crystal structure of Wheat Germ Agglutinin at 2.2A
    resolution, <u>J. molec. Biol.</u>, 111:439.

PEGYLATION OF MEMBRANE SURFACES

Derek Fisher[1,2], Cristina Delgado[2], John Morrison[1],
Gerald Yeung[3] and Colin Tilcock[3]

[1]Department of Biochemistry and [2]Molecular Cell Pathology
Laboratory, Royal Free Hospital School of Medicine
University of London, London NW3 2PF
[3]Department of Radiology, University of British Columbia
Vancouver, BC, Canada, V6T 1W5

INTRODUCTION

Polyethylene glycol (PEG) is a long chain, relatively inert
and non-toxic water-soluble polymer. PEG-modification of
clinically useful proteins makes dramatic improvements in their
biological properties by increasing circulation lifetimes,
conveying resistance to proteolytic attack, reducing antigenicity,
and increasing solubility[1-4]. These advantages arise from the
flexible, hydrophilic shell of the attached PEG which protects the
protein from immune surveillance, serum protein adherence and
proteases.

Liposomes administered i.v are subject to interactions with
plasma proteins (e.g. HDL) and the reticulo-endothelial system
(RES) which results in destabilisation and clearance of the
vesicles from the circulation. Ways are being sought to prolong
liposome half-life in the circulation for purposes such as
controlled drug release, magnetic resonance imaging and targeting
of entrapped pharmaceuticals. For all these applications there is
a need to decrease non-specific removal to enhance specific
delivery and to increase vesicle stability (to prevent loss of
contents). Methods employed to date to improve vesicle survival
include the use of small neutral unilamellar liposomes with a
highly rigid wall[5]. To increase the half life of vesicles even
further and/or to facilitate the slow release of liposomally
entrapped agents into the bloodstream using liposomes with non-
rigid bilayers, other lipid components or surface coatings may be
required[6]. Previously described methods, which can confer a net
increase in hydrophilicity on the liposomal surface include:

incorporation of gangliosides into the lipid bilayer[7], coating the liposomal surface with glycosides[8] or with poloxamers[9]. Our experience of the PEGylation of albumin[10], immunoglobulins[11], granulocyte-macrophage stimulating factor, GM-CSF[12], and proteins in protein/DNA complexes[13] has encouraged us to PEGylate liposomal surfaces with a view to obtaining similar benefits for liposomal performance in the circulation as has been obtained with PEG-proteins[6]. In this article we describe how PEG can be covalently linked to intact lipid vesicles via head groups of phosphatidyl-ethanolamine (PE) or phosphatidylserine (PS) without loss of vesicle integrity. PEGylation confers the twin benefits of decreased serum-vesicle interaction in vitro and increased circulation lifetimes of vesicles in vivo. In addition, the attachment of PEG on the outer surface of lipid bilayers provide a model system with which to examine the influence of water-binding polymer on the surface as well as membrane properties of lipid bilayers.

MATERIALS AND METHODS

Preparation of Lipid Vesicles

Multilamellar vesicles (MLVs) containing 20% (w/w) egg phosphatidylethanolamine (EPE) and 80% (w/w) egg phosphatidylcholine (DPPE) and ^{14}C-EPC were prepared in 0.125M NaCl containing 0.05M sodium phosphate buffer, pH 7.5 (PBS) at 10 mg total lipid/ml. Unilamellar vesicles of dioleyl phosphatidylcholine (DOPC) and dioleyl phosphatidylethanolamine (DOPE) or of dioleoyl phosphatidylserine (DOPS) of defined size were prepared by extrusion of multilamellar vesicles through polycarbonate filters as previously described[14-17]. Small unilamellar vesicles (SUVs) of distearoyl phosphatidylcholine (DSPC), dipalmitoyl phosphatidylethanolamine (DPSE) and cholesterol were prepared as previously described[5,6]. For partitioning studies vesicles were spiked with 1-2 μCi of tritiated dipalmitoyl phosphatidyl choline. To measure liposomal retention of water soluble molecules during the coupling reaction and in subsequent procedures, carboxyfluorescein (6CF) was entrapped.

PEGylation Procedure

PEG was activated with tresyl chloride (2,2,2-trifluoro ethanesulphonyl chloride) as previously described[10]. Activation was confined to one end of the PEG chain, to avoid crosslinking, by using monomethoxyPEG (MPEG) to produce tresylated MPEG(TMPEG). TMPEG was reacted with lipid vesicles at room temperature in buffered solutions (see below).

Partitioning of Vesicles

The covalent attachment of the MPEG to the outer surface of the vesicles was demonstrated by the alteration in the partitioning behaviour of the vesicles in aqueous two-phase

systems of PEG and dextran [18-20]. The composition of the phase
system was adjusted so that the vesicles showed a low partition in
the top PEG-rich phase; vesicles were at the interface or in the
bottom dextran-rich phase[14-15]. Attachment of MPEG to the vesicle
surface makes them more "PEG-like" (increases their wetting by the
PEG-rich phase) and they partition to the top phase. Liposome-
serum interactions were demonstrated by the decrease in top phase
partition that occured when liposomes were exposed to serum
reflecting adsorption of serum components. In these studies phase
composition was selected so that vesicle partition was relatively
high, so that the fall in top phase partition resulting from serum
exposure could be more readily followed. Phase systems of 5% w/w
dextran T500 and 5% w/w PEG 6000 (or PEG 8000) were prepared in
0.15M NaCl buffered with 0.01M sodium phosphate, pH 6.8 from stock
solutions of approximately 20% w/w dextran (calibrated by
polarimetry), 40 % w/w PEG, 0.6M NaCl and 0.44M sodium phosphate
and distilled water. After phase separating at 25°C the top and
bottom phases were removed and stored at 4°C. Vesicles (typically
50 µl) were added to 1ml of top and bottom phases in 50 x10 mm
tubes[6,14] or in other experiments to 1.5ml of both phases in 10 x
70 mm tubes[15] at 25°C. The contents were mixed by repeated
inversion for 1 minute. Triplicate 50 µl samples were removed for
total counts, After the tubes had stood for 20 or 25 min at 25°C
for phase separation to occur triplicate samples
(20-50 µl) were taken from each phase. The difference between
counts in the top and bottom phases was taken to represent counts
at the interface

<u>Measurement of 1/T1 relaxation rates</u>

Vesicles were prepared with entrapped gadolinium-
diethylenetriaminepentaacetic acid and sized to 100nm by extrusion
as described above. Phantoms were images using a Toshiba MR50A
whole-body MR scanner using a standard spin-echo pulse sequence
with a fixed echo time of 20 msec and variable repetition time
(TR). 1/T1 values were calculated from a single exponential fit to
the signal intensity data as a function of TR.

RESULTS AND DISCUSSION

1   Factors influencing the attachment of PEG to liposomal
     surfaces

We have PEGylated three types of lipid vesicles: MLVs,
unilamellar vesicles of defined size, and SUVs[6]. PEGylation
is sensitive to the ratio of TMPEG to PE, the time of reaction,
and the pH.

1.1 Molar ratio of TMPEG to available PE

MLVs of EPC:EPE (8:2) in PBS, pH 7.5 were incubated with
solutions of TMPEG prepared in PBS (final concentrations 0-25

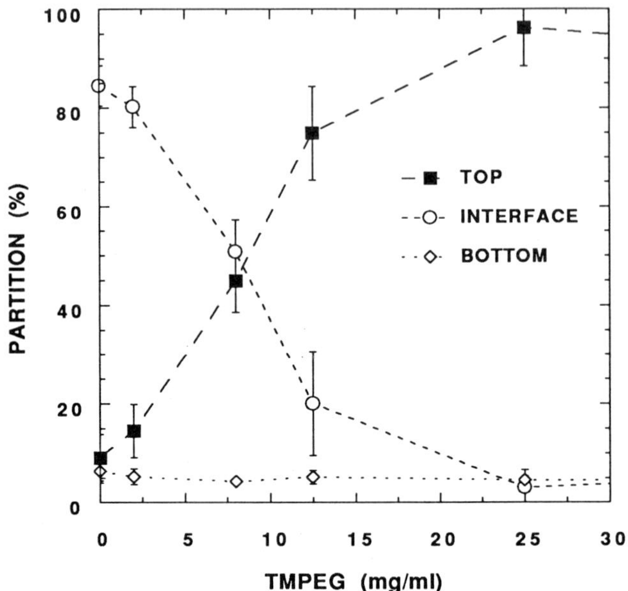

Fig.1.   The effect of TMPEG on the partitioning behaviour of
multilamellar vesicles of EPC:EPE (8:2).

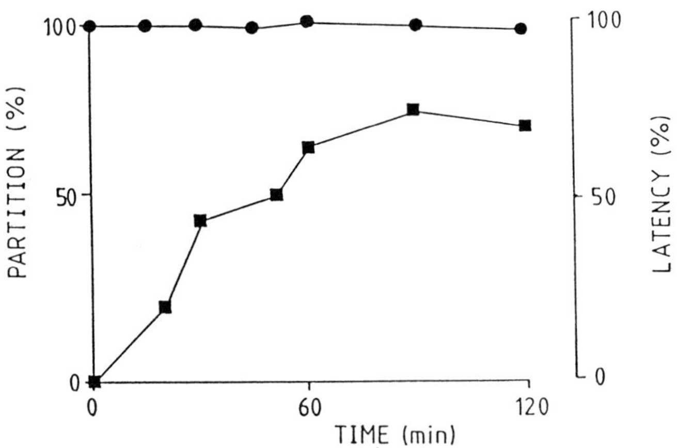

Fig.2.   PEGylation of 100 nm unilamellar vesicles proceeds
without loss of vesicle latency. DOPC:DOPE 8:2)
vesicles were incubated at pH 9 with TMPEG. Top
phase partitioning (■) and latency of entrapped 6CF
(●) were measured at intervals.

mg/ml) for 2 hours at room temperature. Fig. 1 shows that exposure
of the liposomes to TMPEG increases their partition into the PEG-
rich top phase with an associated decrease in partition at the
interface. These changes in partition indicate that PEG has
become attached to the liposome, presumably by the covalent
attachment to the amino group of the EPE.

Treating 100nm unilamellar vesicles of DOPC:DOPE (8:2) at pH
9.0 with varying molar ratios of TMPEG increased partitioning
into the top phase consistent with increasing PEGylation (Fig 3).
There was a marked increase in top phase partitioning between the
molar ratios 1.0 and 1.3 from 20% to 90%.

SUVs composed of DSPC:DPPE:cholesterol (8:2:1) have been
readily PEGylated in PBS (0.125M NaCl buffered with 0.05M sodium
phosphate buffer, pH 8.5) by incubation for 2 hours with a ratio
of TMPEG:total DPPE of 6.25[6]. In this case the vesicles were
separated from unreacted TMPEG by gel filtration on Sepharose 4B-
CL prior to partitioning.

1.2   Reaction time

Both Fig 2 and Fig 3 show that PEGylation increases with
time. In particular Fig 3 shows that the rate of PEGylation
increases with the molar ratio of TMPEG:PE.

1.3   pH Dependency of vesicle PEGylation

The results of PEGylating 100nm vesicles of DOPC:DOPE (8:2)
with a two-fold excess of TMPEG to the DOPE present at the outer
surface is shown in Fig 4. Results are shown for the top and
bottom phases and the interface. Before reaction the vesicles
partitioned between the bottom phase (60%) and the interface
(40%). On exposure to TMPEG there is a transfer of vesicles with
time from the bottom phase to the interface and then to the top
phase as PEGylation precedes.   At pH 9 there is a rapid transfer
of vesicles to the top phase with associated declines in bottom
phase and interface partitions. A similar result at pH 10
suggests that the pH maximum for this reaction lies around pH 9.
At pH 7.2 there was virtually no transfer to the top phase but the
slow decrease in bottom phase partition as vesicles are
transferred to the interface shows that PEGylation is proceeding,
albeit slowly. At pH 8 the partitioning moves from the bottom
phase to the interface and then to the top phase, demonstrating
greater PEGylation than as pH 7.2 but not as rapid as at pH 9 and
above. PEGylation is obtained by a nucleophilic displacement of
the tresyl moiety by the primary amino group of PE. It is
therefore to be expected that the reaction occurs more rapidly in
alkaline conditions. That some PEGylation occurs at pH 7.2 is in
agreement with our studies on proteins, where neutral conditions
are preferred to preserve the activity of labile proteins [10-13].

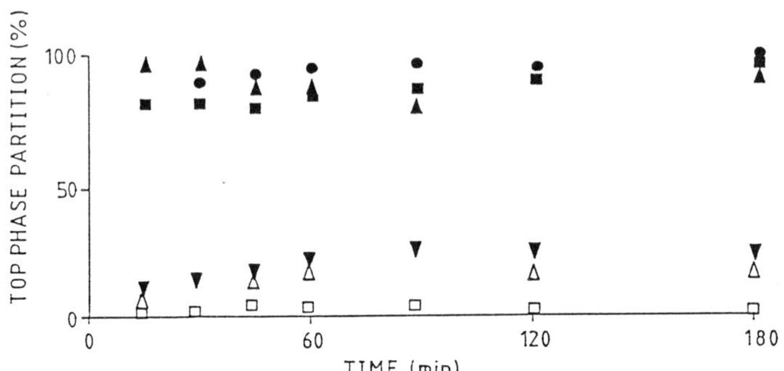

Fig.3.   The effect of TMPEG:PE ratio on PEGylation 100 nm
            unilamellar vesicles of DOPC:DOPE (8:2) were incubated
            with TMPEG in 50 mM HEPES, 100mM NaCl at pH9 with a
            range of molar ratios of TMPEG:DOPE available at outer
            surface: 0.2 (□);0.5 (△); 1.0 (▼); 1.3 (▲);
            1.55 (■); 4.7 (●) and partitioned at intervals.

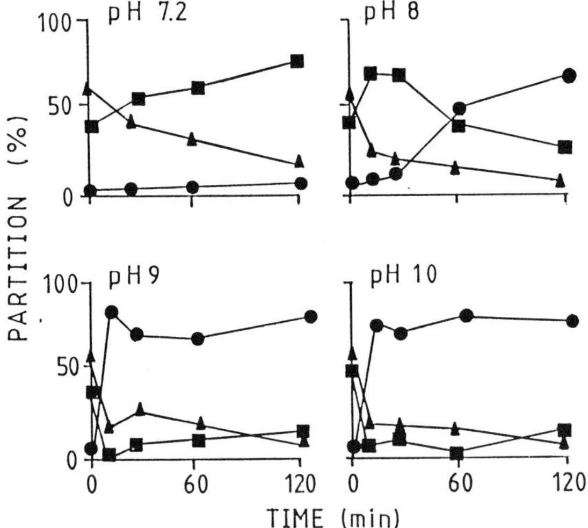

Fig.4.   pH sensitivity of PEGylation of DOPC:DOPE vesicles
            DOPC:DOPE (8:2) 100nm diameter vesicles were incubated
            with a two-fold molar excess of TMPEG to the exterior
            concentration of DOPE at room temperature in 50 mM
            Hepes, 100mM NaCl at pH 7.2, 8, 9 and 10. Partitioning
            into the top phase (●), the interface (■) and the
            bottom phase (▲) are shown.

2. Attachment of PEG to liposomal surfaces occurs via amino
   groups

The presence of PE in the vesicle is required for TMPEG to
have any effect. When MLVs of 100% EPC were treated with TMPEG for
two hours and then partitioned there was no difference compared to
MLVs treated with buffer (Table 1). Samples of TMPEG that had lost
their ability to PEGylate proteins after prolonged storage, were
found to have no effect on the partitioning of liposomes
containing EPE.

Table 1.   Effect of TMPEG on the partition of egg PC
            multilamellar vesicles

FINAL [TMPEG]	PARTITION (%)[*]		
(mg/ml)	TOP	INTERFACE	BOTTOM
0	22.5±13.0	71.6±12.0	5.9±1.0
25	25.8±13.0	67.8±14.0	6.4±1.0

(* ± S.D., n=5)

Treatment of SUVs with MPEG instead of TMPEG produced no
change in the partition of the SUVs, suggesting that adsorption of
polymer was not the cause of increased partition when TMPEG was
used. Taken together, these observations support the view that
TMPEG attaches to PE specifically, and that altered partitioning
does not arise from adsorption of TMPEG to vesicle surfaces.

The covalent attachment of PEG was confirmed by measurement
of the fraction of amino groups (from PE) exposed at the outer
surface of the vesicles with trinitrobenzenesulphonic acid, as
described by Hope and Cullis[21]. Determinations gave values of 47%
for DOPC:DOPE vesicles (8:2), close to the theoretical value of
50% for equal distribution of the PE between the inner and outer
surfaces. PEGylation caused a decrease in the PE content
detectable by this assay, suggesting covalent attachment of the
MPEG to the free amine group of PE. For example when a 3-fold mole
excess of TMPEG to outer PE was added to DOPC:DOPE (7:3) vesicles
for 1 hr, the percentage of outer PE PEGylated was 36%; when a 6
fold molar excess was added, this percentage PEGylation increased
to 45%.

Table 2. Partitioning of PS vesicles

| | PARTITION (%)* | | |
TREATMENT	TOP	INTERFACE	BOTTOM
Untreated	5.3±1.0	4.2±4.4	97.3±16.4
TMPEG-treated	86.6±4.6	10.6±3.8	2.8±1.7
MPEG	5.2±0.8	3.0±5.2	94.7±8.8

(* ± S.D., n=3)

The amino group present in PS was also readily PEGylated. 100nm vesicles of DOPS, prepared at 10 mg/ml in HEPES buffer, pH 9 were incubated with TMPEG (3:1 molar excess to externally exposed DOPS) for 2hr at 25°C. Table 3 shows that TMPEG transferred the vesicles from the bottom phase to the top phase whereas MPEG had no effect.

3   Improved behaviour of PEGylated liposomes for in vivo
      applications

3.1 Vesicle PEGylation is achieved without loss of latency

Unilamellar vesicles of defined size (100 nm diameter uni-lamellar vesicles of DOPC:DOPE 8:2) containing entrapped 6CF were PEGylated with a 3-fold excess of TMPEG to the outer surface DOPE at room temperature in 50 mM HEPES, 100mM NaCl at pH 9. At intervals samples were removed for partitioning and to measure vesicle latency. Fig 2 shows that top phase partitioning increased with time indicating progressive PEGylation, and that PEGylation occurred without the release of the entrapped 6CF. In similar studies with SUV liposomally entrapped 6CF was quantitatively retained in liposomes during procedures indicating that structural integrity of SUVs was maintained during and after the coupling reaction, as well as during phase partitioning.[6]

3.2 PEGylation decreases interactions and destabilisation of
     vesicles with serum

Phase partitioning is a subtle method with which to follow interactions of vesicles with serum[6]. SUVs of composition DSPC:PE:Cholesterol (4:1:5) partitioned about 20% top phase, 60% interface and 20% bottom phase. Treatment with serum caused an immediate (within 1 min) alteration in the vesicle surface properties indicated by their partition changing to 0% top phase, 40% interface and 60% bottom phase. The plasma proteins alone partitioned mainly to the bottom phase (68% bottom, 32% top; partition coefficient, K = 0.47 ± 0.02, n=4). Thus it appears that the SUVs are immediately coated with serum proteins which then cause the vesicles to partition with similar characteristics

to the proteins. PEGylation of the SUVs increased their partition
into the top phase (almost 100%); on exposure to serum there was a
change in their partition towards the interface and the bottom
phase, but importantly this process was very slow compared with
the virtually instantaneous effect of serum on unPEGylated SUVs.
Since the partitioning behaviour relates to the sum of the forces
imposed by the PEGylation and serum binding, and with the former
is not a linear function, it is not simple to determine whether
the effect of serum on partition is equal for the PEGylated and
for the unPEGylated liposomes. This could, however, be determined
with a detailed dose response analysis of the effect of PEGylation
on the partition coefficient so that the influence of serum could
be determined at various parts of the dose response curve in "PEG-
equivalents". This would establish whether serum had different
effects on the PEGylated and unPEGylated liposomes. Nevertheless
the order of magnitude differences in partition behaviour
suggests that PEGylation slows down the adsorption of serum
components onto the vesicles.

Separation of the SUVs exposed to serum by gel chromatography
gave vesicles which showed partitioning behaviour close to that of
the vesicles before exposure. This suggests that plasma
components can alter the partitioning behaviour, but are
effectively removed by gel filtration, and would not therefore be
detected by, for example, gel electrophoresis. Partitioning
offers, therefore, the possibility of studying interactions of
serum components without the need to isolate the vesicles. It
should be noted that partitioning has proved useful in studies of
protein-protein, where even rather weak complexes between enzymes
have been detected, and protein-membrane interactions[18,22,23].

The stability of liposomes can be decreased by serum[9]. Fig 5
shows that PEGylation can decrease this destabilising action of
serum. 100nm diameter vesicles of DOPC:DOPE (7:3; 6:4; and 5:5),
containing entrapped 6CF, were incubated at 37° serum (freshly
hydrated lyophilised human serum, Monitrol-ES, Dade Diagnostics)
to provide a final lipid concentration of approx 1mg/ml, a
concentration corresponding to the maximum in vivo serum
concentrations expected on the basis of the imaging experiments
of Unger et al[24]. Samples were removed at intervals and the 6CF
released was measured fluorimetrically. Vesicles were PEGYlated
with a 3-fold excess of TMPEG to outer surface DOPE overnight at
room temperature, after which time there had been no loss of
latency. Fig 5 shows that vesicles of 7:3 molar ratio DOPC:DOPE
released 6CF on exposure to serum. PEGylation decreased serum
induced loss of latency by a factor of 2.

Increasing the DOPE content to 40 mole% and 50 mole % increased
the stability of the vesicles to serum; nevertheless PEGylation
produced additional stabilisation. It should, however, be noted
that vesicles of DOPC:DOPE (8:2) retained only 35% of entrapped
6CF after 2hrs incubation with serum; loss of latency in vesicles
of this composition was not decreased by PEGylation.

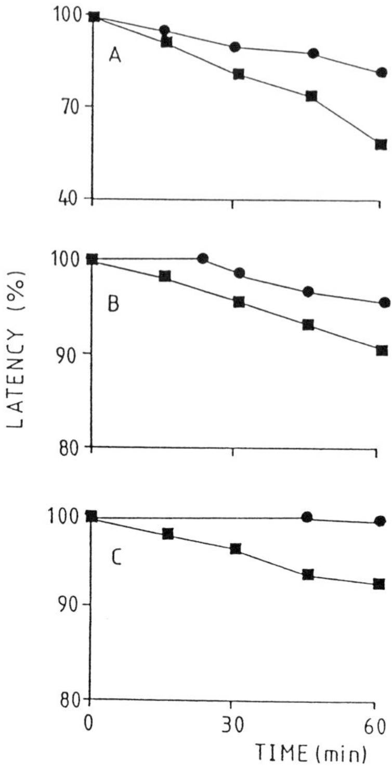

Fig.5.   PEGylation decreases destabilising effect of serum 100nm
           vesicles of (A) DOPC:DOPE (7:3); (B) DOPC:DOPE(6:4);(C)
           DOPC:DOPE (5:5) containing entrapped 6CF, either
           PEGylated overnight at room temperature with a threefold
           molar ratio of TMPEG to DOPE at the outer surface (●)
           or unPEGylated (■) were incubated in 10 volumes of
           serum at 37°C and latency measured.

3.3 PEGylation decreases in vivo clearance

To demonstrate the efficacy of PEGylation in decreasing vesicle
clearance we chose to use SUVs of a composition that has
previously been shown to produce the longest half-lives in the
absence of surface modification: DSPC:DPPE:Cholesterol(4:1:5).[6]
Fig·6 shows that PEGylation produces an increase of about 30% in
half-life.

3.4 PEGylating decreases vesicle-vesicle interactions

PEG modifies water structure. A number of studies have
suggested that approximately two molecules of water bind to each
ether oxygen on the polymer[25,26]. Depending on their concentration, solutions of PEG can precipitate certain proteins and
vesicles,[26,27]; can alter solvent properties[28]; can be fusogenic[29-

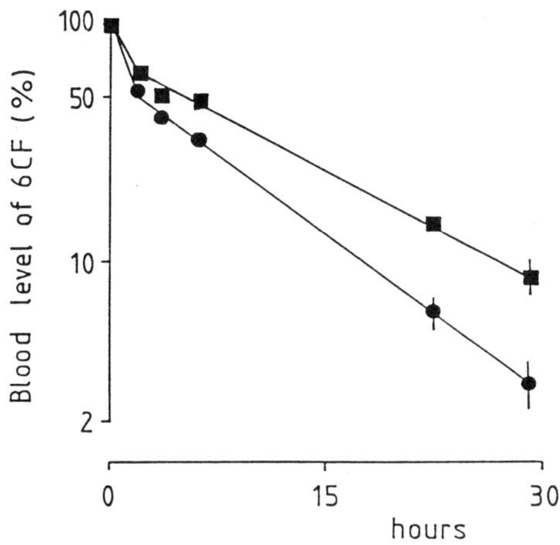

Fig.6.   A comparison of the clearance of PEGylated SUVs and
            unPEGylated SUVs from the circulation in mice. SUVs of
            composition DSPC:PE:Cholesterol(molar ratio 0.4:0.1:0.5)
            containing entrapped 6CF, either PEGylated (■) or
            unPEGylated (●) were injected iv into mice. Blood
            levels of 6CF are shown (± SE,5 animals).
            (Taken from ref 6 with permission)

[32] and membrane permeabilising[33]. It is therefore to be expected
that covalently attached PEG will modify the environment of the
lipid surface; the exclusion of serum components from the surface
is an example of this, described above.

    As a model for membrane-membrane interactions we chose the
Ca^{2+} induced aggregation of phosphatidylserine vesicles[34].
PEGylation of PS vesicles abolishes the vesicle-vesicle
interaction induced by Ca^{2+}.

    Addition of Ca^{2+} to DOPS vesicles caused an increase in
turbidity (Fig 7) which was maximal around 5mM $CaCl_2$ when
flocculation occurred (F in Fig 7); with higher concentrations
precipitation was seen (P in Fig 7) with a decrease in turbidity.
These changes were reversed by EDTA. By contrast PEGylated
vesicles showed no detectable aggregation when $CaCl_2$ was added.
This inhibition appears due to the covalent attachment of the
polymer, rather than simply the presence of polymer in the
solution (8% w/w) since Ca^{2+}-induced aggregation was obtained
with vesicle in the presence of MPEG, used at the same
concentration as TMPEG. It seem likely that this inhibitory
effect of PEG anchored to the membrane surface arises from the
flexible chain preventing the close approach of vesicle required
for Ca^{2+} induced aggregation.

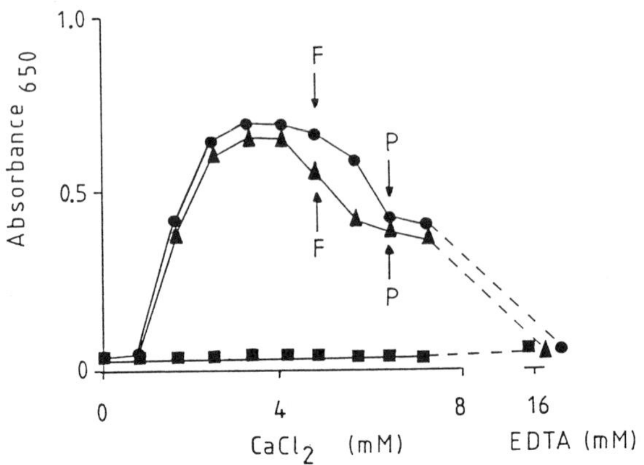

Fig.7.   Aggregation of DOPS 100 nm vesicles with Ca^{2+} is
            inhibited in PEGylated vesicles. Vesicles treated with
            buffer (●); MPEG (▲) or TMPEG (■).

3.5   Application of PEGylated liposomes for Magnetic Resonance Imaging

Liposomes with associated paramagnetic species are of
interest as both vascular and reticuloendothelial contrast agents,
having been shown to improve the detection of hepatic metastases
by magnetic resonance (MR) imaging[24,35]. Vascular enhancement by
liposomal MR agents would be improved if both vesicle stability
and circulation lifetime could be improved.

Because water must cross the lipid bilayer in order to
interact with paramagnetic species either entrapped within the
vesicle interior or on the inner surface of the lipid bilayer, any
factor which decreases the water permeability of the membrane
causes a decrease in the efficacy of the contrast agent. For this
reason the use of saturated lipid species or lipid compositions
with high sterol contents is not the best approach to the design
of liposomal MR contrast agents with extended half-life in the
circulation. To be able to take advantage of PEGylation as a
means to achieve the extended half-life we have investigated
whether covalently attached PEG has any effect upon the relaxivity
of paramagnetic species entrapped within the interior aqueous
space of liposomes. The contrast agent Gd-DTPA was encapsulated in
100 nm diameter vesicles of DOPC:DOPE (7:3)[36]. Half of the
sample was PEGylated with TMPEG (molar ratio of TMPEG: PE on outer
surface of 3:1). Both control and PEGylated samples were diluted
in saline buffer to give four samples with effective Gd
concentrations of 2, 1, 0.5, and 0.25 mM (calculated as described
by Tilcock et al [36] given the known trap volume of the vesicles,
the lipid concentration and assuming the concentration of
entrapped Gd-DTPA was 0.67M). Samples were imaged with a Toshiba

0.5T MRT-50A whole body scanner. Relaxivities are obtained from
linear regressions of 1/T1 (spin lattice relaxation rate) against
the effective Gd-DTPA concentration. Fig 8 shows that the
relaxivities of liposomal entrapped Gd-DTPA were unaffected by
PEGylation of the vesicles.

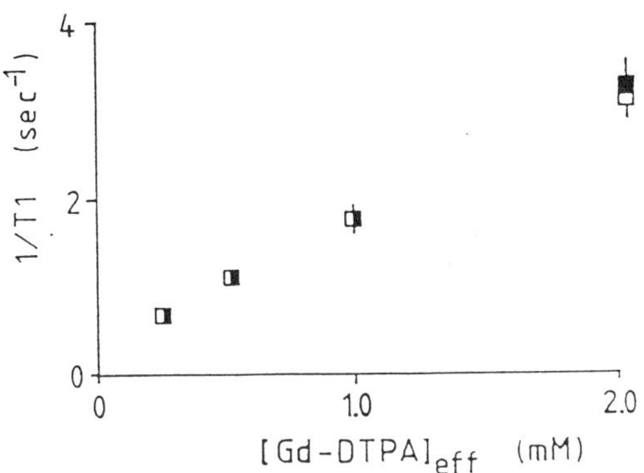

Fig.8.   Relaxivities of liposomal entrapped Gd-DTPA samples are
            unaffected by PEGylation. 1/T1 versus effectiveGd-DTPA
            in vesicles of DOPC:DOPE (7:3) before (■) and after
            PEGylation (□)

Since coupling of PEG to the membrane surface has no measurable
effect on the water permeability, it appears that covalent
attachment of PEG to liposomal MR agents may prove an effective
method for increasing their circulation lifetimes without
decreasing their effectiveness as contrast agents.

CONCLUSIONS

    PEG can easily be covalently linked to the outer surface of
liposomes containing either PE or PS and offers the possibility of
designing slowly cleared, stable (non-leaky), antigenically inert
and highly water permeable liposomes. In other studies a
different approach from our has been taken: PEG-PE derivatives
have been synthesised and used, with other lipids, to form
liposomes. Improved circulation lifetimes etc have also been
observed [37,38].

ACKNOWLEDGEMENTS

    This work was supported by grants from the British
Council(C.D.), NATO and British Council Travel Awards (D.F, C.T.),
Medical Research Council of Canada (C.T.) and the Royal Free
Hospital School of Medicine (D.F.). C.T. is an MRC Scholar.We
thank Dr.G.E.Francis for helpful discussion.

REFERENCES

1   A. Abuchowski, T. van Es, N.C. Palczuk, and F.F. Davis,
        Alteration of immunological properties of bovine serum
        albumin by covalent attachment of polyethylene glycol.
        J.Biol. Chem. 252:3578 (1977)

2   F.F. Davis, A. Abuchowski, T. van Es., N.C. Palczuk,
        K.Sacova, H.-L. Chen and P. Pyatak, Soluble, nonantigenic
        polyethylene glycol-bound enzymes, in "Biomedical Polymers.
        Polymeric Materials and Pharmaceuticals for Biomedical Use",
        (E.P. Goldberg and A. Nakajima eds.) pp 441-452 Academic
        Press, New York (1980)

3   N.V. Katre, Immunogenicity of recombinant IL-2 modified by
        covalent attachment of polyethylene glycol. J.Immunol.
        144:209 (1990)

4   G.E. Francis, C. Delgado and D.Fisher, PEG modified
        proteins, in "Pharmaceutical Biotechnology. Vol3: Stability
        of Protein Pharmaceuticals: In Vivo Pathways of Degradations
        andStrategies for Stabilization" (T.J. Ahern and M.C. Manning
        eds.) Plenum Press. In press

5   J. Senior, J.C.W. Crawley and G. Gregoriadis. Tissue
        distribution of liposomes exhibiting half-lives in the
        circulation after intravenous injection. Biochim. Biophys.
        Acta 839:1 (1985)

6   J. Senior, C. Delgado, D. Fisher, C. Tilcock and G.
        Gregoriadis. Influence of surface hydrophilicity of liposome
        on their interaction with plasma proteins and clearance from
        the circulation: studies with polyethylene glycol-coated
        vesicles. Biochim. Biophys. Acta, 1062: 77 (1991)

7   T.M. Allen, J. Ryan and D. Papahadjopoulos. Gangliosides
        reduce leakage of aqueous space markers from liposomes in the
        presence of human plasma. Biochim. Biophys. Acta 818:205
        (1985)

8   P.Ghosh and K.Bachhawat. Grafting of different glycosides on
        the surface of liposomes and its effect on the tissue
        distribution of ^{125}I-labelled-globulin encapsulated in
        liposomes.Biochim. Biophys. Acta 633: 562 (1980)

9   J. H. Senior. Fate and behaviour of liposomes in vivo: a
        review of controlling factors. CRC Critical Reviews in
        Therapeutic Drug Carriers 3: 123 (1987)

10   C.Delgado, J.N. Patel, G.E. Francis and D. Fisher. Coupling
        of poly(ethylene glycol) to albumin under very mild
        conditions by activation with tresyl chloride:
        characterization of the conjugate by partitioning in aqueous
        two-phase systems. Biotechnol. Appl. Biochem. 12:119 (1990)

11   C.Delgado, R.J. Anderson, G.E. Francis and D. Fisher.
        Separation of cell mixtures by immunoaffinity cell
        partitioning:strategies for low abundance cells. Anal.
        Biochem. 192: 322 (1990)

12   F. Malik, C. Delgado, C. Knusli, S.A. Irvine, D. Fisher and
        G.E. Francis. Construction of polyethylene glycol (PEG)

modified granulocyte-macrophage colony stimulating factor (GM-CSF) with conserved biological activity. Exp. Haematol. 18: 624 (Abstract) (1990)

13   R.J. Anderson, C.Delgado, D. Fisher, J.M. Cunningham, and G.E. Francis. A method for the purification of DNA/protein complexes applied to DNA topoisomerase II cleavage sites. Anal. Biochem. 193: 101 (1991)

14   C. Tilcock, P.Cullis, T. Dempsey, B.N. Youens and D. Fisher. Aqueous two-phase polymer partitioning of lipid vesicles of defined size and composition. Biochim. Biophys. Acta 979:208 (1989)

15   C. Tilcock, R. Chin, J. Veiro, P.Cullis, and D. Fisher. Detection of surface charge-related properties in model membrane systems by aqueous two-phase partition. Biochim. Biophys. Acta 986:167 (1989)

16   C. Tilcock, S. Eastman and D. Fisher. Induction of lipid asymmetry and exchange in model membrane systems. Biocolloids and Biosurfaces, in press

17   C. Tilcock, S. Eastman and D. Fisher. Determination of lipid asymmetry and exchange in model membrane systems. This volume

18   P.-Å.Albertsson. "Partition of Cell Particles and Macromolecules" pp 227-250, John Wiley and Sons, New York (1986)

19   H. Walter, D.E. Brooks and D. Fisher (eds.). "Partitioning in Aqueous Two-Phase Systems. Theory, Methods, Uses, and Applications to Biotechnology," Academic Press, Orlando (1985)

20   D. Fisher and I. A. Sutherland (eds.). "Separations Using Aqueous Phase Systems. Applications In Cell Biology and Biotechnology," Plenum Press, New York (1989)

21   M.J. Hope and P.R. Cullis. Lipid asymmetry induced by transmembrane pH gradients in large unilamellar vesicles. J. Biol. Chem. 262: 4360 (1987)

22   L. Backman. Interacting systems and binding studied by partitioning. In "Partitioning in Aqueous Two-Phase Systrems. Theory, Methods, Uses, and Applications to Biotechnology". (H. Walter, D.E. Brooks and D. Fisher, eds.) pp 267-314, Academic Press (1985)

23   P.E. Andreasen. Streoid receptors and steroid-binding plasma proteins studied by partitioning. In "Partitioning in Aqueous Two-Phase Systrems. Theory, Methods, Uses, and Applications to Biotechnology". (H. Walter, D.E. Brooks and D. Fisher, eds.) pp 315-326, Academic Press (1985)

24   E.C. Unger, T. Winokur, P. MacDougall, J. Rosenblum, M. Clair, R. Gatenby and C. Tilcock. Hepatic metastases: liposomal Gd-DTPA-enhanced MR imaging. Radiology 171: 81 (1989)

25   P. Molyneux. Synthetic polymers. In "Water, a Comprehensive Treatise, vol 4, Aqueous Solutions of Amphiphiles and Macromolecules", (F. Franks,ed.) pp 569-757. Plenum Press, New York and London (1975)

26   C.P.S. Tilcock and D. Fisher. The interaction of phospholipid membranes with poly(ethylene glycol). Vesicle aggregation and lipid exchange. Biochim. Biophys. Acta 688: 645 (1982)

27   K.C.Ingham. Protein precipitation with polyethylene glycol. Methods in Enzymology 104: 351 (1984)

28   K. Arnold, A. Herrmann, L. Pratsch and K. Gawrisch. The dielectric properties of aqueous solutions of poly(ethylene glycol) and their influence on membrane structure. Biochim. Biophys Acta 815: 515 (1985)

29   D.Fisher. Mechanisms of cell fusion induced by polyethylene glycol. Royal Swedish Academy of Engineering Sciences (IVA)- Rapport 330 pp 67-102 (1987)

30   D. Fisher and A.H. Goodall. Membrane fusion by viruses and chemical agents. Techniques in Cellular Physiology P115: 1 (1981)

31   A.M.J. Blow, G.M. Botham, D. Fisher, A.H.Goodall, C.P.S. Tilcock and J.A. Lucy. Water and calcium in cell fusion induced by poly(ethylene glycol). FEBS Lett 94: 305 (1978)

32   Q.F. Ahkong, D. Fisher, W. Tampion and J.A. Lucy. Mechanisms of cell fusion. Nature 253:194 (1975)

33   T.J. Aldwinckle, Q.F. Ahkong, A.D. Bangham, D. Fisher and J.A. Lucy. Effects of poly(ethylene glycol) on liposomes and erythrocytes. Permeability changes and membrane fusion. Biochim. Biophys. Acta 689: 548 (1982)

34   D.Papahadjopoulos, W.J. Vail, W.A. Pangborn and G. Poste. Studies on membrane fusion II. Induction of fusion in pure phospholipid membranes by calcium ions and other divalent metals. Biochim. Biophys. Acta 448: 265.(1976)

35   E.Unger, P. MacDougall, P. Cullis and C. Tilcock. Liposomal Gd-DTPA: effects of encapsulation on enhancement of hepatoma model by MRI. Magn. Res. Imaging 7: 417 (1989)

36   C.Tilcock, E. Unger, P. Cullis and P MacDougall. Liposomal Gd-DTPA: preparation and characterization of relaxivity. Radiology 171: 77-80 (1989)

37   A.L. Klibanov, K. Muruyama, V.P. Torchilin and L. Huang. Amphipathic polyethyleneglycols effectively prolong the circulation time of liposomes. FEBS Lett. 268: 235 (1990)

38   G. Blume and G. Cevc. Liposomes for the sustained drug release in vivo. Biochim. Biophys. Acta 1029: 91 (1990)

INFLUENCE OF POLAR POLYMERS ON THE AGGREGATION AND FUSION OF MEMBRANES

Klaus Arnold, Mathias Krumbiegel, Olaf Zschörnig,
Dieter Barthel, and Shinpei Ohki[*]

Institute of Biophysics, Medical School, University of
Leipzig, Leipzig, Germany
[*]Department of Biophysical Sciences, SUNY, Buffalo, USA

INTRODUCTION

Fusion and aggregation of liposomal membranes have attached much
attention in conjunction with the great importance of fusion processes in
cell physiology. It has been reported that several polar polymers are able
to induce the aggregation and fusion of liposomes. Poly(ethylene glycol)
(PEG) is known to induce the fusion at relatively high concentrations
(Tilcock & Fisher, 1982). This polymer has also been used frequently to
fuse biological cells for practical purposes, e.g. production of hybridoma
cells. It has been shown that some polysaccharides as dextran induce the
aggregation and fusion of liposomes (Sunamoto et al., 1980). The polysaccharides are important components of cell surfaces and studies of the
interactions with phospholipid surfaces can contribute to the understanding of their role in membrane-membrane interactions.

Charged polymers seem to be good candidates for electrostatic interactions with phospholipid surfaces. So far fusion of liposomes was mainly
described for cationic polymers such as polylysine and polyamine. Such
studies could contribute to the understanding of the protein-mediated
fusion which can involve both hydrophobic and electrostatic interaction
between the protein and the participating membranes. First reports about
fusion in the presence of anionic polymers were given by Beigel et al.,
1988, and Keren-Zur et al., 1989. Glycosaminoglycans (GAG) comprise
negatively charged polymers such as heparin and chondroitin sulfate which
are important components of the extracellular matrix. Dextran sulfate has
a structure very similar to GAG and this polymer was used instead of GAG
in many investigations. The interest in this polymer increased because it
was used as an anti-atherosclerotic drug (Radhakrishnamurthy et al., 1978)
and potent agent against HIV infection (Mitsuya et al., 1988). Studies of
the fusion of virions with liposomes and erythrocyte membranes have given
some insights into the mechanism of the action of dextran sulfate (Arnold
et al., 1990c, Ohki et al., 1991).

The effects of polymers on liposome aggregation are strongly dependent on the mode of interaction of the polymer with the phospholipid
surface. Therefore, mainly the molecular mechanisms of the adsorption or
exclusion processes of polymers are to be discussed in this contribution.
Microelectrophoretic measurements were used for the detection of the

polymer adsorption. Special attention was given to the contribution of the hydration forces to the interaction between polymers and phospholipid surfaces. From the behaviour of polymer adsorption mechanisms of the aggregation process were derived.

Fusion of liposomes is only possible after a strong approaching of membranes is realized in the aggregation state. The mechanisms which lead to fusion from strongly approaching membranes are not well understood. In fusion processes induced by both charged and uncharged polymers, lipid vesicles are brought into close contact by quite different interactions of polymers with phospholipids. Therefore, different mechanisms of polymer-induced fusion should be expected.

INTERACTIONS OF POLAR POLYMERS AND MECHANISMS OF PARTICLE AGGREGATION

Polar Interaction

The following discussions are restricted to the effects of polar polymers on the aggregation and fusion of vesicles and the virus fusion with vesicles and cells. The term "polar" is used for polymers that have appreciable dipole moments or/and a high tendency to form hydrogen bonds. Of course, no material is completely polar and apolar interactions such as Lifshitz-van der Waals interactions also occur. There is experimental evidence that strong repulsive forces can occur for the interaction between such polymers (e.g. PEG (Van Oss et al., 1990a)) and between polymers and phospholipid membrane surfaces (Arnold et al., 1988a). The physical nature of these repulsive forces can be compared with the hydration repulsion observed between surfaces of phospholipid bilayers (Rand & Parsegian, 1989). By use of X-ray diffraction and the osmotic stress technique the existence of such forces was directly shown for the interaction of biopolymers such as DNA (Rand et al., 1985) and some polysaccharides (Rau & Parsegian, 1990). At short distances these forces are quantitatively much stronger than the Lifshitz-van der Waals attraction and the electrostatic repulsion.

It was found that the surfaces of many water-soluble polymers are very strong proton acceptors and weak proton donors (Van Oss et al., 1987a, 1988). Such molecules were termed as monopolar. This behaviour has a strong influence on the intermolecular interaction in water. The water molecules of hydration are almost equally oriented on the surface of more extended polymers due to the proton acceptor property of the polymer: The water molecules are bound to the polymer surface with the hydrogen atoms pointing to the polymer strand. This orientation is mediated to the next layers of water and the orientation decreases in layers which are more distant from the polymer surface. A repulsive interaction between the polymer molecules results which decays exponentially with distance and the decay length is in the order of 0.2 nm. Despite the short decay length the effects are still measurable at large distances due to their high magnitude.

Very recently, a theory for a quantitative description of the total interaction energy between strongly polar molecules in water was developed (van Oss et al., 1987, 1988b). In this theory the surface tension components of the apolar and polar interactions are used. These parameters are determined by contact angle measurements of a material using different apolar and polar liquids. From the application of this theory and the use of experimental data results a strong repulsion of PEG molecules dissolved in water (van Oss et al., 1987). It was found that many other polar polymers and biological macromolecules exhibit the same property.

The recognition of the existence of such forces for polar polymers allows an understanding of several processes observed with polar polymers, such as the phase separation of different polymers in aqueous solutions (Van Oss et al., 1987b), anomalously high osmotic pressure (Van Oss et al., 1990a), pronounced solubility in water (Van Oss & Good, 1989), steric stabilization of particles (see below) and the exclusion of polymers from polar surfaces of phospholipid bilayers (Arnold et al., 1988, 1990b).

Monopolarity is also a property of phospholipid bilayer surfaces. Vesicles and cells repel each other owing to the existence of the repulsive hydration force (polar repulsion), even in the absence of an electrostatic repulsion. To achieve particle destabilization or membrane fusion this repulsion must be overcome. Ca^{2+} is a very effective agent for the enhancement of fusion processes. Very recently it was shown that Ca^{2+} very strongly reduces the proton acceptor capacity or monopolarity of the phospholipid surface (Van Oss, 1988). The reduction of the monopolarity simultaneously changes the mode of hydration of the surface. In other words, Ca^{2+} increases the hydrophobicity of the bilayer surface. Thus the hydration repulsion is sufficiently reduced for membranes to make a close molecular contact. It is one of the most important properties of polar interactions that the magnitude and sign of the forces can be strongly influenced by small changes of surface properties, such as the binding of Ca^{2+}.

Influence of Polar Polymers on Particle Aggregation

The effects of polymers on the interaction of particles is strongly dependent on the mode of interaction of the polymer with the particle surface and the polymer-polymer interaction. Different possibilities of polymer-particle interaction are compared in Fig. 1 and the influences on particle-particle interaction are discussed.

Steric Stabilization: The adsorption or covalent attachment of polar polymers in such a manner that the polymer chains extend into the aqueous phase results in a repulsion of the particles due to steric hindrance, changes in the conformational entropy and excluded volume effects. For charged polymers an additional contribution results from the electrostatic repulsion. These effects were extensively discussed by Napper (1983) and the term "Steric Stabilization" was introduced (Fig. 1a). More recently the importance of polar (hydrogen bonding) repulsion between the polymer layers of the surfaces was realized for the explanation of steric stabilization (Van Oss, 1991 a; Van Oss et al., 1990a). A strong repulsion was recently reported between surfaces coated with poly(ethylene glycol) immersed in water using an Israelachvili device for the measurement of forces (Claesson & Golander, 1987).

As expected from this model the attachment of polyoxyethylene-containing surfactants to phospholipid bilayers results in an increase of the bilayer repulsion (Arnold et al., 1986). These surfactants consist of a hydrocarbon (HC) group (e.g. alkane chain) and a polyethylene oxide (PEO) chain $((OC_2H_4)_nOH)$. The surfactants are incorporated into the bilayer system in such a manner that the HC part is imbedded in the phospholipid bilayer and the PEO part extends into the water phase (Arnold et al., 1990b). In a similar manner polysaccharides that form glycocalices on cell surfaces contribute to the stability of the cell suspension. Adsorption of blood cells and platelets is also significantly reduced on surfaces covered with anchored PEO chains. Such surfaces are expected to have potential applications as blood-compatible materials. It should be emphasized that the stabilizing effect of attached polymers can only occur if the polymer has a high solubility in the liquid medium.

    <u>Bridging Mechanism</u>. The behaviour is completely changed if the polymer is capable of binding to the particle surface with different binding sites along the polymer chain (Fig. 1b). Cross-binding of particles occurs readily under appropriate circumstances. The important point here is that even for a relatively small adsorption energy per each binding site of the polymer, the cumulative action will result in a substantial energy of adsorption, especially for polymers of high molecular weight.

    As a rule this bridging mechanism can only occur at low polymer concentrations where the surface of the particles is covered to a lesser extent. At high polymer concentrations strong electrostatic, steric and polar repulsive forces due to the coverage of the particle surface prevent the mutual approach of the particles. Dextran at low concentrations is able to form bridges between erythrocytes (Van Oss et al., 1990c). As will be shown below ionic polymers are very potent candidates for the aggregation of phospholipid vesicles by formation of polymer bridges. While in the general case the interactions of polar polymers with polar surfaces are repulsive, under special circumstances electrostatic attractions between the charged groups of the polymer and the surface can overcome the repulsion in cases of opposite signs of charges. Another type of electrostatic interaction between sites of equal sign, especially negatively charged surfaces, is the formation of cation bridges between the polymer and the particle surface.

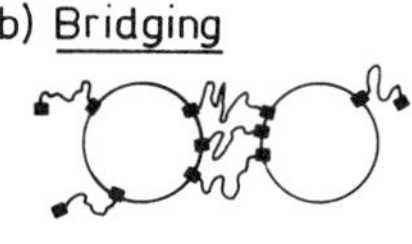
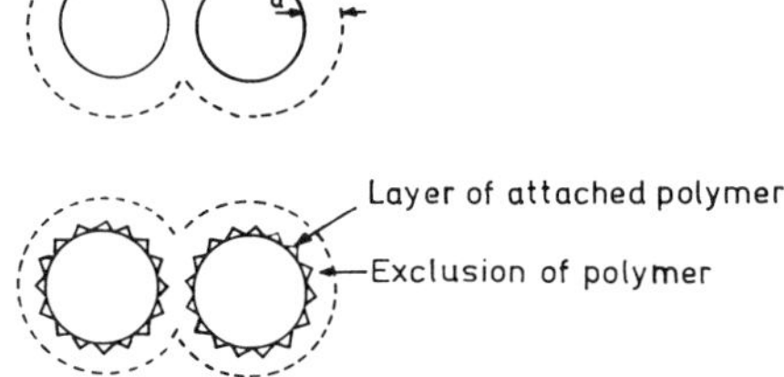

Fig. 1. Polymer-induced steric stabilization (a), bridging formation (b) and depletion flocculation (c) of particles.

    <u>Depletion Flocculation</u>: In those cases where no polymer molecules adsorb on the particle surface depletion flocculation can result (Fig. 1c). If a mutual polar repulsion between the free polymer and the particle surface occurs, polymer molecules are excluded from the particle surface and a depletion layer is formed. Particles are aggregated due to the osmotic pressure of the dissolved polymer (depletion flocculation). This mechanism is effective if the repulsion between the polymer molecules as well as between the polymer molecules and the particle surface is higher than the particle-particle repulsion. A critical polymer concentration is necessary to bring particles together close enough for destabilization.

    Depletion flocculation can also occur if part of the polymer molecules attach themselves to the particle surface (Fig. 1c). A depletion layer resulting from the repulsion between free and attached polymers occurs between the polymer surface layer and the bulk polymer. Such phenomena were observed for the effect of dextran on the stability of erythrocytes (Bäumler and Donath, 1987; Van Oss et al., 1990c). In some cases depletion flocculation was followed by depletion stabilization at higher polymer concentrations. It is assumed that this mechanism appears when the surface concentration of the polymer is higher than its bulk concentration (Van Oss et al., 1990c).

In a discussion of the depletion interaction Van Oss (1991) concluded that the existence of a depletion layer is only an accompanying effect. The primary reason for flocculation is the phase separation of polymers and particles driven by the polar repulsion between particles and polymers.

AGGREGATION AND FUSION INDUCED BY UNCHARGED POLYMERS

Physicochemical Properties of Aqueous Solutions of Neutral Polymers

PEG is one of the simplest synthetic polymers and its physicochemical properties were reviewed in Arnold et. al., 1988. From the chemical point of view the ethylene oxide group can interact with two water molecules on the basis of hydrogen bonding between the water molecules and the ether oxygens. It has been demonstrated that PEG may exercise a structuring effect on several layers of water around the polymer. This can involve up to 16 water molecules per monomer unit (Baran et al., 1972). The influence of PEG on the rotational mobility of water molecules is confirmed by a drastic decrease of the dielectric constant from about 80 for water to about 50 for 50 wt% PEG (Arnold et. al., 1985) and by the decrease of the polarity of the solution (Herrmann et. al., 1983).

We have measured the osmotic properties of aqueous PEG solutions over a wide range of concentrations (Arnold et. al., 1988). Unusually high osmotic pressures are observed, which are remarkably independent on the molecular weight. It was shown that this behaviour can be explained by a strong polar repulsion between polymer moieties dissolved in water. With increasing concentrations of PEG the average distance between the polymer molecules decreases which leads to an increase of the polar repulsion (exponential dependence on distance) and a high tendency to include water. This effect occurs with PEG to a very pronounced degree. For dextran weaker effects are observed. A theory based on the polar forces was developed which can explain the influence of the molecular weight and the concentration of the polymer on the osmotic pressure (van Oss et al., 1990a).

Exclusion of Polar Polymers from Vesicle Surfaces

A partial or complete exclusion of PEG from the water layer around water-soluble proteins was recently established (Atha & Ingham, 1981). The solvent composition is perturbed from that in the bulk in a region near the protein and the protein surface is preferentially hydrated. An exclusion of PEG from the water layer between interacting multilamellar systems has been described (Arnold et al., 1983). Evans & Needham (1988) developed a depletion based theory to explain the PEG-induced aggregation of giant phospholipid vesicles. As discussed above Van Oss (1988) found from the discussion of the polar interactions of PEG and phospholipid vesicles that the two repel each other in water.

Measurements of the electrophoretic mobility of charged phospholipid vesicles and erythrocytes have supported the idea that PEG has a lower concentration near the vesicle surface than in the bulk phase (Arnold et al., 1987). In the convential application of the Smolukhovski equation to the determination of the zeta potential the viscosity of the suspension is used. This viscosity is increased because of the addition of the polymer.

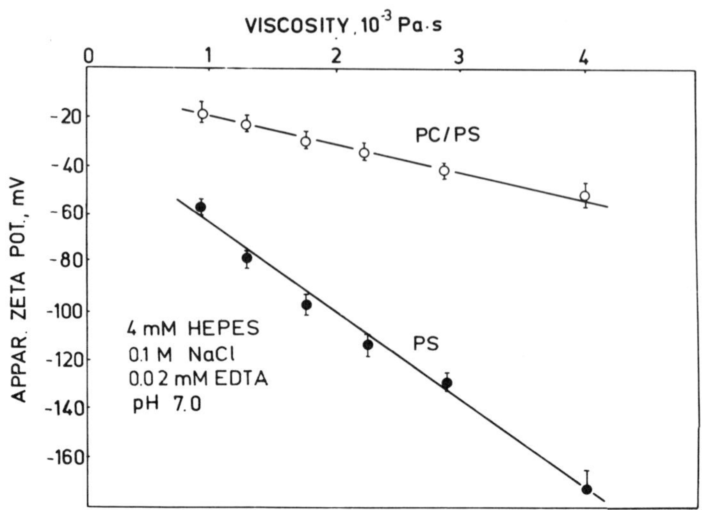

Fig. 2.   Influence of PEG 6000 on the apparent zeta potential of PS and
          PC/PS vesicles (1:1). The zeta potential calculated from equ. 2
          is plotted as a function of the viscosity of the suspension
          medium.

However, if an exclusion of the polymer from the surface of the vesicle
occurs the local viscosity near the surface is lower than the viscosity in
the bulk phase.

A position- dependent viscosity has been introduced (Arnold et al., 1987,
1990b):

$$\eta(x) = \eta_\infty \left[1 + \left(\frac{\eta_\infty}{\eta_0} - 1 \right) \exp\left(-\frac{x - x_s}{d} \right) \right]^{-1} \qquad (1)$$

     In this equation η_∞ is the bulk viscosity, η_0 is the viscosity of
the pure solvent, x is the distance from the surface, x_s is the distance
of the shear plane from the surface and d is an estimate of the width of
the depletion layer. Obviously η_∞ is the higher the higher the bulk concentration of PEG. The function $\eta(x)$ describes the decrease of the viscosity
from η_∞ in the bulk phase to η_0 near the shear plane. The Smolukhovski
equation can be formally used to define an apparent zeta potential

$$\zeta_{app} = \frac{\eta_\infty u}{\varepsilon_0 \varepsilon_r} \qquad (2)$$

where u is the electrophoretic mobility, ε_r the relative dielectric
constant and ε_0 the electric field constant. In accordance with the theory
developed by Arnold et al., 1990b, a linear dependence of the apparent
zeta potential on the bulk viscosity occurs

$$\zeta_{app} = \psi(x_s) + \left(\frac{\eta_\infty}{\eta_0} - 1 \right) \int_0^{\psi(x_s)} \exp\left(-\frac{x(\psi) - x_s}{d} \right) d\psi \qquad (3)$$

$\psi(x)$ is the potential profile, $x(\psi)$ the corresponding inverse
function, $\psi(x_s)$ is identical with the real zeta potential. $\psi(x)$ follows
from theory. Linear regression analysis gives an empirical linear function

$$\zeta_{app} = A + B\eta_\infty \qquad\qquad (4)$$

which fits the experimental data (η_∞, ζ_{app}). A comparison with eq. (3) leads to an equation for the width of the depletion layer, d. Fig. 2 shows the linear relationship between bulk viscosity and apparent zeta potential for two sets of data, as predicted by theory.

The interpretation of the experimental data developed above relies on the assumption that the real electrostatic potential ψ (x_s) remains uncharged when PEG is added to the solvent. Independent measurements of the surface potential of the vesicles by use of charged spin probes and pH-sensitive fluorescence probes did not show an influence of PEG on the surface potential (Arnold et al., 1990b; Ohki & Arnold, 1989). This finding allows the conclusion that $\psi(x_s)$ is not influenced by PEG and equ. 3 can be used for the determination of the thickness d of the exclusion layer from the measurement of the electrophoretic mobility and the bulk viscosity. The values determined for mixtures of phosphatidylcholines with egg-PA or egg-PS are in the order of 1 nm for PEG 6000.

A similar behaviour of the electrophoretic mobility was observed for erythrocytes in the presence of dextran (Bäumler & Donath, 1987) and PEG (Arnold et al., 1988a,c, Pratsch & Donath, 1988). The finding about the exclusion of the polymers from phospholipid surfaces and biological membranes is of basic importance for the interpretation of the molecular mechanisms of polymer-induced aggregation of vesicles, cells and particles in terms of depletion flocculation (Fig. 1c). For erythrocytes depletion stabilization was also observed for dextran concentrations higher than 10 wt% (Van Oss et al., 1990) which is an indication for the existence of a thin layer, which is depleted of dextran, between a dextran layer adsorbed onto the erythrocytes and the bulk dextran solution.

It is suggested that the depletion effect is significantly influenced by the hydration of the phospholipid surface as well as the hydration of the polymer molecules. Therefore, it is expected that the depletion is decreased if the phospholipid surface becomes more hydrophobic or the hydration of the polymer is reduced. Indeed, experimental evidence for a binding of PEG 20000 to didodecylphosphate vesicles at higher temperatures was given (Hoekstra et al., 1988). It was assumed that the hydration shell around PEG is disrupted due to conditions of the system close to the clouding point of PEG.

<u>Aggregation of Vesicles</u>

Experiments have shown that PEG is excluded from phospholipid vesicle and erythrocyte surfaces. As discussed above this behaviour of polymers results in depletion flocculation (Fig. 1c) of such particles. First systematic studies on the aggregation of vesicles induced by PEG were reported by Tilcock and Fisher (1982) and Boni et al. (1984). Aggregation was also observed for low density lipoproteins (LDL) and high density lipoproteins (HDL) in the presence of PEG (Arnold & Zschörnig, 1988b). Because the surface composition of lipoproteins is very similar to the surface of cell membranes these experiments show the comparable behaviour of PEG at such surfaces. Several polysaccharides, such as dextran, pullulan, hydroxyethyl starch and amylopectin were also found to aggregate phospholipid vesicles (Sunamoto et al., 1980). Whereas a complete exclusion of PEG was shown for different surfaces, the polysaccharides could form a thin layer of attached polymer with the bulk polymer separated by a depletion layer from the surface layer as shown in Fig. 1c.

The critical concentration (wt.%) of the polymer for aggregation of

the particles depends on the molecular weight of the polymer and is higher
for lower molecular weights. This concentration is also influenced by the
concentration of the particles, the pH and the surface potential of the
particles. These influences result from the properties of the energies of
polymer-polymer, particle-particle and polymer-particle interaction which
determine the process of depletion flocculation: The size of polymers
varies the polymer-polymer and polymer-particle repulsion because the
interaction energy depends on the cross sectional area of contact. The
concentrations of polymers and particles influence the average distance of
the interacting surfaces and surface potentials change the particle-
particle repulsion. The last contribution is influenced by the surface
charge of particles, ionic strength of monovalent ions, binding of multivalent ions and pH. This agrees with findings that the higher the particle
concentration the lower the critical polymer concentration for aggregation
and that the smaller particles are aggregated at higher polymer concentrations. Higher surface potentials result in an increase of the critical
polymer concentration independent of the sign of the surface potential.

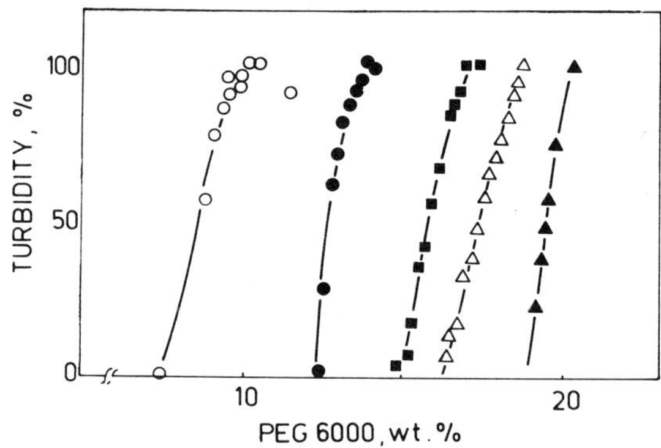

Fig. 3. Aggregation of LDL induced by PEG 6000 as a function
of LDL modification by formaldehyde. Values of the normalized
turbidity of LDL in the presence of PEG are given.
LDL concentration: 0.1 mg/ml protein
Formaldehyde concentrations: 0 ($\bigcirc$), 0.045 ($\bullet$), 0.09 ($\blacksquare$), 0.22
($\triangle$), 0.45 ($\blacktriangle$) mol/l

    As an example studies of the influence of the surface charge of LDL
on the aggregation induced by PEG 6000 are given in Fig.3. The formal-
dehyde-treated lipoproteins have an increased mobility in agarose gel
electrophoresis compared to native LDL (Arnold et al., 1989). The effect
increases with increasing concentrations of formaldehyde. These changes of
the electrophoretic mobility are interpreted as the result of an increase
of the negative surface charge of LDL due to a reduction of the number of
positively charged amino groups on the LDL surface caused by these
chemical modifications. A PEG concentration of about 8 wt% is necessary to
induce a marked increase of the turbidity. Modified LDL shows higher
critical concentrations of PEG for aggregation. In the experiments it was
found that the higher the extent of modification, the higher is the
critical concentration of PEG for aggregation. This result is in agreement
with the theoretical prediction that a higher electrostatic repulsion of
particles results in increased polymer concentration for aggregation.

The aggregation of small PS vesicles by PEG starts at about 10 wt% PEG 6000. This concentration is shifted to lower values in the presence of Ca^{2+} due to the binding of this cation, reducing the surface potential (Ohki & Arnold, 1989). Similar influences of surface charge and ions were found for many other charged vesicles (Zschörnig et al., 1990a, Yamazaki et al., 1989).

At the critical concentration of PEG the particles are aggregated as a result of the phase separation process. Once the vesicles are aggregated, the mutual approaching of membranes increases with the increase of the PEG concentration. If it is assumed that a depletion of PEG in the gap between particle surfaces occurs, a compression of particles results from the osmotic effect of PEG outside the gap. The thickness of the water layer between PC bilayers was measured by use of ^{2}H-NMR and a reduction to about 10 water molecules per phospholipid molecule was found for 30 wt.% PEG 6000 (Arnold et al., 1983; Arnold et al., 1988a; Arnold & Gawrisch, 1991). An exponential relation between the thickness of the water layer and the osmotic pressure exerted by PEG was found for all lamellar systems studied. Thus, the highly positive value for the interaction of PEG in water and the high osmotic pressure are the real reason for the flocculating power of PEG for vesicles and cells.

Fusion of Vesicles and Surface Dielectric Constant

Unlike other nonionic polymers, PEG itself is able to induce fusion of lipid vesicles to a greater extent at concentrations larger than about 15 wt% for molecular weights of PEG higher than 1000 (Blow et al., 1978; Tilcock & Fisher 1979; Boni et al., 1981, 1984; Mac Donald, 1985; Parente & Lentz 1986; Hoekstra 1982; Ohki & Arnold, 1989). Dextran can induce limited fusion of large vesicles, but at a much lower rate than PEG (Boni et al., 1981; Parente & Lentz, 1986). Fig.4 shows the phospatidylserine vesicle fusion induced by PEG and Ca^{2+}. In the absence of PEG a remarkable fusion starts at concentrations of Ca^{2+} of about 1 mM. Without Ca^{2+} the fusion occurs for PEG 6000 concentrations higher than 25 wt%. In the presence of both fusogens fusion is observed for subthreshold concentrations of PEG and Ca^{2+}. For 0.6 mM Ca^{2+}, the fusion already starts between 5 and 10 wt% PEG, and in the presence of 15 wt% PEG, a strong fusion occurs at a concentration of Ca^{2+} as low as 0.2 mM. These results are in agreement with measurements of the fusion of PS and PS/PE vesicles under the influence of Ca^{2+}, Mg^{2+} and PEG (Hoekstra, 1982). A facilitation of the Ca^{2+}-induced fusion by PEG was also found for didodecylphosphate vesicles (Rupert et al., 1988).

The aggregation of the PS vesicles under the same experimental conditions as used in the fusion experiments occurs for PEG concentrations larger than 10 wt%. As stated above, addition of Ca^{2+} reduces the threshold concentration of PEG for aggregation. Compared to the threshold concentration for the fusion, higher PEG concentrations are necessary to bring the vesicles from the aggregated to the fused state. For the relatively high concentrations of PEG used for the fusion of vesicles the membranes of apposed vesicles are intensely compressed due to the high osmotic pressure of the PEG molecules. It is assumed that the close contact of the membranes is a necessary condition for two membranes to fuse. The combined effect of Ca^{2+} and PEG results from the decrease of the electrostatic repulsion by Ca^{2+} and the increase of the membrane approach from depletion flocculation of PEG. Measurements of the surface potential have shown that the approach of vesicles creates improved conditions for a stronger binding of Ca^{2+} to the bilayer surface (Ohki & Arnold, 1989). This stronger binding is directly related to the increase of the hydrophobicity of the membrane surface as an important step in the fusion events (Ohki & Ohshima, 1984; Ohki & Arnold, 1989, 1990).

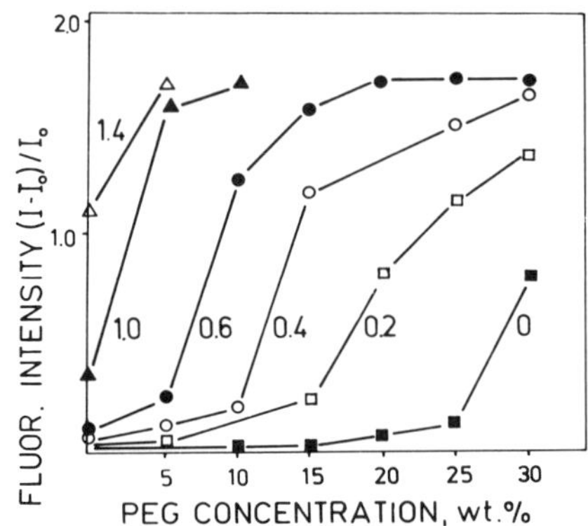

Fig. 4.   Fluorescence   intensity   of   the   N-NBD-PE   probe   in   a
mixture of 0.05 mM PS vesicles labelled with N-NBD-PE and N-Rh-PE
and   0.1   mM   unlabelled   PS   vesicles   as   a   function   of   the   concentration   of   PEG   6000   in   the   presence   of   different Ca^{2+} concentrations   (the   numbers   given   in   the   figure   correspond   to   millimolar Ca^{2+} concentrations).   The   increase   of   the   NBD   fluorescence
given   in   the   figure   results   from   a   reduction   of   the   resonance
energy   transfer   from   NBD   to   Rh   fluorophore   due   to   a   dilution   of
the   fluorophores   in   the   vesicles   after   fusion   of   labelled   and
unlabelled   vesicles   (Struck   et   al.,   1981).

The   energy   of   adhesion   of   water   to   the   membrane   surface   depends   on
the   nature   of   the   membrane   surface   (Ohki,   1988).   Many   biological   membrane
interfaces,   including   lipid   membrane   surfaces,   are   usually   strongly   hydrophilic   or   monopolar   in   nature   and   their   repulsion   is   much   greater   than   the
other   attractive   interaction   energies   as   discussed   before.   Binding   of
divalent   and   polyvalent   cations   makes   the   membrane   surface   strongly   hydrophobic   which   was   demonstrated   by   the   measurements   of   the   increase   of
surface   tension   and   the   decrease   of   the   surface   dielectric   constant   (Ohki
& Arnold, 1990).

A   good   correlation   between   the   extent   of   PEG-induced   fusion   and   the
decrease   in   dielectric   constant   was   found   (Ohki   &   Arnold,   1990).   In   Fig.5
the   change   of   the   dielectric   constant   is   given   as   a   function   of   the   PEG
concentration.   These   experiments   can   be   directly   compared   with   the   fusion
experiments   given   in   Fig.   4   because   the   same   vesicle   preparations   were
used.   A   good   correlation   of   the   decrease   in   the   dielectric   constant   with
the   fusion   properties   is   observed.   In   the   presence   of Ca^{2+} the   fusion
occurs   for   dielectric   constants   lower   than   8.   However,   PEG-induced   fusion
appears   at   higher   dielectric   constants   (12-14)   in   the Ca^{2+}-free   system,
indicating   that   the   membrane   surfaces   are   still   more   hydrophilic   than   the
PS   membrane   in   the   presence   of Ca^{2+}.   This   suggests   that   additional
mechanisms   are   necessary   for   membrane   fusion   in   the   PEG   system   where   PEG
destabilizes   the   region   of   contacting   membranes.   The   vesicles   are
flattened   due   to   the   withdrawel   of   free   water   from   the   vesicles   by   PEG   and
fusion   could   occur   at   the   highly   curved   boundaries   as   proposed   by   Mac
Donald   (1985)   and   Parente   &   Lentz   (1986).   The   detergent- like   properties
of   PEG   could   also   contribute   to   the   bilayer   destabilization   (Arnold   et
al.,   1988a).   In   addition,   a   transient   leakage   of   internal   contents   from
the   vesicles   was   observed   at   the   concentrations   that   caused   fusion
(Yamazaki   et   al.,   1989,   1990).   These   authors   proposed   a   mechanical   stress
model   for   the   mechanism   of   PEG-induced   fusion.   The   actions   of Ca^{2+} on   the

membrane surface dielectric constant are in agreement with the effect of Ca^{2+} on the monopolarity of the surface. It was found that Ca^{2+} reduces the proton acceptor capacity of the surface (van Oss, 1988) and the surface tension is increased (Ohki, 1988).

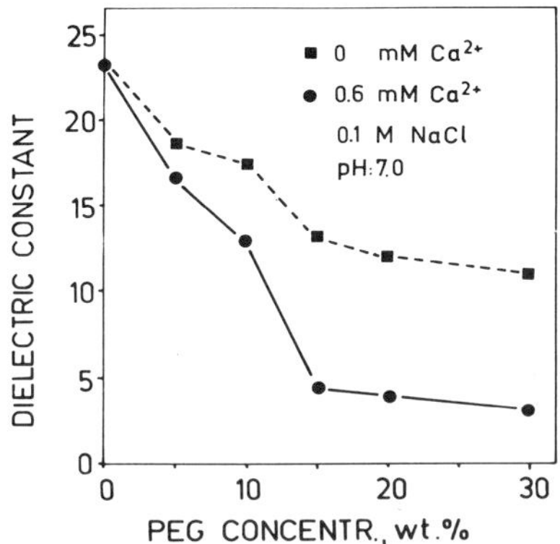

Fig. 5. Dielectric constant of PS vesicles in the presence and absence of 0.6 mM Ca^{2+} as a function of the concentration of PEG 6000. The dielectric constant was determined from the position of the fluorescence signal of Dansylphosphatidylethanolamine (DPE) incorporated in the PS vesicles at a molar ratio PS/DPE = 300, suspended in 0.1 M NaCl/ 4 mM TES/0.02 mM EDTA, pH 7.0 (details of the method are given in Ohki & Arnold, 1990).

AGGREGATION AND FUSION OF LIPOSOMES INDUCED BY CHARGED POLYMERS

Adsorption

The charged macromolecules with fusogenic properties (listed in Tables I. and II.) adsorb on liposomes via electrostatic binding of charges of their polymeric subunits with opposite charges on the membrane surface. A binding to zwitterionic, but neutral phospholipids does not occur. Polyanions bind only to vesicles with a positive net charge whereas polycations adsorb only on acidic vesicles. The positive net charge can be reached by incorporation of charged amphiphilic molecules in the bilayer or by adsorption of divalent ions on the liposome surface. For the polypeptide polylysine a hydrophobic contribution to the binding at high polylysine/lipid ratios is discussed (Carrier and Pezolet, 1984) but the main binding force is of ionic nature.

For the high molecular weight polymers as dextran sulfate 500 with a high charge density a typical high affinity binding can be observed (Krumbiegel and Arnold, 1990 and Fig. 6). At relatively low concentrations of the polymer a strong saturation of the binding occurs. This is in agreement with theoretical considerations and experimental results of adsorption processes of other polymers on colloidal particles (van der Schee, 1984). The binding energy increases due to the concerted binding of the great number of subunits interacting with the particle surface. The saturation may result from electrostatic and steric repulsion forces between polymers free in solution and bound macromolecules. The tendency exists that the shorter polymers are adsorbed to a lesser extent and the saturation is reached at higher concentrations.

The adsorption is connected with some structural effects on the
bilayer. The polyanionic DNA, dextran sulfate and glycosaminoglycans
induce an elevation of the main phase transition temperature (see Fig. 7
and the refs. Grudzev et al., 1982; Bichenkov et al., 1988). As shown in
Fig. 7 an unperturbed and a shifted peak appear arising from distinct
membrane regions that do not interact or interact, respectively, with the
polyanions dextran sulfate and heparin. The position of the shifted peak
does not depend on the polymer concentration, but the intensity ratio of
the peaks is influenced. Because of the constant temperature of the
shifted peak it can be assumed that, once they are bound to a membrane
region, a constant stoichiometric relation between the polyanions and the
other components involved (lipid, calcium) occurs. From Kim & Nishida
(1977) it is known that a very high amount of calcium is included in the
lipid-dextran sulfate complex. Possibly, the dehydrating effect of calcium
combined with a lateral fixation of the lipid molecules by the polymer
causes the elevation of the phase transition temperature. Comparable to
the results shown in Fig. 7 dextran sulfate with its high molecular
weight and charge has a greater influence, which is indicated by a higher
shift of the phase transition temperature. At high dextran sulfate concentration the undisturbed peak fully disappeared, i.e. all lipid is involved
in the interaction. Differential scanning calorimetry and fluorescence
experiments with unilamellar dipalmitoylphosphatidylcholine liposomes in
the presence of Mg^{2+} gave very similar results after addition of dextran
sulfate and heparin (Grudzev et al., 1982; Budker et al. 1989). The
occurrence of two phase transitions in unilamellar as well as multilamellar liposomes indicates a coexistence of domains with bound polymers
and uncovered regions.

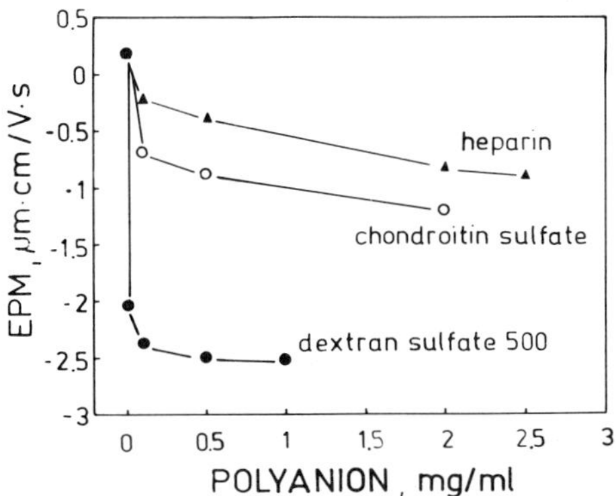

Fig. 6.   The binding of dextran sulfate (MW 500 000), heparin and chondroitin sulfate to 5 mol% stearylamine-containing multilamellar
dimyristoylphosphatidylcholine (DMPC) liposomes was investigated
by microelectrophoresis measurements. In the absence of the polyanions the vesicles have a positive electrophoretic mobility
(EPM) due to their positive surface potential. Binding of the
negatively charged polymers renders the EPM strongly negative. It
is clearly shown that dextran sulfate with its much higher chain
length and charge per subunit binds with very high affinity. The
saturation is reached at low polymer concentrations. In the same
experiments done with pure DMPC liposomes the polyanions did not
effect the EPM, i.e. no binding occurred. (Lipid concentration:
0.1 mg/ml, temperature: $20^{\circ}C$, buffer: 10 mM Tris, 145 mM NaCl,
0.3 mM EDTA, pH 7.4)

Direct evidence was given that polylysine induces a domain formation in mixtures of negatively charged and neutral phospholipids (Hartmann et al., 1977; Hartmann and Galla, 1978). This effect is caused by a condensation of the negative lipids associated to the polycation that leads to a lateral phase separation. Also in liposomes containing one negative

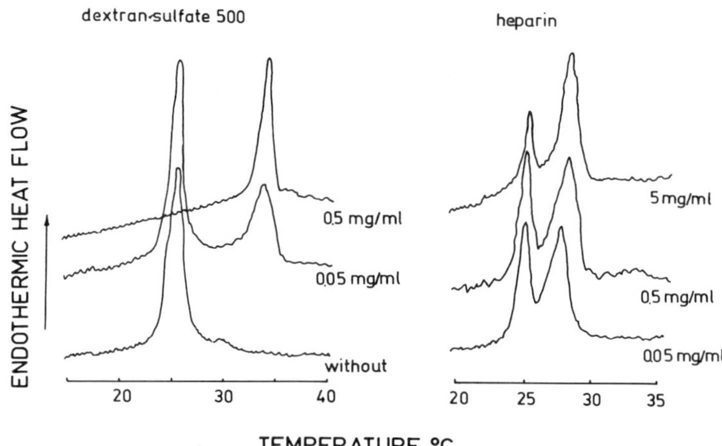

Fig. 7. Differential scanning calorimetry experiments reveal the influence of dextran sulfate 500 and heparin on the main phase transition temperature of multilamellar DMPC liposomes in the presence of 10 mM calcium (lipid concentration: 0.5 mg/ml, heating rate: 0.5 grd/min, buffer: 10 mM Tris, 145 mM NaCl, pH 7.4).

phospholipid a lateral phase separation induced by polylysine can be observed (Carrier et al., 1985; Carrier and Pezolet 1986). The phase transition temperature in the perturbed regions is shifted to higher values (Carrier et al., 1985; Houbre et al., 1988) which is connected with a stabilization and closer packing in the acyl chain matrix of the bilayer (Carrier and Pezolet, 1986; Laroche et al., 1988). On the other side, the structure of the polymer itself is changed with binding which is indicated by an altered ratio of fractions occurring in α-helices, ß-sheets or random coils (Fukushima et al., 1988; Laroche et al., 1988; Carrier and Pezolet, 1986). These structural changes seem in turn to increase the binding affinity of the long polylysines to the lipid membranes.

Another structural event that correlates well with the fusion process (see below) is an increase of the membrane permeability induced by dextran sulfate, heparin (Budker et al., 1990) and the polycations listed in Table II (Gad et al., 1982, 1986; Wang and Huang, 1984; Oku et al., 1986).

<u>Aggregation</u> <u>and</u> <u>Fusion</u>

In Tables I and II and Figs. 8 and 9 the fusions of some liposomal systems by polyanions and polycations are given.

Table I.   Polyanion induced fusion of liposomes

Polyanion	Lipid and liposome type		References
Dextran sulfate	PC, PC/PE (+ calcium)	SUV	Arnold et al. (1990a)
	PC (+ calcium)	SUV	
		LUV	Budker et al. (1990)
	PC/Stearylamine	SUV	Zschörnig et al. (1991)
	PC/chol/DEBDA[OH]	REV	Keren-Zur et al. (1989)
Heparin	PC (+ calcium)	SUV	
		LUV	Budker et al. (1990)
DNA	PC/Chol/DEBDA[OH]	REV	Keren-Zur et al. (1989)
PASP	PC/chol/DEBDA[OH]	REV	Beigel et al. (1988) Keren-Zur et al. (1989)

Abbreviations: PC, phosphatidylcholine; Chol, cholesterol; SUV, small unilamellar vesicles; LUV, large unilamellar vesicles; REV, reverse-phase evaporated vesicles; PASP, poly(aspartic acid);DEBDA[OH], a quaternary ammonium detergent (see Keren-Zur et al., 1989).

Table II.   Polycation induced fusion of liposomes

Polycation	Lipid and liposome type		References
Polylysine	PC/PS	SUV	Walter et al. (1986)
	PC/PE/CL	SUV	Gad et al. (1982) Gad (1983)
	PE/PC/PS	LUV	Uster & Deamer (1985)
	PC/CL (+ mannitol)	LUV	Gad et al. (1986)
Poly-histidine	PE/PA, PE/CL, PE/PG, PC/PS, PC/PG, PC/CL, Sph/PS, PS, PG, CL PE/PS	SUV	Wang & Huang (1984)
	PE/PC/PS(/Chol)	LUV	Uster & Deamer (1985)
Polyallyl-amine	PC/PS, PS	SUV	Oku et al. (1986)
Poly-ethyleneimine	PC/PS, PS	SUV	Oku et al. (1986)

Abbreviations: PS, phosphatidylserine; PE, phosphatidylethanolamine; PG, phosphatidylglycerol; CL, cardiolipin; PA, phosphatidic acid; Sph, sphingomyelin; for other abbreviations see Table I.

A prerequisite for the fusion process is an aggregation that brings the vesicles into close contact. Polyion induced aggregation is explained by a bridge formation of the polymers between adjacent particles as shown in Fig. 1b. This mechanism could be clearly evidenced by experiments with aspartic acid (Keren-Zur et al., 1989) and polyhistidine (Wang and Huang, 1984) in its polymeric and monomeric forms. In contrast to the polymer the monomer with only one binding site was not able to aggregate and fuse vesicles. In agreement with the adsorption process the high molecular weight polyions often cause a stronger aggregation and fusion than smaller ones (Budker et al., 1990; Gad et al., 1986) or lower concentration of the polymer (mass per volume) is necessary to induce similar effects (Gad et al., 1982, Fig. 9). For large polyions as high molecular weight dextran sulfate and polylysine a stabilization, i.e. decreased aggregation, and inhibition of fusion can be observed at high polymer concentrations (Budker et al., 1990; Zschörnig et al., 1991; Gad et al., 1986 and Figs. 8 and 9).

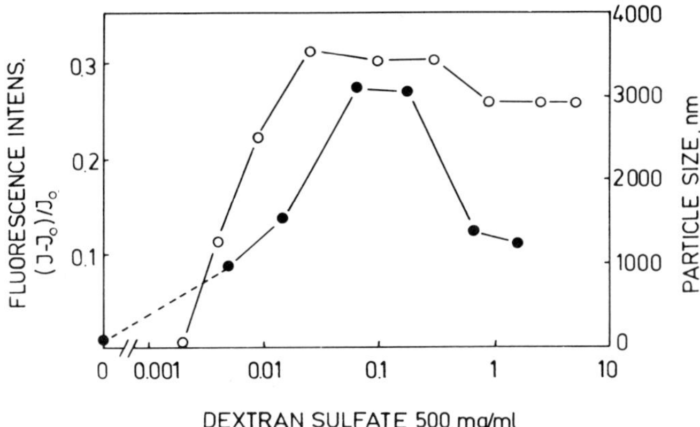

Fig. 8. The aggregation (●) and fusion (○) of small unilamellar egg
PE/DMPC (10 mol% egg PE) vesicles by dextran sulfate 500 in the
presence of 2 mM calcium were measured by QELS (●) and a
fluorescence fusion assay (○). The particle size is a direct
indicator for the aggregation process. As a fusion assay the
"probe dilution" method (Struck et al., 1981) was used. The
increase of the fluorescence intensity of an N-NBD-PE probe in a
mixture of vesicles labelled with N-NBD-PE and N-Rh-PE and unlabelled vesicles monitors the fusion process (J, J_o: N-NBD-PE
fluorescence in the presence and absence of dextran sulfate,
respectively). In this experiment the polymer was added subsequently. Once it has occurred the fusion is irreversible. In
contrast the size of the aggregates is reduced again after
addition of excess amounts of the polymer due to strong repulsion
between the polymer coated surfaces of the (partly fused)
liposomes. (Fusion assay: 0.16 mg/ml lipid (labelled), 0.3 mg/ml
lipid (unlabelled vesicles), 37°C; QELS: 0.3 mg/ml lipid, 30°C;
buffers: 10 mM HEPES, 50 mM NaCl, pH 7.4)

This effect is obviously due to electrostatic and steric repulsion forces
between the liposomes that are tightly covered with the polymers.

The influence of pH and ionic strength in the medium on aggregation
and fusion confirms the electrostatic nature of the processes of polyion
adsorption and inter-vesicle forces. For instance for polylysine and
polyhistidine it was shown that fusion occurs only in the range where the
polymers are charged (Uster and Deamer, 1985).

Comparable to other fusion processes described in the literature
small unilamellar vesicles are more easily fused by polyions than large
unilamellar vesicles. For instance polylysine induces fusion of large
unilamellar liposomes composed of phosphatidylcholine and cardiolipin only
in the presence of the (probably dehydrating) agent mannitol (Gad et al.,
1986) or promotes only its calcium-triggered fusion (Gad et al., 1985).

In most experiments fusion promoting substances as phosphatidylethanolamine and detergents were incorporated in the liposomes.
For the polyanion dextran sulfate it was shown that the fusion occurs with
pure phosphatidylcholine vesicles (in the presence of calcium) without

such destabilizing agents (Fig. 9 and Arnold et al., 1990a; Budker et al., 1990).

The detailed mechanism of polyion induced vesicle fusion is unclear. It may be assumed that vesicle crosslinking and membrane destabilization are the crucial points in this process. Aggregation brings the membranes of the vesicles into close contact. In all cases polyion induced fusion is accompanied by a leakage of the liposome content indicating the occurrence of defects in the membranes. Such defects may arise at the boundaries between domains formed by the adsorption. A phase separation or domain formation was also observed for all polyions. According to the theory of Papahadjopoulos et al. (1977) such defects or local destabilizations may serve as initiation sites for the fusion process.

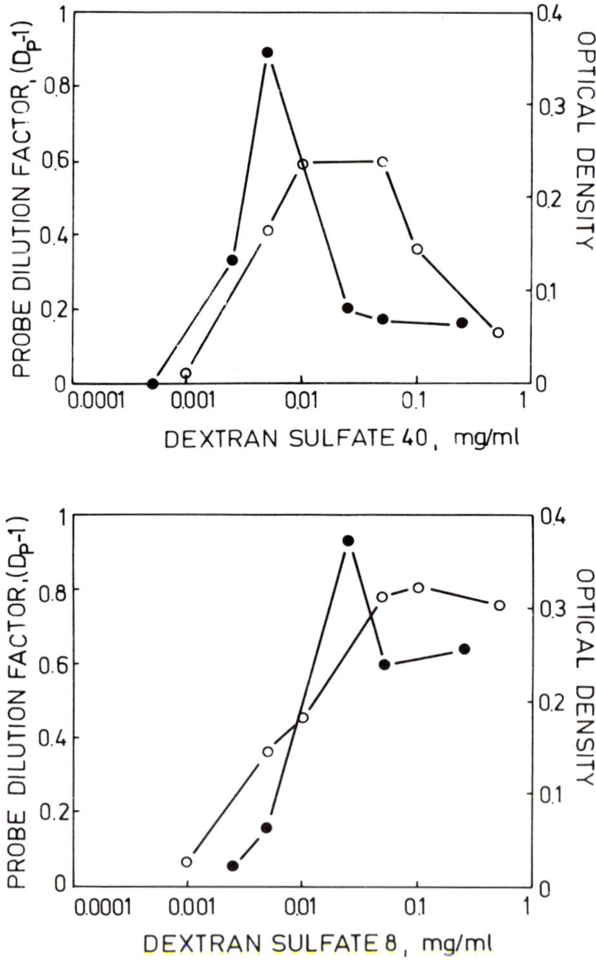

Fig. 9. Aggregation and fusion of small unilamellar DMPC vesicles in the presence of 2 mM calcium by dextran sulfate 8 (MW 8000) and dextran sulfate 40 (MW 40 000). The aggregation was observed by

INFLUENCE OF POLAR POLYMERS ON THE FUSION OF VIRUSES WITH LIPOSOMES AND ERYTHROCYTES

The elucidation of virus-membrane fusion processes is of special interest because of their clinical relevance. The fusion of viruses with liposomes depends on the solution properties (temperature, pH, ionic strength) as well as on the types of viruses and the composition of the liposomes (Haywood & Boyer, 1984). In contrast to biological membranes, the fusion of Sendai viruses with negatively charged phospholipid membranes is an unspecific process (Klappe et al., 1986). However, the use of liposomes instead of host cells provides a system which is simple enough for experimental manipulations and for experiments under definite conditions. The Sendai virus contains two major surface proteins: HN plays a role in virion adsorption on the cell surface and F is the fusion protein. For the fusion of acidic phospholipid membranes and Sendai virions, only the presence of HN in viral envelope seems to be sufficient to induce fusion of viral envelope-lipid membranes (Amselem et al., 1986). On the other hand, for virus-cell fusion systems the presence of both proteins (HN and F) is necessary.

Dextran sulfate influences the adsorption of viruses on cells (Bengtsson, 1965). It is also able to suppress the binding of HIV to CD 4 cells (Mitsuya et al., 1988) and it is used against HIV-1 infection in vitro (Ueno & Kuno, 1987). For Sendai virus - phosphatidylserine SUV fusion it was found that the fusion inhibition by dextran sulfate, heparin, chondroitin sulfate is higher at lower pH (pH = 5.0) (Zschörnig et al., 1991). The extent of fusion inhibition for the polymers is changed in the order dextran sulfate 40000 > heparin > dextran sulfate 8000 > chondroitin sulfate (Zschörnig et al., 1991). For Influenza virus-DOPC/Chol-LUV fusion no influence of pH on the inhibition by DS (MW 500000) could be measured (Lüscher-Mattli & Glück, 1990). The effect of dextran sulfate 40000 (1 mg/ml) on the fusion between Sendai virus and phosphatidylserine SUV is demonstrated in Fig. 10. Compared to the control sample the E/M ratio in the presence of dextran sulfate is decreased, indicating a less extent of fusion.

Ohki et al. (1991) found that Sendai virus - erythrocyte ghost fusion is suppressed by the polymers in the order dextran sulfate > heparin > chondroitin sulfate. The lower the pH of the solution, the more effective were the polymers in suppressing Sendai virus - erythrocyte ghost fusion.

---

the increase of the optical density at 500 nm ($\bullet$). For the fusion assay with Pyr-PC (O) the probe concentration dependence of the excimer/ monomer fluorescence ratio (E/M) was used to detect bilayer intermixing (Amselem et al., 1986). When unlabelled and labelled membranes fuse the probe concentration and thus E/M is reduced. The probe dilution factor that increases parallel to the fusion is defined by $(D_p-1)=[(E/M)_o-(E/M)]/(E/M)$ where $(E/M)_o$ is the value before fusion. Each point represents a separately prepared sample. The maximum of aggregation coincides quite well with the maximum of fusion. For dextran sulfate 8 only a small inhibition of fusion and no significant reduction of the aggregation can be found at excess polymer concentrations. At high concentrations the larger dextran sulfate 40 stabilizes the suspension which is connected with a low extent of fusion. (Fusion assay: 0.02 mg/ml DMPC, 10 mol% 10-Pyr-PC (labelled vesicles), 0.04 mg/ml DMPC (unlabelled vesicles), 35°C; optical density measurements: 0.1 mg/ml, 20°C; buffers: 10 mM Tris, 100 mM NaCl, pH 7.4)

In general, the suppression of virus - membrane fusion is concentration dependent. The effect tends to saturate at higher concentrations for each polymer. The suppression of the virus-membrane fusion depends also on the ionic strength of the solution; the lower the ionic concentration, the higher the inhibition by dextran sulfate, heparin and chondroitin sulfate (Zschörnig et al., 1991).

It is obvious that the mere existence of anionic groups such as $-SO_3^-$ does not guarantee inhibition. Measurements done with Na_2SO_4 (Zschörnig et al., 1991) or glucosamin 1,6-disulfate (Lüscher-Mattli & Glück, 1990) did not lead to an inhibition of fusion. Probably, the existence of anionic polymeric structures is a prerequisite. The higher the concentration and molecular weight of the anionic polymer, the higher is the inhibitory effect (Lüscher-Mattli & Glück, 1990, Ohki et al., 1991, Arnold et al., 1990c, Zschörnig et al., 1991).

Sendai virus exhibits a negative surface charge under physiological conditions as was measured by microelectrophoresis (Haywood, 1974). Decreasing the pH to 5.0 reduces the average negative surface charge and an increased binding of anionic macromolecules is possible. Taking into account that the isoelectric points of the viral HN and F glycoproteins are expected to be 6.5 and 4.9, respectively (Shimuzu et al.,1974), the ave. ʒe surface charge at pH 5 tends to be zero or positive. Because of greater electrostatic attractions the binding of anionic macromolecules is then higher.

In fact an aggregation of Sendai virus by dextran sulfate, heparin and chondroitin sulfate was found for pH values lower than 7, which can be explained by a binding of the polymer to the virus surface and the formation of polymer bridges between virus particles (Ohki et al., 1991). The interaction of the polymers with Sendai virus was measured by turbidity (Ohki et al., 1991). An increase in turbidity of a Sendai virus sample was found at concentrations higher than 1 μg/ml.

From all these experiments it can be concluded that these anionic polymers bind electrostatically to the virus surface. The binding of these polymers to the target membranes (PS-SUV, PC-Chol LUV, Ery ghosts) used can be excluded (see previous section).

It was observed that the rate and the extent of fusion of Sendai virus with PS membranes are increased in the presence of PEG (Nir et al., 1986), especially the rate constant of fusion. The function of PEG in this process is seen in a close approach of the virus and liposome membrane enabling a better penetration of the virus fusion protein into the liposome membrane (Hoekstra et al., 1989), probably due to the ability of PEG to dehydrate the local contact area between viral and target membranes (Hoekstra et al., 1989, Nir et al., 1986). It is known that not only PEG causes dehydration of phospholipid surfaces, also dextran, polyvinylpyrrolidin and polyvinyl alcohol. Interestingly enough, Lüscher- Mattli and Glück (1990) did not measure a facilitated fusion of Influenza virus and DOPC/Chol LUV in the presence of up to 30 μg/ml Dextran (MW: 70000). Further investigations are necessary to clarify this point.

Incubation of the Sendai virus with PEG 6000 in the presence of the erythrocyte ghost membranes also leads to a doubling of the fusion extent (Hoekstra et. al., 1989). These results are interpreted as due to PEG mimicking the hydrophobic environment, which exists in the stage of close contact of the viral and target membranes.

Our experiments show (Fig.10) that the presence of 8 wt.% PEG 6000 in the solution increases the extent of fusion between Sendai virus and phosphatidylserine SUV. The additional presence of dextran sulfate (1 mg/ml) still has a strong inhibiting effect on the fusion process in the presence of PEG. However, the fusion is higher by about the same factor as PEG increases the fusion in the system free of dextran sulfate. The observation that PEG is able to reduce the inhibition effect of dextran sulfate may be explained by opposite actions of these polymers: PEG increases whereas dextran sulfate prevents the close approach of the virus and liposome membrane due to the adherence of this negatively charged and hydrophilic molecule to the virus surface.

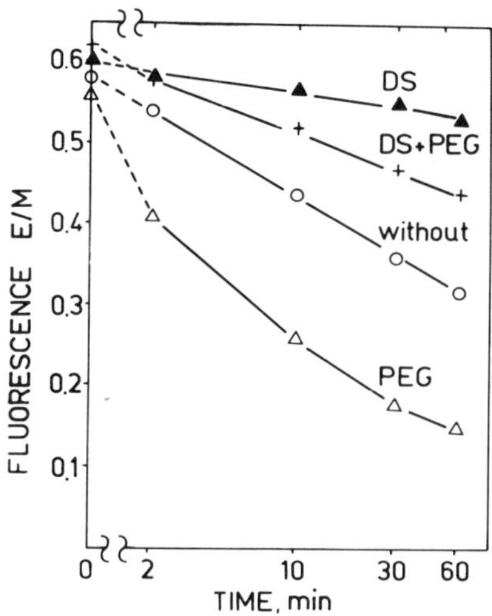

Fig. 10. Effect of PEG 6000 and DS 40k on the fusion of Sendai virus with phosphatidylserine liposomes. Buffer: 5 mM HEPES, 0.1 M NaCl, pH 6.0. Fusion was monitored by measurement of the excimer to monomer fluorescence ratio without polymer (O), with 8 wt.% PEG 6000 ($\triangle$), with 1 mg/ ml DS 40k ($\blacktriangle$), with 8 wt.% PEG 6000 and 1 mg/ml DS 40k (+).

CONCLUSIONS

The interaction of uncharged polar polymers with phospholipid surfaces is strongly influenced by the polar repulsion resulting from the hydration of polymers and phospholipids. A complete exclusion from the surface was observed for PEG due to a high monopolarity and a low Lifshitz-van der Waals attraction. The aggregation of liposomes is driven by the phase separation of polymer molecules and liposomes (depletion flocculation).

Charged polymers are strongly adsorbed on liposome surfaces of opposite charge. The aggregation of liposomes results from the formation of polymer bridges between the surfaces. In the presence of Ca^{2+} a binding of anionic polymers to uncharged liposomes can also occur.

The diversity of interactions of charged and uncharged polymers with phospholipid bilayers results in different fusion mechanisms. For Ca^{2+}-induced fusion the importance of a "hydrophobic" contact between contacting membranes was demonstrated. PEG-induced fusion has suggested that the high ability of PEG to aggregate membranes could be a sufficient requirement for the initiation of fusion without a need for a drastic increase of the hydrophobicity of bilayer surfaces. The process could be supported by the formation of defect structures. The combination of the effects of both fusogens, i. e. the compression of bilayers by PEG and the increase of surface hydrophobicity by Ca^{2+}, increases the extent of fusion. Compared to polysaccharides the effect of PEG on liposome fusion is larger due to the high tendency of this polymer to be excluded from phospholipid surfaces and the relatively high osmotic pressures. That is the reason for the wide application of this polymer to the artificial fusion of cells.

The fusion of anionic and cationic liposomes by cationic and anionic polymers, respectively, could have properties in common with Ca^{2+}-induced fusion. The polymers are sandwiched between the bilayers. Therefore, it seems to be clear that a major function of charged polymers in liposome fusion is the promotion of liposome aggregation by neutralization of the surface charge. Further studies are required to elucidate the additional molecular events of membrane destabilization. It is possible that lipid reorganization leads to the fusion or mixing of membranes at points where a close contact of phospholipid membranes is mediated by the polymer. In some cases the addition of membrane active agents is necessary.

Anionic polymers are able to fuse neutral liposomes in the presence of Ca^{2+} concentrations comparable to extracellular concentrations. The cation forms bridges between the phospholipid head group and the polymer. An optimum polymer concentration for fusion occurs.

Anionic polymers can induce the fusion of liposomes under definite circumstances whereas most biological fusion processes are inhibited by these polymers. This was shown for virus-membrane fusion. It is suspected that the anionic polymer is adsorbed on positively charged domains of virus proteins influencing the adsorption process of the virus on the membrane surface.

REFERENCES

Amselem, S., Barenholz, Y., Loyter, A., Nir, S., and Lichtenberg, D., 1986, Fusion of Sendai virus with negatively charged liposomes as studied by pyrene-labelled phospholipid liposomes, Biochim. Biophys. Acta, 860:301-313.
Arnold, K., Pratsch, L., and Gawrisch, K., 1983, Effect of poly(ethylene glycol) on phospholipid hydration and polarity of the external phase, Biochim. Biophys. Acta, 728:121-128.
Arnold, K., Herrmann, A., Pratsch, L., and Gawrisch, K., 1985, The dielectric properties of aqueous solutions of poly (ethylene glycol) and their influence on membrane structure, Biochim. Biophys. Acta, 815:515-518.
Arnold, K., Lvov, Y.M., Szögyi, M., and Gyorgyi, S., 1986, Effect of poly(ethylene oxide)-containing surfactants on membrane-membrane interaction, studia biophys., 113:7-14.
Arnold, K., Zschörnig, O., Herold, W., and Barthel, D. 1987, Effect of PEG on the electrophoretic mobility of liposomes, Mol. Cryst. Liq. Cryst., 152:357-362.
Arnold, K., Herrmann, A., Gawrisch, K., and Pratsch, L., 1988a, Water-mediated effects of PEG on membrane properties and fusion, in "Molecular Mechanism of Membrane Fusion", Ohki, S., Doyle, D., Flanagan, T.D., Hui, S. W., Mayhew, E., eds., Plenum Press, New York, pp 255-272.
Arnold, K., and Zschörnig, O., 1988b, Aggregation of human plasma low density lipoproteins by means of poly(ethylene glycol), Biomed. Biochim. Acta, 47:949-954.
Arnold, K., Zschörnig, O., Pratsch, L., Donath, E., Barthel, D., and Herrmann, A., 1988c, Exclusion of PEG from membrane surfaces and

aggregation of liposomes and lipoproteins, <u>studia biophys.</u>, 127: 113-120.

Arnold, K., Arnhold, J., Zschörnig, O., Wiegel, D., and Krumbiegel, M., 1989, Characterization of chemical modifications of surface properties of low density lipoproteins, <u>Biomed. Biochim. Acta</u>, 48:735-742.

Arnold, K., Ohki, S., and Krumbiegel, M., 1990a, Interaction of dextran sulfate with phospholipid surfaces and liposome aggregation and fusion, <u>Chem. Phys. Lipids</u>, 55:301-307.

Arnold, K., Zschörnig, O., Barthel, D., and Herold, W., 1990b, Exclusion of poly(ethylene glycol) from liposome surfaces, <u>Biochim. Biophys. Acta</u>, 1022:303-310.

Arnold, K., Flanagan, T.D., and Ohki, S., 1990c, Influence of dextran sulfate on the fusion of sendai virus with liposomes, <u>Biomed. Biochim. Acta</u>, 49: 633-635

Arnold, K., and Gawrisch, K., 1991, Effect of fusogenic agents on membrane hydration, <u>Meth. Enzymol.</u>, in press.

Atha, D.H., and Ingham, K.C., 1981, Mechanism of precipitation of proteins by PEG, <u>J. Biol. Chem.</u>, 256:12108-12117.

Baran, A.A., Solomentseva, I.M., Mank, V.V., and Kurilenko, O.D., 1972, Role of the solution factor in stabilizing, disperse systems containing water soluble polymers, <u>Dokl. Akad. Nauk USSR</u>, 207:363-366.

Bäumler, H., and Donath, E., 1987, Does dextran indeed significantly increase the surface potential of human red blood cells, <u>studia biophys.</u>, 120:113-122.

Beigel, M., Keren-Zur, M., Laster, Y., and Loyter, A., 1988, Poly(aspartic acid)-dependent fusion of liposomes bearing the quarternary ammonium detergent [[[(1,1,3,3-Tetramethylbutyl)cresoxy]ethoxy]ethyl]dimethylbenzylammonium hydroxide, <u>Biochemistry</u>, 27: 660-666.

Bengtsson, S., 1965, <u>Proc. Soc. Exp. Biol. Med.</u>, 118: 47-53

Bichenkov, E.E., Budker, V.G., Korobeinicheva, I.K., Savchenko, E.V., and Filimonov, V.V., 1988, DNA interaction with phosphatidylcholine liposomes. Melting of DNA and phase transition of the lipid membrane in the complex (Russian), <u>Biol. Membrany</u> (Moscow), 5:843-851.

Blow, A.M.J., Botham, G.M., Fisher, D., Goodall, A.H., Tilcock, C.P.S., and Lucy, J.A., 1978, Water and calcium ions in cell fusion induced by PEG, <u>FEBS Lett.</u> 94: 305-310

Boni, L.T., Hah, J.S., Hui, S.W., Mukherjee, P., Ho, J.T., and Jung, C.Y., 1984, Aggregation and fusion of unilamellar vesicles by poly(ethylene glycol), <u>Biochim. Biophys. Acta</u>, 775:409-418.

Boni, L.T., Stewart, T.P., Alderfer, J.L. and Hui, S.W., 1981, Lipid-polyethylene glycol interactions: Induction of fusion between liposomes, <u>J. Membrane Biol.</u> 62: 65-70

Budker, V.G., Markushin, Yu.Ya., Vakrusheva, T.E., Kiseleva, E.V., Maltseva, T.V., and Sidorov, V.N., 1990, Ca^{2+}-mediated interaction of negatively charged polysaccharides with phosphatidylcholine vesicles (Russian), <u>Biol. Membrany</u> (Moscow), 7:419-427.

Carrier, D., and Pezelot, M., 1984, Raman spectroscopic study of the interaction of poly-L-lysine with dipalmitoylphosphatidylglycerol bilayers, <u>Biophys. J.</u>, 46:497-506.

Carrier, D., Dufourcq, J., Faucon, J.-F., and Pezolet, M., 1985, A fluorescence investigation of the effects of polylysine on dipalmitoylphosphatidylglycerol bilayers, <u>Biochim. Biophys. Acta</u>, 820:131-139.

Carrier, D., and Pezelot, M., 1986, Investigation of polylysine-dipalmitoylphosphatidylglycerol interactions in model membranes, <u>Biochemistry</u>, 25:4167-4174.

Claesson, P.M., and Golander, C.G., 1987, Direct measurements of steric interactions between mica surfaces covered with electrostatically bound low-molecular-weight polyethylene oxide, <u>J. Colloid Interface Sci.</u> 117, 366-374

Eckert, R. & Randall, D., 1978, Animal Physiology, Freeman, E.H. Co., San Francisco

Evans, E., and Needham, D., 1988, Intrinsic colloidal attraction/repulsion between lipid bilayers and strong attraction induced by non - adsorbing polymers, _in_ "Molecular Mechanisms of Membrane Fusion", Ohki, S., Doyle, D., Flanagan, T.D., Hui, S.W. and Mayhew, E., eds., Plenum Press, New York, pp 83-99.

Fukushima, K., Muraoka, Y., Inoue, T., and Shimozawa, 1988, Conformational change of poly(L-lysine) induced by lipid vesicles of dilauroylphosphatic acid, _Biophys. Chem._, 30:237-244.

Gad, A.E., Silver, B.L., and Eytan, G.D., 1982, Polycation-induced fusion of negatively charged vesicles, _Biochim. Biophys. Acta_, 690:124-132.

Gad, A.E., 1983, Cationic polypeptide-induced fusion of acidic liposomes, _Biochim. Biophys. Acta_, 728:377-382.

Gad, A.E., Bental, M., Elyashiv, G., and Weinberg, H, 1985, Promotion and inhibition of vesicle fusion by polylysine, _Biochemistry_, 24:6277-6282.

Gad, E.A., Elyashiv, G., and Rosenberg, N., 1986, The induction of large unilamellar vesicle fusion by cationic polypeptides: the effect of mannitol, size, charge density and hydrophobizity of the cationic polypeptides, _Biochim. Biophys. Acta_, 860:314-324.

Grudzev, A.D., Khramtsov, V.V., Weiner, L.M., and Budker,V.G., 1982, Fluorescence polarization study of the interaction of biopolymers with liposomes, _FEBS Letters_, 137:227-230.

Hartmann, W., Galla, H.-J., and Sackmann, E., 1977, Direct evidence of charged-induced lipid domain structure in model membranes, _FEBS Letters_, 78:169-172.

Hartmann, W., and Galla, H.-J., 1978, Binding of polylysine to charged bilayer membranes - Molecular organization of a lipid-peptide complex, _Biochim. Biophys. Acta_ 509:474-490.

Haywood, A.M., 1974, Characteristics of Sendai virus receptors in a model membrane, _J. Mol. Biol._ 83: 427-436

Haywood, A.M. and Boyer, B.P., 1984, Effect of lipid composition upon fusion of liposomes with Sendai virus membranes, _Biochemistry_ 23 4161-4166

Herrmann, A., Pratsch, L., Arnold, K., and Laßmann, G., 1983, Effect of PEG on the polarity of aqueous solutions and on the structure of vesicle membrane, _Biochim. Biophys. Acta_, 733:87-94.

Hoekstra, D., 1982, Role of lipid phase separation and membrane hydration in phospholipid vesicle fusion, _Biochemistry_ 21: 2833-2840

Hoekstra, D., Rupert, L.A.M., Engberts, J.B.F.N., Nir, S., Hoff, H., Klappe, K., and Novick, S.L., 1988, On the role of hydrophobic interactions in membrane fusion effects of poly(ethylene glycol), _studia biophys._, 127:105-112.

Hoekstra, D., Klappe, K., Hoff, H., and Nir, S., 1989, Mechanism of fusion of Sendai virus: Role of hydrophobic interactions and mobility constraints of viral membrane proteins, _J. Biol. Chem._, 264: 6786-6792

Houbre, D., Kuhry, J.-G., and Duportail, G., 1988, Effects of random copolymers on the thermotropic behaviour of dipalmitoylphosphatidylglycerol vesicles, _Biophys. Chem._, 30:245-255.

Keren-Zur, M., Beigel, M., and Loyter, A., 1989, Induction of fusion in aggregated and nonaggregated liposomes bearing cationic detergents, _Biochim. Biophys. Acta_, 983:253-258.

Kim, Y.C., and Nishida, T., 1977, Nature of interaction of dextran sulfate with lecithin dispersions and lysolecithin micelles, _J. Biol. Chem._, 252:1243-1249.

Klappe, K., Wilschut, J., Nir, S. and Hoekstra, D., 1986, Parameters affecting fusion between Sendai virus and liposomes. Role of viral proteins, liposome composition, and pH. _Biochemistry_, 25: 8252-8260

Krumbiegel, M., and Arnold, K., 1990, Microelectrophoresis studies of the binding of glycosaminoglycans to phosphatidylcholine liposomes, _Chem. Phys. Lipids_, 54:1-7.

Laroche, G., Carrier, D., and Pezolet, M., 1988, Study of the effect of poly(L-lysine) on phosphatidic acid and phosphatidylcholine/ phosphatidic acid bilayers by Raman spectroscopy, <u>Biochemistry</u>, 27:6220-6228.

Lüscher-Mattli, M., and Glück, R., 1990, Dextran sulfate inhibits the fusion of influenza virus with model membranes, and suppresses influenza virus replication in vivo, <u>Antiviral Research</u>, 14: 39-50

Mac Donald, R.I., 1985, Membrane fusion due to dehydration by polyethylene glycol, dextran or sucrose, <u>Biochemistry</u>, 24: 4058-4066

Mitsuya, H., Looney, D.J., Kuno, S., Ueno, R., Wong-Staal, F., and Broder, S., 1988, Dextran sulfate suppression of viruses in the HIV family: Inhibition of virion binding to $CD4^+$ cells, <u>Science</u>, 240:646-649

Napper, D.H., 1983, "Polymeric stabilization of colloid dispersions", Academic Press, London.

Nir, S., Klappe, K., and Hoekstra, D., 1986, Mass action analysis of kinetics and extent of fusion between sendai virus and phospholipid vesicles, <u>Biochemistry</u>, 25: 8261-8266

Ohki, S., 1988, Surface tension, hydration energy and membrane fusion, <u>in</u> "Molecular mechanisms of membrane fusion", Ohki S., Doyle, D., Flanagan, T.D., Hui, S.W. & Mayhew, E., eds., Plenum Press, New York, pp. 123-138

Ohki, S., and Arnold, K., 1989, Phospholipid vesicle fusion induced by cations and poly(ethylene glycol), <u>in</u> "Biophysics of the Cell Surface", Glaser, R. and Gingell, D., eds., Springer-Verlag, Berlin, Heidelberg, London, New York, pp 193-219.

Ohki, S., and Arnold, K., 1990, Surface dielectric constant, surface hydrophobicity and membrane fusion, <u>J. Membrane Biol.</u>, 114:195-203.

Ohki, S., Arnold, K., Srinivasakumar, N., and Flanagan, T.D., 1991, Effect of dextran sulfate on fusion of sendai virus with human erythrocyte ghosts, <u>Biomed. Biochim. Acta</u>, 50: 199-206

Oku, N., Yamaguchi, N., Yamaguchi, N., Shibamoto, S., Ito, F., and Nango, M., 1986, The fusogenic effect of synthetic polycations on negatively charged lipid bilayers, <u>J. Biochem.</u> (Tokyo), 100:935-944.

Papahadjopoulos, D., Vail, W.J., Newton, C., Nir, S., Jacobson, K., Poste, G., and Lazo, R., 1977, Studies on membrane fusion. III. The role of calcium-induced phase changes, <u>Biochim. Biophys. Acta</u>, 465:579-598.

Parente, R.A. & Lentz, B.R., 1986, Rate and extent of poly (ethylene glycol)-induced large vesicle fusion monitored by bilayer and internal content mixing, <u>Biochemistry</u> 25: 6678-6688

Pratsch, L., and Donath, E., 1988, Poly(ethylene glycol) depletion layers on human red cell surfaces measured by electrophoresis, <u>studia biophys.</u>, 123:101-108.

Radhakrishnamurthy, B., Ruitz, H.A., Srinivasan, S.R., Preau, W., Dalferes,E.R., and Berenson, G.S., 1978, Studies of glycosaminoglycan composition and biological activity of $Vessel^R$, a hyperlipidemic agent, <u>Atherosclerosis</u>, 31: 217-224

Rand, R.P., Das, S., and Parsegian, V. A., 1985, The hydration force, its character, universality and application, <u>Chemica Scripta</u>, 25:15-21.

Rand, R.P., and Parsegian, V.A., 1989, Hydration forces between phospholipid bilayers, <u>Biochim. Biophys. Acta</u>, 988:351-376.

Rau, D.C., and Parsegian, V.A., 1990, Direct measurement of forces between linear polysaccharides xanthan and schizophyllan, <u>Science</u>, in press.

Rupert, L.A.M., Engberts, J.B.F.N., and Hoekstra, D., 1988, Effect of poly (ethylene glycol) on the Ca^{2+}-induced fusion of didodecyl phosphate vesicles, <u>Biochemistry</u>, 27: 8232-8239

Shimuzu, K., Shimuzu, Y.K., Kohama, T. and Ishida, N., 1974, <u>Virology</u>, 62: 90-101

Struck, D.K., Hoekstra, D., and Pageno, R.E., 1981, Use of resonance energy transfer to monitor membrane fusion, <u>Biochemistry</u>, 20:4093-4099.

Sunamoto, J., Iwamoto, K., Kondo, H., and Shinkai, S., 1980, Liposomal membranes. VI. polysaccharide - induced aggregation of multilamellar liposomes of egg lecithin, J. Biochem., 88:1219-1226.

Tilcock, C.P.S. & Fisher, D., 1979, Interaction of phospholipid membranes with PEG, Biochim. Biophys. Acta, 577: 53-61

Tilcock, C.P.S., and Fisher, D., 1982, The interaction of phospholipid membranes with poly(ethylene glycol). Vesicle aggregation and lipid exchange, Biochim. Biophys. Acta, 688:645-652.

Ueno, R. and Kuno, S., 1987, Dextran sulphate, a potent anti-HIV agent in vitro having synergism with zidovudine, Lancet i, 1379

Uster, P.S., and Deamer, D.W., 1985, pH-Dependent fusion of liposomes using titratable polycations, Biochemistry, 24:1-8.

Van der Schee, H.A., 1984, "An experimental and theoretical study of oligo- and polyelectrolyte adsorption", thesis, Agricultural University Wageningen (Netherlands).

Van Oss, C.J., Chaudhury, M.K., and Good, R.J., 1987a, Monopolar surfaces, Advan. Colloid Interface Sci., 28:35-64.

Van Oss, C.J., Chaudhury, M.K., and Good, R.J. , 1987b, The mechanism of partition in aqueous media, Sep. Sci. Technol., 22:1515-1526.

Van Oss, C.J., 1988a, "Polar interfacial interactions, hydration pressure and membrane fusion" in "Molecular Mechanism of Membrane Fusion", Ohki, S., Doyle, D., Flanagan, T.D., Hui, S. W., Mayhew, E., eds., Plenum Press, New York, pp 113-122

Van Oss, C.J., Chaudhury, M.K., and Good, R.J., 1988b, Interfacial Lifshitz-van der Waals and polar interactions in macroscopic systems, Chem. Rev., 88:927-941.

Van Oss, C.J., and Good, R.J., 1989, Surface tension and the solubility of polymers and biopolymers, J. Macromol.Sci.Chem., A 26:1183-1203.

Van Oss, C.J., Arnold, K., Good, R.J., Gawrisch, K., and Ohki, S., 1990a, Interfacial tension and the osmotic pressure of solutions of polar polymers, J. Macromol. Sci.-Chem., A 27: 563-580.

Van Oss, C.J., 1990b, Surface free energy contribution to cell interactions, in "Biophysics of the cell surface", Glaser, R., and Gingell, D., eds., Springer-Verlag, Berlin, Heidelberg, London, New York, pp 131-152.

Van Oss, C.J., Arnold, K., and Coakley, W. T., 1990c, Depletion flocculation and depletion stabilization of erythrocytes, Cell Biophys., 17:1-10.

Van Oss, C.J., 1991, Interaction forces between biological and other polar entities in water: How many different primary forces are there?, Biocoll. Biosurf., in press.

Walter, A., Steer, C.J., and Blumenthal, R., 1986, Polylysine induces pH-dependent fusion of acidic phospholipid vesicles: a model for polycation-induced fusion, Biochim. Biophys. Acta, 861:319-330.

Wang, C.-Y., and Huang, L., 1984, Polyhistidine mediates an acid-dependent fusion of negatively charged liposomes, Biochemistry, 23:4409-4416.

Yamazaki, M., Ohnishi, S. & Ito, T., 1989 Osmoelastic coupling in biological structures: decrease in membrane fluidity and osmophobic association of phospholipid vesicles in response to osmotic stress, Biochemistry, 28: 3710-3715

Yamazaki, M. & Ito, T., 1990, Deformation and instability in membrane structure of phospolipid vesicles caused by osmophobic association: mechanical stress model for the mechanisms of poly(ethylene glycol)-induced membrane fusion, Biochemistry, 29: 1309-1314

Zschörnig, O., Arnold, K., Machill, H., Lachmann, U., and Herold, W., 1990a, Influence of surface charge on aggregation of phospholipid vesicles induced by PEG, studia biophys., 137:153-160.

Zschörnig, O., Machill, H., Wiegel, D., Arnhold, J., and Arnold, K., 1990b, Aggregation of human plasma high density lipoproteins induced by PEG, Biomed. Biochim. Acta, submitted.

Zschörnig, O., Arnold, K., and Ohki, S., 1991a, Dextran sulfate-dependent fusion of liposomes containing cationic stearylamine, in preparation.

Zschörnig, O., Arnold, K., and Ohki, S., 1991b, Effect of glycosamino-
    glycans and PEG on fusion of Sendai virus with phosphatidylserine
    liposomes, in preparation.

CONTROL OF FUSION OF BIOLOGICAL MEMBRANES BY PHOSPHOLIPID ASYMMETRY

Andreas Herrmann, Alain Zachowski[†], Phillipe F. Devaux[†] and
Robert Blumenthal[‡]

Humboldt-Universität, Fachbereich Biologie, Institut für
Biophysik, Invalidenstr. 42, D-O-1040 Berlin, Germany
[†]Institut de Biologie Physico-Chimique, 13, rue P. et M.
Curie, F-75005 Paris, France
[‡]National Institutes of Health, NCI, Bldg. 10 Rm 4B56
Bethesda, MD, 20892, USA

INTRODUCTION

Membrane fusion is one of the most fascinating properties of cellular membranes important for the homeostasis of the cell (organism). Cell-division, sperm-egg fusion and polykaryon formation in muscle and bone are typical examples of intercellular fusion reactions. Those natural cell-cell fusion processes lead to significant physiological and developmental changes. Intracellular fusion events are involved in endocytosis, exocytosis, intracellular transport and targeting. Viral infection and parasite invasion are other meaningful biological phenomenons depending on membrane fusion processes.

Fortunately, fusion of membranes is an energetically unfavorable process, otherwise heterogenous organization and functional specifity of organisms, cells and organelles would be impossible. To maintain biological specifity fusion events have to occur at desired locations and to develop in highly regulated way in order to keep specifity of composition, structure and function. Many barriers prevent membranes from fusing, e.g. steric constraints (for instance due to the presence of a cell glycocalix), electro-static and dehydration repulsion, packing of membrane lipids, and resistance to deformation. In order to overcome those restrictions distinct mechanisms must exist triggering membrane fusion in a controlled way both spatially and temporally (Blumenthal, 1987; 1988).

In an attempt to understand the complex phenomenon of membrane fusion, systems have been investigated which are more simple than real biological membranes. Examples for those rather simple systems are the electrofusion or poly(ethylene glycol) induced fusion of cell membranes, or even more simple, the Ca^{2+} -induced fusion of negatively charged phospholipid vesicles. It turned out that these artificially induced membrane fusion processes are often uncontrolled events. In addition, these investigations clearly showed that nature has developed gentle mechanism(s) of membrane fusion in order to avoid membrane and cell damage. An essential reason for using more simple membrane systems as lipid vesicles arised from the early idea that proteins (e.g. glycoproteins of plasma membranes) were considered as a barrier for membrane approach, and fusion was a result of lipid-lipid interaction. This was sustained by the following observations: (i) As deduced from electron microscopy artificially induced fusion of cell membranes was accompanied by formation of protein-depleted (lipid-rich) membrane areas which were regarded as fusing points. (ii) Often it was found that artificially induced fusion of pure lipid vesicles was rapid. From these observations the idea was developed that fusion originated from lipid-lipid interaction between protein-free patches of attached membranes.

However, from investigation of fusion of enveloped viruses with appropriate target membranes, exocytosis and, more recently, intracellular transport and targeting a different concept on the role of proteins in membrane fusion has been emerged. It has been clearly shown that specific viral spike proteins of enveloped viruses are necessary to induce fusion with the target membrane. Removal or distinct modification in structure and arrangement of those 'fusogenic' spike proteins inhibits virus fusion. Intracellular fusion along the biosynthetic pathways exhibits an absolute dependence on the presence of proteins on the surface of the Golgi membranes (for a review see Wilschut, 1989). Moreover, several proteins have been implicated in natural cell-cell fusion reactions, although their exact role in the overall process in merging of plasma membranes remains to establish (for a review see White and Blobel, 1989). Consequently, research on intra- and intercellular membrane fusion processes is mainly directed to understand the role of proteins as a fusogen.

However, although fusogenic proteins are required for most, if not all, natural occurring fusion one can assume that lipid composition and arrangement are important for effectiveness and kinetic of that process. White and Blobel (1989) proposed that in addition to proteins responsible for the specifity of cell-cell fusion events, specific proteins facilitate the bilayer destabilization necessary for fusion by interacting with lipid components of two apposed bilayers. Therefore, one may ask how lipids and their properties can effect and modulate protein-mediated membrane fusion. In particular, one may ask whether lipids in plasma and subcellular membranes may serve as a barrier for spontaneous inter- and intracellular fusion, or, alternatively to facilitate those events. It is generally assumed that those fusogenic proteins disturb bilayer structure by interacting with the lipid phase of the host membrane. One may speculate that hydrophobic sequences of those proteins induce (i) defects in lipid packing and/or (ii) nonlamellar structures resulting in fusion of apposed membranes. The relevance of the latter structure for fusion has been demonstrated by Ellen et al. (1989) in that fusion of lipid vesicles consisting of phosphatidylethanolamine (PE) or PE derivatives was caused by inverted micellar structures which were temporally formed due to the inherent properties of PE to adopt nonbilayer arrangements.

To elucidate the role of lipid composition on fusion of biological membranes, principally, two strategies exist: (i) utilization of liposomes with well defined composition, (ii) modification of lipid phase of real biological target membranes.

In the present article we will discuss only very shortly the advantage and disadvantage in using liposomes as a model system for biological membranes. Mainly we will focus on how composition and arrangement of phospholipids in biological (eukaryotic) membranes can affect artificially induced cell-cell fusion as well as natural (protein-mediated) fusion events. We will give some evidence that the biological significance of the asymmetric transversal lipid distribution of cellular membranes and its maintenance may also be justified in the inhibition (barrier) or, vice versa, support of natural fusion processes. The heterogenous asymmetric organization of membrane lipids might be an essential factor to preserve this heterogeneity. Furthermore, we want to see whether disturbance of asymmetric transversal lipid distribution affects both protein-mediated (natural) fusion processes and artificially induced fusion processes of cells. In this respect we will show that specific modifications of the transversal distribution of endogenous phospholipids offer a worthwhile way to elucidate the role of lipid composition of the target membrane in fusion. To do that one has to know how asymmetry can be affected in a definite way. For that purpose we will shortly review the present knowledge of phospholipid asymmetry, its regulation, and mechanisms of transmembrane motion of lipids in eukaryotic membranes.

LIPOSOMES AS A MODEL FOR BIOLOGICAL TARGET MEMBRANES TO STUDY THE ROLE OF LIPID COMPOSITION ON FUSION

Model membranes are still extensively used as a target to study the role of membrane lipid composition on fusion. E.g., several methods have been developed to investigate and to probe fusion of enveloped viruses and phospholipid vesicles. As early as 1974 Haywood has shown that Sendai virions are able to fuse with lipid vesicles. Those model systems provide a transparent way to elucidate the influence of the lipid state on fusion. The chemical composition of those lipid systems differing in their headgroups and hydrocarbon chains can be varied easily. Thus, it is possible to define the headgroup specifity, the role of physical properties of the bilayer on fusion, e.g. charge, lipid packing, phase transition. Indeed, studies using liposomes as a target have suggested that physical parameters, such as lipid packing and the surface hydration, may important factors modulating the ability of viruses to fuse (Nir et al., 1986, Klappe et al., 1986).

Incorporation of other components as glycolipids and cholesterol allows to approach more closely the properties of natural (plasma) membranes. Those studies have implied that some viruses require the presence of cholesterol for fusion. However, the relevance of cholesterol in enveloped virus fusion is still controversial. It has been claimed that fusion of Semliki Forest Virus with phospholipid vesicles requires cholesterol (Kielian and White, 1984). Similar observations were reported for fusion of Sendai virions with liposomes. Only very little fusion activity was observed with liposomes consisting of neutral phospholipids such as phosphatidylcholine (PC). Incorporation of cholesterol into PC liposomes renders them susceptible to fusion with Sendai virions (Citovsky et al., 1985; Chejanovsky et al., 1988). In contrast, Haywood and Boyer (1984) did not observed any effect of cholesterol on Sendai virus fusion with liposomes containing zwitterionic phospholipids and gangliosides.

However, some doubts might exist as to the relevance of results based on liposomes to fusion of natural membranes. An important criticism concerns the size of liposomes. Commonly used small unilamellar vesicles (< 300Å) are not recommended to resemble fusion behavior of natural membranes. In particular, problems arises with the high degree of curvature accompanied by lipid packing constraints. Those vesicles can fuse spontaneously and under conditions in which large liposomes do not show any fusion (Wilschut and Hoekstra, 1986). Therefore large unilamellar liposomes should be used for fusion studies resembling more closely the properties of natural membranes. A crucial factor is also lipid composition. For instance, it has been clearly shown for influenza as well as Sendai virus that fusion with liposomes consisting of negatively charged lipid is unrelated to their biological fusion activity. Fusion of Sendai virions with biological membranes is maximal between pH 7.0 and 9.0, whereas fusion with negatively charged liposomes is maximal at low pH values (Chejanovsky et al., 1986). Influenza virus fusion with cardiolipin liposomes does not reflect its biological fusion behavior. Even under conditions of a complete loss of biological fusion activity, fusion was observed when using cardiolipin (Stegmann et al.,

1986). One reason for this unusual behavior might be a surface pH several
units lower than in the bulk medium which leads to protein denaturation
(Stegmann et al., 1989).

PROPERTIES OF LIPIDS INFLUENCING MEMBRANE FUSION

From studies using liposomes as a target several properties of lipids
have been elucidated influencing membrane fusion processes. These are mainly
(a) shape of the lipids; (b) saturation degree of the fatty acid carbon
chains; (c) hydration and (d) electrical charge of the polar headgroup; and
related (e) packing properties and phase behavior of lipids.

Shape of lipid molecules: The effective shape of phospholipids is
determined by the polar headgroup and the hydrocarbon chains and their
properties (e.g. hydration of polar headgroup and degree of saturation of
fatty acyl chains, respectively). The shape can be characterized by the
packing parameter f depending on the molecular volume V, the area per polar
head a, and the maximum length h of the hydrocarbon tail: $f=V/(a*h)$
(Israelachvili et al., 1980). Those lipids as lysoforms in which the area of
polar heads in plane is greater than that of the hydrophobic parts are shaped
like a cone ($f<1$). If both areas are similar as in the case of diacyl-PC the
shape can be described by a cylinder ($f=1$). Diacyl-PE is a typical example
for an inverted cone ($f>1$) because the hydration of the polar head is lower
in comparison to PC. Dehydration of polar lipids and/or a decrease in packing
of hydrocarbon chain (unsaturated fatty acid residues) tend the lipid to
approach an inverted cone. The effective shape of lipid molecules is
important for the bilayer structure, in particular the spontaneous curvature
of its monolayers. Inverted cone shaped lipids as PE tend to form nonbilayer
structures such as the inverse hexagonal phase (H_{II}) or inverted micelles
(for recent review see Seddon, 1990). This arises from an imbalance between
lateral stresses from the headgroup region, the polar/nonpolar interface, and
the hydrocarbon tail of the bilayer. The important observation that most
biological membranes contain large amounts of lipids which have the inherent
property to adopt nonlamellar phases leads to the suggestion that nonbilayer
structures are of fundamental relevance for many processes connected to
biomembranes, in particular membrane fusion (Verkleij et al., 1984; Ellens
et al., 1989). It is already known that PE which is shaped like an inverted
cone is more preferable for membrane fusion than lipids of cylindrical or
cone shape as PC and lyso-PC, respectively (Nir et al., 1983). It has been
shown that fusion is hindered by the presence of lyso-PC not only in model
membranes (Chernomordik et al., 1987) but also in biological membranes, for
instance in case of myoblasts (Reporter and Norris, 1973) or Sendai virus
mediated-fusion (Poste and Pasternak, C.A., 1978).

Saturation degree of fatty acid carbon chain: The degree of saturation
of the fatty acid carbon chains of lipids effects the packing properties of
membrane lipids which can be rationalized on the basis of the conformation
of fatty acid residues. Typical characteristics imposed by unsaturated fatty
acids to the membrane structure reside in the *cis* double bonds (Brenner,
1984). When the hydrocarbon chain is in a fully extended configuration the
appearance of a *cis* bond creates a kink because of the restricted motion
about the double bound. As a consequence the molecules are prevented from
packing close together with other fatty acid chains in particular if the
double bonds are located at the same depth of the bilayer. This is
accompanied by an increased average cross-sectional area of the hydrocarbon
chain which leads for a given cross-sectional area of the polar headgroup to
an enhancement of the effective hydrophobicity of the membrane surface. As
pointed out above membrane hydrophobicity is a critical factor in membrane
fusion.Although the increase of membrane fluidity by *cis* unsaturated acids

has been shown, there is no strict correlation between the unsaturation and membrane fluidity and ordering (Stubbs and Smith, 1984; Quinn et al., 1989).

Hydration properties: The hydration of phospholipid headgroups is one of the main barrier of membrane fusion. In particular, at very close apposition of membranes in the order of several angströms the hydration forces increase exponentially and dominate the interaction of phospholipid membranes as they approach contact. The hydration of phospholipids depends mainly on the headgroup. It has been shown that the hydration of PC is much higher than that of PE (Lis et al., 1982) and phosphatidylserine (PS). Recently, it has been suggested that the non-electrostatic double layer repulsion of PS bilayers is more like that between PE bilayers than between PC bilayers (Rand and Parsegian, 1989). Dehydration of phospholipid headgroups by polymers such as poly(ethylene glycol) or by binding of cations as Ca^{2+} to negatively charged phospholipids (see below) facilitates membrane fusion.

Electrical charge of headgroups: Charged headgroups can increase electrostatic repulsion between apposing membranes if the overall charge of the headgroups in both membranes is of the same sign. Electrostatic repulsion between membranes composed of negatively charged lipids can be screened by cations. Most efficient screening by cations was achieved with Ca^{2+}. It has been shown that binding of cations, in particular Ca^{2+}, to negatively charged headgroups as phosphorylserine can trigger fusion of liposomes containing such lipids. Besides a screening of the negative charges of polar headgroups by Ca^{2+}, binding of this cation reduces significantly the hydration of the phospholipid headgroup. Even at micromolar concentrations of Ca^{2+} surface water is displaced by the bound cation which allows to completely overcome hydration repulsion. Bilayers precipitate to virtually anhydrous contact often accompanied with crystallization of hydrocarbon chains changing the phase state of the membrane (Rand and Parsegian, 1989). Binding of Ca^{2+} is accompanied by an increase of interfacial tension which is solely ascribed to the change in nature of the surface hydrophilic bilayer. It was suggested that this increase in interfacial tension arises mainly from a conformational change in the interfacial polar headgroups upon divalent cation chelation binding which can be regarded as equivalent to a change of the effective shape of lipid molecules, namely a decrease in area of polar lipid heads. This is accompanied (a) by a partial exposure of the hydrocarbon portion of the membrane molecule to the water phase (Ohshima and Ohki, 1985; Ohki, 1988) increasing the effective hydrophobicity. The degree of fusion of liposomes correlates well with the degree of increased interfacial tension or increased hydrophobicity of the membrane caused by cations (Ohki, 1984, 1988). It has been suggested that the hydrophobicity of the membrane surface is an important determinant in controlling membrane fusion processes. (b) If the area of the hydrophobic part of PS-molecules is assumed to remain unchanged in this case (Ohki, 1982), then it is equivalent to a decrease in the spontaneous curvature of the layer since the effective shape of lipids is more like an inverted cone. Another relevant factor of interaction Ca^{2+} with negatively charged phospholipids as PS in liposomes made of mixtures of lipids is the phase separation of PS with the formation of crystalline domains in membranes. Local defects of the membrane structure at the boundaries between such domains and the region of the bilayer which are in the liquid state may trigger or facilitate membrane fusion (Wilschut et al., 1985).

Another important aspect of charged headgroups is the modulation of the membrane surface pH with respect to the bulk pH (see above).

In conclusion, basically the aminophospholipids PS and PE are lipids which support membrane fusion processes. Model membranes as liposomes containing these aminophospholipids are known to undergo fusion in the presence of Ca^{2+} (Düzgünes et al., 1981, 1985; Ohki, 1982). This was not observed for membranes containing exclusively zwitterionic lipids as PC.

TRANSLOCATION AND TRANSVERSAL DISTRIBUTION OF PHOSPHOLIPIDS ACROSS EUKARYOTIC MEMBRANES

Phospholipid asymmetry in eukaryotic membranes

Earlier, intrinsic and extrinsic membrane proteins embedded or superficially anchored, respectively, to the lipid bilayer were considered as responsible for the functional properties of biological membranes. Much attention was given to their inhomogeneous lateral and asymmetric transverse arrangement in membranes. However, increasing evidence for the involvement of membrane lipids in complex cellular processes in recent years (Hanahan and Nelson, 1984; Yeagle 1989) has given important impetus to lipid biophysics and biochemistry. Much information on the lateral and transversal inhomogeneous organization and dynamics of the lipid phase has been accumulated in the past decade. Now it is well established that also phospholipids can display a transversal asymmetric distribution in biological membranes. The human erythrocyte plasma membrane is the most highly characterized system with respect to lipid asymmetry. This is mainly attributed to the fact that mammalian erythrocytes comprise only a single membrane, and difficult membrane isolation and resealing steps are not necessary. Usually, eukaryotic cells posse intracellular membranes corresponding up to 85% of total cell membranes.

In the human erythrocyte membrane, the aminophospholipids PS and PE are preferentially located on the inner monolayer while zwitterionic lipids as PC and sphingomyelin (SM) are on the outer layer. The inner leaflet of human erythrocyte membranes contains about 96% of PS molecules, 80% of the PE, 30% of the PC and 10% of SM. With few exceptions, a similar asymmetric distribution with a majority of aminophospholipids localized on the inner leaflet was found in erythrocytes from many animals, in the plasma membrane of fibroblasts, platelets, ascites cells, synaptosomes from brain, brush borders from intestine and kidney (for a review see Op den Kamp, 1979, Zachowski and Devaux, 1990; Herrmann et al., 1990b). The only exception so far reported are chick myoblasts, where PS is symmetrically distributed and PE preferentially localized on the outer leaflet (Session and Horwitz, 1983; see below).

The headgroup asymmetry in human erythrocyte membranes is accompanied by an asymmetrical distribution of fatty acid hydrocarbon chains with respect to their saturation degree. The double bound index is 1.54 for the inner leaflet and 0.78 for the outer leaflet (Middelkoop, 1989). For instance, polyunsaturated fatty acid chains of PE are preferentially localized in the inner layer whereas saturated and monoene species of PE are enriched on the outer leaflet. About 88% of PS has 18:0 fatty acyl chain in position 1 and 18:1 (4.7%), 20:3, 20:4 (61.4%), 22:4 (7.5%), 22:5, 22:6 (6.5%) in position 2 (Hullin et al., 1990). In contrast, 66% of SM has 18:1 in position 1, and 16:1, 24:0 or 24:1 in position 2 (Middelkoop 1989).

Phospholipid asymmetry in eukaryotic cell organelles is less well established. For instance, contradictory data have been reported for the transversal phospholipid distribution in the endoplasmic reticulum (ER) and nuclear membranes (see Zachowski and Devaux, 1990). Recent results suggest that phospholipids are randomly distributed in the ER (Herrmann et al., 1990a). Consistent data were reported for the sarcoplasmic reticulum where the distribution of PC is symmetric, PE is preferentially oriented outside and PS inside (Herbette et al., 1984). Evidence was given that in membranes of chromaffin granules and in synaptic vesicles PE appears to localized

preferentially on the outer (cytoplasmic) leaflet (Buckland et al., 1978;
Michaelson et al., 1983). No data are available on transversal distribution
of fatty acid chains in intracellular membranes of eukaryotic cells.

Transverse motion of phospholipids in eukaryotic membranes

An important question concerns the origin and maintenance of
phospholipid asymmetry in plasma and intracellular membranes. So far, there
is no report that such a pronounced and stable asymmetric distribution as
shown for natural membranes can be established in vesicles of lipid mixtures.
The phenomenon of phospholipid asymmetry is also related to the problem
whether asymmetric lipid distribution is a static or dynamic equilibrium,
that means, whether phospholipids are able to move between the outer and
inner leaflet of the membrane. Transverse motion of lipids in artificial and
biological membranes has been investigated extensively. Here, we consider
only the transmembrane movement of phospholipids in biological membranes. As
we will see understanding the mechanisms of this movement enables the control
of the transbilayer distribution of phospholipids in biological membranes,
in particular in plasma membranes.

Several mechanisms involved in transmembrane movement of lipids in
biological membranes have been established (Zachowski and Devaux, 1990): (i)
simple diffusion; (ii) facilitated diffusion and (iii) active transport. As
shown recently only the latter process created an asymmetric distribution of
phospholipids between the inner and outer leaflet of biological membranes.
In contrast, simple and facilitated diffusion creates a random phospholipid
arrangement in the absence of any other driving forces such as an electric
field, a pH gradient or high curvature of the membrane. With a few exceptions
spontaneous transmembrane diffusion of phospholipids (*simple diffusion*) is
slow with half-times of several hours or days; half-times of 2 -10 h for PC
transmembrane diffusion in human erythrocytes have been reported (Zachowski
and Devaux, 1990). However, it has been shown that this slow phospholipid
diffusion in erythrocyte membranes is still faster than in liposomes from the
lipid extract (Zachowski et al., 1985) presumably due to the presence of
integral membrane proteins. Assuming that the rate limiting step of
transverse diffusion is the occurrence of corresponding defects in both
leaflets (Homan and Pownall, 1987), it is likely that the probability of
those defects can be enhanced at boundaries of integral membrane proteins (De
Kruijff et al., 1978). *Facilitated diffusion* has been suggested to occur in
membranes of the endoplasmic reticulum. Bishop and Bell (1985) postulated the
existence of a PC transporter in membranes of the rat liver microsomes, the
location of phospholipid synthesis. Recently, we have shown that in such
membranes almost all phospholipids can diffuse rapidly by a protein-dependent
mechanism, which is independent on ATP (Herrmann et al., 1990b). This result
suggests that the phospholipid flippase in the ER is probably a nonspecific
transporter, which tends, in the absence of any other metabolic events, to
equilibrate rapidly the phospholipid composition of both halves.

Active phospholipid transport: An active, ATP-dependent, translocation
of phospholipids was first described by Seigneuret and Devaux (1984) in human
erythrocyte membranes. This translocase is specific for aminophospholipids
which are transported continuously from the outer to the inner leaflet on
expense of ATP. This outside-inside migration is much faster in comparison
to the slow, ATP independent, motion of choline-containing lipids, PC and SM,
respectively. When appropriate analogues of PS and PE were incorporated in
the outer membrane leaflet at 37°C, they rapidly translocate to the inner
monolayer with half-times of 3-5 min for PS and 30-45 min for PE, respectively (Morrot et al., 1989; Calvez et al., 1988) (Fig. 1). This fast, ATP-
dependent transmembrane movement of aminophospholipids, which was demonstrated originally by Electron-Spin-Resonance (ESR), was confirmed by different

laboratories using independent techniques (see Zachowski and Devaux, 1990). The rapid translocation of aminophospholipids requires 1 mM cytosolic Mg^{2+}-ATP. Micromolar concentrations of intracellular Ca^{2+} and vanadate inhibit the active transport. In ATP-depleted cells aminophospholipids display the same slow transverse diffusion observed for PC and SM (Seigneuret and Devaux, 1984). Competition experiments have shown that PS and PE are recognized and transported by the same protein. Systematic investigation of the substrate specifity of the flippase established a high sensitivity to all chemical modifications of the aminophospholipid headgroup (Morrot et al., 1989). Moreover, the glycerol backbone and at least a short ß-chain are required for recognition and fast inward motion via the aminophospholipid translocase. Important to note, that the chain lengths and the degree of unsaturation are not essential properties for the active transport. Recently, a PS stimulated Mg^{2+}-ATPase of molecular weight 120 kDa was enriched from solubilized human erythrocyte membranes which share precisely the same properties as the aminophospholipid translocase. Therefore, it was hypothesized that the same protein is responsible for the translocase and Mg^{2+}-ATPase activity (Morrot et al., 1990).

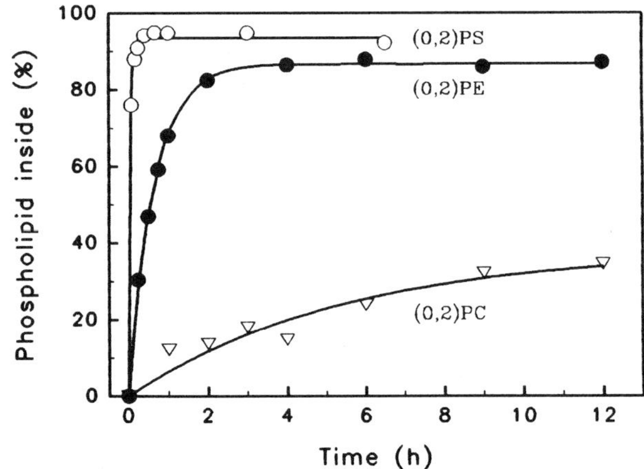

Figure 1. Kinetics of transmembrane reorientation of spin-labeled phospholipids (0,2)PC, (0,2)PE and (0,2)PS (1-palmitoyl-2-(4-doxylpentanoyl)-PC, PE, PS), respectively, initially (t=0) incorporated in the outer leaflet of human erythrocytes at 37°C. The percentage of labeled lipids on the inner monolayer is presented. The transmembrane distribution of those spin-labeled phospholipids is monitored easily by back exchange using bovine serum albumin to extract the labeled lipids remaining on the outer monolayer (for details see Calvez etal).

There is now compelling evidence that the physiological significance of the aminophospholipid translocase in several plasma und subcellular membranes relies in the formation and maintenance of an transversal asymmetric arrangement of phospholipids in the membrane. Several years ago, it was suggested that the interactions between the membrane skeleton proteins (spectrin) and the aminophospholipids could be responsible for the asymmetric distribution. However, biophysical studies established only a weak

interaction between spectrin and PS of the order of the thermal energy
(Maksymiw et al., 1987). It has been shown recently, that in spectrin-
depleted vesicles from human erythrocytes, aminophospholipids are accumulated
in the inner leaflet, as in native cells, provided that the vesicles contain
ATP (Calvez et al., 1988). Another argument in favor of a minor role of the
cytoskeleton in maintaining lipid asymmetry is the fact that the lipid
transmembrane equilibrium is not a fixed state with a fraction of lipids
anchored to spectrin but rather a steady-state with continuous inward and
outward motion of aminophospholipids across the cell membrane. As suspected
from theoretical consideration (Herrmann and Müller, 1986) the equilibrium
distribution for *all* phospholipids are in fact steady-state corresponding to
the balance between continuous inward and outward translocation of lipids
(Bitbol and Devaux, 1988; Sune et al., 1998). The accumulation of choline-
containing phospholipids in the outer layer can be simple explained by a
counter diffusion mechanism assuming that the lipid sites on each leaflet are
fixed.

An aminophospholipid translocase activity was found in red blood cell
membranes of other mammals: cows, sheep, rats (Devaux et al., 1988) and
guinea pigs (Sune et al, 1988). Selective translocation of aminophospholipids
across plasma membranes was also established for platelets, reticulocytes,
lymphocytes, and synaptosome membranes (for review see Zachowski and Devaux,
1990). Recently, it has been demonstrated that such a translocase activity
is not confined only to plasma membranes, but can be found also in subcellular membranes. It has been shown for adrenal medulla chromaffin granule,
that PS is accumulated on the external (cytosolic) leaflet by an ATP-
dependent phospholipid carrier (Zachowski et al., 1989). However, the
presence of a translocase in inner membranes seems to be not universal. As
mentioned a nonspecific, ATP-independent flippase was found in microsomal
membranes.

Physiological significance of the lipid asymmetry

The maintenance of the pronounced phospholipid asymmetry at the expense
of ATP consumption suggests some physiological significance for cells. There
are several, nonexclusive indications that the preservation of nonsymmetrical
lipid distribution is important for the homeostasis of the cells. This is
sustained by observations where the disturbance of the nonsymmetrical distribution of phospholipids is accompanied by alterations of cell properties and
functioning.

Platelet stimulation: In platelets, the transbilayer asymmetry is
rapidly lost upon activation by platelet agonists. This phenomenon could play
an important role in local blood-clotting reactions, since part of
coagulation cascade reactions are accelerated upon exposure of phosphatidylserine at the outer surface (Zwaal, 1988). One may speculate that the lipid
redistribution could be due to a concomitant fusion of granules with the
plasma membrane.

Cell-cell recognition: It has been shown that red blood cells with PS
on the outer leaflet were selectively taken up rapidly by macrophages (Tanaka
and Schroit, 1983). There is some experimental support for the hypothesis,
that recognition and removal of aged red blood cells by macrophages is
related to a partial loss of lipid asymmetry and the exposure of aminophospholipids on the exoplasmic leaflet (Herrmann and Devaux, 1990).

Membrane fluidity: There is evidence that the asymmetric fluidity of
plasma membranes is based on the asymmetric distribution of phospholipids
(Cribier et al., 1990, Herrmann et al., 1990c). Yet, the role of asymmetric
membrane fluidity remains to establish. It has been suggested that the
asymmetric viscosity of the lipid phase of red blood cell may contribute to
the determination of cell shape (Zachowski and Devaux, 1990).

Endocytosis: Recently, Devaux (1990) proposed that the aminophospholipid translocase could be involved in early stages of endocytosis. Endocytosis requires membrane invagination which must be accompanied with an accumulation of lipids on the inner leaflet. The activity of the aminophospholipid translocase could be involved in creating an excess of aminophospholipids in the inner leaflet. High levels of ATP in red blood cells can induce endocytosis (Ben-Bassat et al., 1972).

Cell shape: Randomization of lipid distribution in ATP-depleted red blood cells is accompanied by echinocyte formation. Progressive restoration of lipid asymmetry by enrichment of intracellular ATP results in a continuous shape transformation to discocytes and stomatocytes. Incorporation of exogenous phospholipids in the outer leaflet of erythrocytes also induces echinocytes. When PS or PE are incorporated in the exoplasmic monolayer, erythrocytes return to their initial shape with a time scale corresponding to the translocation of aminophospholipids from the outer to the inner leaflet (Seigneuret and Devaux, 1984).

Activity of membrane bound enzymes: It has been shown that the activity of the membrane-bound enzyme protein kinase C is dependent on the presence of PS and its concentration (Rando, 1988; Newton and Koshland, 1990). In principle, the control of the transversal distribution of PS offers a way to regulate such enzyme activities.

In the following section we will discuss the very important role of phospholipid asymmetry in membrane fusion.

EFFECT OF TRANSBILAYER PHOSPHOLIPID DISTRIBUTION IN FUSION OF BIOLOGICAL MEMBRANES

Fusion of enveloped viruses with erythrocyte membranes

The viral nucleocapsid of enveloped viruses is surrounded by a membrane, the 'envelope', which consists of viral glycoproteins and lipids forming a bilayer. For many viruses, the protein composition of the envelope is rather simple consisting as few as one or two different types of proteins. Enveloped viruses deliver their nucleocapsids into a host cell by attachment to cell surface receptors followed by fusion of the viral membrane with the cellular plasma membrane at neutral pH (e.g. Sendai virus) or at low pH in an endocytic vesicle as in the case of orthomyxoviruses (e.g. influenza) (for a review see White 1990) or rhabdoviruses (e.g. Vesicular Stomatitis virus (VSV)) (Puri et al., 1988). There is convincing evidence that specific viral proteins are not only responsible for attachment to target membrane receptors, but also for mediating fusion. It is assumed that fusion activity resides in hydrophobic domains of those viral fusion glycoproteins. Exposure of these hydrophobic sequences enable them to penetrate into the target membrane which finally results in the merging of the viral and target membrane.

To understand the mechanism of virus fusion much attention is currently given to conformational changes of viral glycoproteins leading to the exposure of the fusion sequence as well as to the interaction (incl. attachment) of the fusion protein (sequence) with the target membrane. We were particularly interested in the role of phospholipid arrangement in the target membrane in fusion of Vesicular Stomatitis virus and influenza virus (Herrmann et al., 1990c, 1991). As the biological target membrane we choose the human erythrocyte membrane, whose phospholipid arrangement can readily be modified as shown above. To monitor fusion we have used a fluorescence assay based on lipid mixing of fusing membranes. The fluorophore octadecylrhodamine (R18) is inserted into intact viral membranes at self-quenching concentrations. Relief of R18 self-quenching occurs as result of fusion with target membrane (Hoekstra et al., 1984).

Vesicular stomatitis virus: The envelope of VSV consists of a bilayer membrane with a single type of spike glycoprotein, the G protein, which is responsible for attachment to the cell surface and for induction of fusion between the viral and the target membrane (Pal et al., 1987). pH-dependent fusion of VSV with cells has been demonstrated by a variety of methods. For instance, low pH-induced fusion of VSV has been shown using Vero cells to be maximal at about pH 5.6 (Puri et al., 1988) while no fusion was observed at neutral pH.

Although the G protein has been very well studied and characterized, it is not clear what components are necessary in the target membranes to render them susceptible to VSV fusion. A number of studies have been undertaken to classify receptor activity for VSV by identifying those components on the target membrane which inhibit binding and infectivity. Such studies have implicated phosphatidylserine (Schlegel et al., 1983) and phosphatidylinositol or the ganglioside GM3 (Mastromarino et al., 1988) as significant components involved in VSV attachment and fusion. However, an "antagonist", which inhibits the activity of a given protein, is not necessarily part of the functional activity. To study the involvement of a given membrane component in the fusion reaction, that component needs to be incorporated into the target membrane, and the fusion reaction with the modified target should be examined directly. Since the transversal distribution of aminophospholipids PS and PE can be regulated by the aminophospholipid translocase activity, human red blood cell membranes are very useful for those investigations.

Even at low pH, no fusion of VSV was observed with intact red blood cell membranes having an asymmetrical arrangement of phospholipids with PS and PE predominantly exposed to the inner leaflet (Grimaldi et al., 1989). However, erythrocyte membranes with a lipid-symmetric phospholipid bilayer distribution are susceptible to low pH-induced fusion with VSV with a maximum at pH 5.6 similar as that seen for fusion of VSV with Vero cells (Puri et al., 1988). Lipid symmetrization of erythrocyte membranes can be readily achieved by preparation of ghosts in the presence of Ca^{2+}. The procedure is mainly based on that of Williamson et al. (1985), who reported that the ghosts which were lysed and resealed in the presence of Ca^{2+} lose the characteristic asymmetric distribution of membrane lipids (Ca^{2+}-ghosts), whereas ghosts which were prepared in the presence of Mg^{2+} (absence of Ca^{2+}) retain membrane asymmetry (Mg^{2+}-ghosts). Figure 1 shows rapid fusion of VSV with lipid-symmetric ghosts (Ca^{2+}-ghosts) at pH 5.5, whereas little fusion activity was observed with Mg^{2+}-ghosts or intact RBC with an asymmetric lipid distribution (not shown). Little fusion activity was also seen with ghosts prepared in the presence of Ca^{2+} and an ATP-generating system (Ca^{2+}-ATP-ghosts) (Fig. 2). Since it has been shown that the aminophospholipid asymmetry of Ca^{2+}-ghosts prepared under those conditions is restored (Verhoeven er al., 1990) and the aminophospholipid translocase is functioning in the presence of intracellular ATP and Ca^{2+} (Bitbol et al., 1987) we infer that fusion of VSV with erythrocytes is modulated by lipid asymmetry and that the fusion activity of VSV was not due to effects of intracellular Ca^{2+}.

Since PS and PE are expressed on the outer layer of the erythrocyte membrane after symmetrization, it might be that either headgroups operate as a necessary component for VSV fusion. However, there was no phospholipid headgroup specifity for VSV fusion as indicated by a variety of independent experiments involving either incorporation of phospholipid analogues into lipid-asymmetric erythrocyte membranes or modification of phospholipid headgroups in lipid-symmetric ghosts. For that purpose we have used spin labeled phospholipid analogues with a short ß-chain (0,2)PC, (0,2)PE and (0,2)PS [1-palmitoyl-2-(4-doxylpentanoyl) PC, PE, PS], respectively, which immediately incorporate into the outer leaflet of human erythrocyte membranes

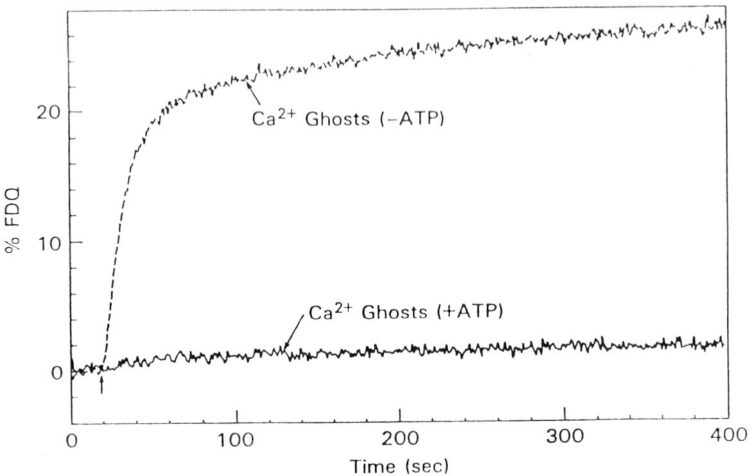

Figure 2   Role of phospholipid distribution in fusion of VSV with erythrocyte ghosts. Ghosts were prepared in the presence of Ca^{2+} (1 mM) and ATP (1.5 mM) [**Ca^{2+}-ATP-ghosts**, lipid-asymmetric] or Ca^{2+} (1mM) alone [**Ca^{2+}-ghosts**, lipid-symmetric] (for details see Herrmann et al., 1990c). R18 labeled VSV was bound to ghosts at pH 7.4 in ice-cold phosphate buffered saline (PBS). After washing R18 VSV-ghost complexes were transferred to prewarmed (37°C) PBS, pH 7.4 (t=0 sec). After 20 sec fusion was triggered by lowering the pH to 5.5 (arrow) at 37°C. The percentage of fluorescence dequenching (%FDQ) was calculated according to %FDQ=$(F-F_0)/(F_t-F_0)*100$ where F, F_0, and F_t are fluorescence values at a given time, at zero time, and after disruption of the VSV-erythrocyte ghost complexes with Triton X-100, respectively.

(Seigneuret and Devaux, 1984). By means of ESR it is very easy to establish their transversal distribution (Herrmann et al., 1990c). In order to prevent fast outside-inside translocation of PS and PE we have choosen those conditions where aminophospholipid translocase is inhibited. Incorporation of 5 mol% of spin-labeled phospholipid analogues with respect to the total membrane phospholipid into the outer leaflet of red blood cells with an intact asymmetry of endogenous phospholipids or lipid-asymmetric ghosts did not render those membrane fusogenic to VSV. The amount of spin-labeled lipids inserted corresponds to the % endogenous PS in lipid-symmetric membranes. Even in the presence of (0,2)PS no enhancement of fusion was observed. An inhibitory effect of these spin labels on fusion can be excluded since fusion of VSV with lipid-symmetric ghosts was not affected in the presence of (0,2)PS. The lack of headgroup specifity in fusion of VSV with target membranes was confirmed by treatment of lipid-symmetric ghosts with PS-decarboxylase which converts PS to PE. This conversion did not affect VSV fusion suggesting that (a) the serine headgroup is not a necessary determinant in the fusion of VSV with the target membrane and (b) the ethanolamine group is not a critical factor in VSV fusion, otherwise one would expect an enhancement of VSV fusion with more PE in the target membrane.

In conclusion, VSV fusion is not dependent on any particular phospholipid headgroup (summary of the data see Table) and the presence of sialoglycolipids on the outer surface of the target membrane appears not to

Table. 1. Effect of red blood cell (RBC) preparation and treatment on low
pH induced fusion with VSV at 37°C. *PS refers to the spin
labeled phospholipid analogue (0,2)PS [1-palmitoyl-2-(4-
doxylpentanoyl) PS] which was incorporated into the outer leaflet
of intact RBC membranes (5 mol% of total membrane phospholipid).
PSD - PS decarboxylase.

RBC prepara-tion	Asymmetry of endogenous phospholipids			Treatment of erythro-cyte membranes	Fusion with VSV
	lost	main-tained	restored		
intact RBC		+		-	no
intact RBC		+		insertion of *PS into the outer leaflet	no
Mg^{2+}-ghosts		+		-	no
Ca^{2+}-ghosts	+			-	yes
Ca^{2+}-ATP-ghosts			+	-	no
Ca^{2+}-ghosts	+			PSD-treat-ment: conversion of PS to PE	yes

be sufficient for VSV fusion. This is consistent with results of Yamada and
Ohnishi (1986) who showed that fusion of VSV with liposomes was rather
independent on any phospholipid headgroup specifity, but cis-unsaturated
acyl chains were required for efficient fusion. The requirement of
unsaturated acyl chains suggests that lipid packing properties of the target
membrane might be important for VSV fusion. This is consistent with our
observation of a correlation between acyl chain mobility in the outer leaflet
of erythrocyte membranes and fusion activity of VSV. Recently, we have shown
by ESR measurement an increase of acyl chain mobility of the outer leaflet
after lipid-symmetrization of erythrocyte membranes using phospholipid spin
labels (0,2)PC distributed exclusively in the outer leaflet (Herrmann et al.,
1990c). Therefore, we attribute the differences of fusion activity between
lipid-symmetric and lipid-asymmetric erythrocyte membranes to lipid packing
depending on the transversal distribution of phospholipids. As discussed
above the two leaflets of the normal erythrocyte differ in lipid packing,
with the outer leaflet being more tightly packed than the inner leaflet
(Tanaka and Ohnishi, 1976; Seigneuret et al., 1984), presumably as a result
of higher levels of unsaturated fatty acid chains in the aminophospholipids
which are preferentially on the inner leaflet. In the lipid-symmetric ghosts,
those unsaturated acyl chains have presumably moved to the outer monolayer
leading to a decrease of lipid packing which is accompanied by an enhanced

acyl chain mobility. Recently, it was shown by photobleaching technique that the difference in lipid diffusivity of the two leaflets of intact (asymmetric) erythrocyte membranes may be accounted for solely by a difference in phospholipid composition, and may be independent on cholesterol distribution and protein asymmetry (Cribier et al., 1990). Moreover, a large increase in staining of lipid-symmetric ghosts with the fluorescent dye merocyanin 540 indicates a decrease in lipid packing of the outer leaflet (Williamson et al., 1985).

Influenza virus: Influenza virus delivers its nucleocapsid into the cytosol of the host cell by attachment of its envelope to cell surface receptors followed by endocytosis and fusion of the viral membrane with the endosomal membrane at acidic pH (White et al., 1983). Both attachment and fusion are mediated by influenza virus spike protein, the hemagglutinin (HA). The final fusion is preceded by a series of activation steps. It has been shown that the pH-dependence of influenza virus fusion correlates with that of a conformational change of the HA (Doms et al., 1985). However, the molecular steps leading to fusion are unclear at the present time. It is assumed that one important step of the fusion process is the exposure of a hydrophobic amino acid sequence of the HA_2 subunit and its penetration into the lipid phase of the target membrane (Brunner 1989). Similar to VSV we have examined by the R18 assay the influence of phospholipid arrangement of the target membrane on influenza virus (A/PR 8/34) fusion choosing erythrocyte membranes as the biological target.

Although influenza virus fusion was not exclusively for lipid-symmetric membranes, and was observed also for lipid asymmetric membranes, we found that lipid symmetrization modulates the fusion characteristics of influenza virus prebound to the target at pH 7.4. Independent on the erythrocyte membrane preparation used fusion was maximal for the given target at a pH of about 5.0, 37°C. Rates of fusion are enhanced if the outer leaflet of target membrane is enriched in phospholipids with unsaturated fatty acyl chains. Fusion with lipid-symmetric ghosts was almost completed 80 sec after lowering the pH to 5.0 at 37°C. A slower, but still fast fluorescence dequenching was established for influenza with lipid asymmetric ghosts reaching the final level of fluorescence dequenching after about 200 sec under these conditions. The initial rates show the sequence lipid-symmetric (Ca^{2+}-ghosts) > lipid-asymmetric (Mg^{2+}-ghosts) > intact RBC over the whole pH range investigated (pH 4.8 – 6.4). The initial rate of influenza virus fusion with Ca^{2+}-ghosts was decreased after restoration of phospholipid asymmetry by including ATP and an ATP-generating system in the internal medium of ghosts (Fig. 3). The rates were similar to those of lipid-asymmetric ghosts.

To elucidate the role of phospholipid headgroup specifity in modulating influenza virus fusion we have undertaken experiments similar to those in case of VSV: (i) incorporation of spin-labeled phospholipid analogues into lipid-asymmetric ghosts and (ii) conversion of PS to PE by PS-decarboxylase treatment of lipid-symmetric ghosts. The results parallel our observation for VSV fusion with erythrocyte membranes that fusion of influenza virus was affected by the presence of unsaturated fatty acyl chains in the outer leaflet, but not by specific phospholipids. This suggests that lipid packing properties of the target membrane may influence influenza virus fusion as well.

One might speculate whether interaction of influenza virus with the target membrane (perhaps early fusion events) causes a disturbance of the target membrane which results in an enhanced (local) flip-flop of phospholipids. The (local) symmetrized lipid arrangement could then sustain later fusion events. It has been shown that influenza virus induces flip of lysophosphatidylcholine from the outer leaflet of erythrocyte membranes (Bashford, 1988). Hoekstra (1983) observed a transbilayer movement of

fluorescent phospholipid analogues in plasma membranes of Chinese hamster
fibroblasts upon insertion of viral components of Sendai virus, presumably
the F-protein. This viral fusion protein may facilitate the flip-flop of
lipids similarly as observed in case of glycophorin enhanced transbilayer
movement of PC in liposomes (De Kruijff et al., 1978).

Our results with natural target membranes are basically in agreement
with those using liposomes, namely, that no specific phospholipid headgroup
is required for influenza fusion. It has been shown that fusion between
influenza virus and liposomes are fairly independent of the lipid composition
of liposomes (Maeda et al., 1981; White et al., 1982). White et al. (1982)
established a stimulating effect in the presence of unsaturated soy PE. This
is consistent with the observation of van Meer et al. (1985) that fusion of
HA-expressing Madin-Darby canine kidney cells with liposomes was dependent
on the portion of unsaturated phospholipids present. Liposomes containing a
high level of unsaturated PE fuse two times faster than PC liposomes (no PE).
Incorporation of PE and PC liposomes did not influence fusion.

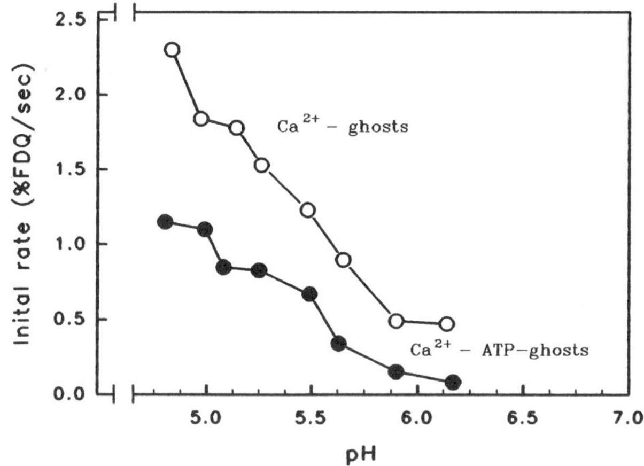

Figure 3. Effect of phospholipid distribution on the initial rate of
influenza virus fusion (A/PR 8/34) with erythrocyte ghosts.
R18 labeled virus was prebound either to lipid-symmetric (Ca^{2+})
or to lipid-asymmetric (Ca^{2+} + ATP) ghosts at pH 7.4, PBS.
Fusion was triggered at 37°C by lowering the pH to the desired
value.

Our results on fusion of VSV and influenza imply that lipid packing
might play an important role in natural membrane fusion processes. Ohki
(1988) suggested that increased "hydrophobicity" of the membrane surface,
which is related to membrane interfacial tension and phospholipid packing,
is a critical factor in lowering those barriers preventing membranes from
fusing. Cell partioning in a two-phase aqueous polymer system indicates that
the surface of lipid-symmetric erythrocytes is more hydrophobic than the
surface of lipid-asymmetric ghosts (McEvoy et al., 1986). The penetration of
hydrophobic sequences into the target membrane might be facilitated (a) by

a lowered phospholipid packing of the outer leaflet of the target *per se* and
(b) by the accompanied enhancement of surface hydrophobicity (see below).

Cell-cell fusion

Natural cell-cell fusion - Myoblast fusion: Myoblast fusion is a key event
in the differentiation of skeletal muscle. During the embryonic phase
mononucleated myoblasts align up and adhere before fusing into multinucleate
myotubes. These myotubes are the precursor of skeletal muscle fibres.

Fusion of myoblasts has been extensively used as a naturally occurring
system to understand the mechanism of cell-cell fusion (see Wakelam 1988).
Fusion proceeds via several stages: acquisition of fusion competence,
apposition of plasma membranes, and reorganization of lipids and proteins in
the plasma membrane at restricted areas. However, despite many efforts the
process of myoblast fusion, its stimulation and regulation, is not
understood. Many changes in myoblast membranes which are claimed to be
essential for fusion have been reported (Wakelam, 1988). It is generally
assumed that (glyco)proteins are involved in cell-cell recognition and
adhesion as well as in the final fusion of myoblast membranes. The formation
of myotubes coincides with a biochemical differentiation indicated by an
increase of various muscle-specific proteins and enzyme activities. Several
proteins have been implicated as regulators in myoblast fusion. However,
there precise role is unknown so far (White and Blobel, 1989; Wakelam, 1988).

Much attention has been given to role of plasma-membrane lipid
composition in myoblast fusion. It has been shown that the relative
composition of the major phospholipids remains for the most part unchanged
during myoblast fusion (Kent et al., 1974, Boland et al., 1977). However,
examination of minor phospholipids such as polyphosphoinositides and
gangliosides suggested that gangliosides may contribute to the adhesion
process in concert with gylcoproteins (Whatley et al., 1981) and that
breakdown of polyphosphoinositides mediates calcium gating which may be an
important signal in myoblast fusion (Wakelam, 1983).

However, gross investigation of lipid composition of myoblasts does not
provide detailed information on transversal lipid distribution in myoblast
plasma membranes. It has been shown that enzymatic modification of plasma
membrane lipids by nonpenetrating phospholipase inhibit myoblast fusion
(Schudt and Pette, 1976). This is consistent with the observation that
myoblast fusion was also inhibited by addition (i) of lysolecithin which is
formed by phospholipase (Reporter and Norris, 1973) and (ii)
dipalmitoylphosphatidylcholine (van der Bosch et al., 1973), respectively.
Moreover, it has been demonstrated that plasma membrane fluidity increases
rapidly before the onset of fusion (Prives and Shinitzky, 1977). Areas of
cell contact between fusing cells have a higher fluidity than other regions
of the plasma membrane (Herman and Fernandez, 1978, 1982). These results
indicate that indeed myoblast cell-cell fusion is affected by lipid structure
of the plasma membrane. It has been shown recently that synthesis of all
major phospholipids, predominantly of phosphatidylcholine and phospha-
tidylinositol-phosphatidylserine, increased significantly during myogenesis
(Sauro et al., 1988). In chick and quail myoblasts a transversal distribution
of aminophospholipids unusual for eukaryotic plasma membranes has been
established by utilizing two nonpenetrating amidating reagents (Session and
Horwitz, 1981, 1983). 65 % of PE and 45 % of PS were present in the
exoplasmic layer. The enrichment of these aminophospholipids in the outer
leaflet does not itself confer on myoblast surface the fusogenic property,
since the same transversal arrangement was observed beyond the fusion-
susceptible period of myoblasts. However, the preferential localization of
PS and PE in the external leaflet may facilitate the fusion process. Both

lipids may serve as a binding site for Ca^{2+} enhancing the fusion competence of apposed membranes. It has been shown that myoblast fusion requires the presence in the medium of Ca^{2+} concentrations in the order of millimolar (van der Bosch et al., 1973). Isolated plasma membranes from myoblasts undergo fusion in Ca^{2+}-specific fashion at physiological concentrations (Schudt et al., 1976). The unusual distribution of aminophospholipids, their fusogenic properties proved in model membrane studies (Düzgünes, 1985) on the one side and the essential step of myoblast fusion in myogenesis on the other side suggest that lipid asymmetry might be an important factor in regulating natural events of membrane fusion. In this respect it will be interesting to see whether myoblasts exhibit (i) no aminophospholipid translocase activity, or (ii) an aminophospholipid translocase activity depending on the development of mononucleated myoblasts.

Artificially induced cell-cell fusion - PEG induced fusion: Another argument in favor of the regulatory role of phospholipid asymmetry in cell-cell fusion came from recent observations on poly(ethylene glycol)-induced fusion of human erythrocytes (Tullius et al., 1989).

Poly(ethylene glycol) (PEG) of a molecular weight of about 6000 is commonly used to induce aggregation and, subsequently, fusion of liposomes as well as of a wide variety of plant protoplasts and animal cells (for review Arnold et al., 1988). The fusogenic property of higher molecular weight PEG and similar polymers can be explained by the high osmotic pressure of aqueous solutions of those polymers and by the excluded volume effect or depletion of polymers in the vicinity of (membrane) surfaces (Arnold et al., 1988, Evans and Needham, 1988, Pratsch et al., 1989). The thermodynamically unfavored interaction of PEG with polar surfaces, i.e, the steric exclusion and the competition of water and PEG molecules for interaction with polar groups on the surface are the basis of the depletion layer. The reduction of concentration of the polymer in the contact region between aggregated membranes relative to the exterior bulk solution gives rise to an osmotic imbalance and an osmotic stress that acts to remove water overcoming a main barrier of membrane fusion, the dehydration forces.

Erythrocytes have been extensively used as a model system for investigating PEG-mediated cell-cell fusion (e.g. Knutton 1979; Hui et al., 1985). Recently, Tullius et al. (1989) has shown that PEG-induced fusion of lipid-symmetric ghosts (Ca^{2+}-ghosts) is 5-fold more efficient than fusion of lipid-asymmetric ghosts (**Mg^{2+}-ghosts**). Moreover, fusion of mixtures of lipid-symmetric and lipid-asymmetric ghosts cells suggested that both fusing partners must have a symmetric distribution for fusion to be enhanced. Ca^{2+} was required in the medium since fusion was negligible in the absence of Ca^{2+} independent on the modification of the erythrocyte membrane. The enhanced PEG-induced fusion of lipid-symmetric ghosts was attributed to a decrease in packing of the exterior lipids as judged by a large increase in staining with fluorescent dye merocyanin 540 (Tullius et al., 1989). Several non-exclusive reasons have been discussed to explain membrane alterations caused by a loss of lipid asymmetry and leading to an increase of fusion suscepti-bility: **(a)** Loose lipid packing of the outer leaflet of lipid-symmetric ghosts enhances its surface hydrophobicity (McEvoy et al., 1986), presumably as a result of higher levels of unsaturated fatty in the exoplasmic leaflet after symmetrization (see above). Ross and Chopin (1985) have been able to show that the PEG-induced fusion of mouse LM cells can be controlled by the fatty acyl chain composition of membrane phospholipids. The fusion yield was significantly enhanced by increasing the level of unsaturated fatty acids. As already mentioned the hydrophobicity of the membrane surface might be an important determinant in membrane fusion by reducing dehydration energy of the membrane surface. For example, an increased hydrophobicity may enhance fusion by facilitating the close apposition of membranes. Recently, Dimitrov and Sowers (1990) attributed the delay in electrofusion of erythrocytes in part to the time required for movement of two membranes into an apposition appropriate for fusion which may be determined by the degree of the hydrophobicity of the fusing membranes. Ohki (1985, 1988) has proposed recently that repulsive interaction forces due to hydration water on the membrane surface ("hydration pressure") can be reduced by altering membrane surfaces to a more hydrophobic nature. Enhancement of hydrophobicity of membrane surface may not only be caused by a lower lipid packing, but also by lowering the hydration barrier due to the increased level of PE in the outer leaflet after symmetrization. Lamellar phases of PE are less strongly hydrated than are bilayers of PC (Lis et al., 1982), and aqueous dispersions of PE's can adopt nonlamellar configurations much more readily than PC's with the same acyl chain configuration (Verkleij et al., 1984) (see above). Dressler et al. (1983) discussed a link between enhancement of flip-flop and fusion of human erythrocytes by dielectric breakdown. These authors observed an increased accessibility of PE on the outer monolayer and an enhanced transbilayer mobility of exogenous lyso-PC after dielectric breakdown. An enhanced mobility and a redistribution of PE may sustain the merging of two closely apposed erythrocyte membranes. The relevance of hydrophobicity in membrane-membrane interaction is supported by the long been known and verified correlation between membrane hydrophobicity of particles and bacteria and the degree to which they get engulfed by phagocytic cells (Van Oss and Gillman, 1972; Van Oss, 1986). **(b)** An enhanced membrane fluidity in the outer leaflet of ghosts upon lipid symmetrization which has indeed shown (Herrmann et al., 1990c) may enhance the ability of intramembraneous particles to cluster. A correlation between clustering of IMP's and fusion of human erythrocytes has been established (Knutton, 1979; Hui et al., 1985).

   (c)   Another explanation for the increased efficiency of PEG-induced fusion of erythrocytes upon lipid symmetrization might be the exposure of PS on the outer leaflet. The occurrence of PS in the outer leaflet of erythro-cytes may thereby render the membrane more fusible in the presence of Ca^{2+}. This is consistent with requirement of Ca^{2+} in the external medium for PEG-mediated fusion of lipid-symmetric erythrocytes. As mentioned, liposomes containing PS can be fused in the presence of Ca^{2+}. However, Ca^{2+} was also required in case of lipid-asymmetric cells. One may speculate that the osmotic stress exerted by PEG or early events of cell-cell interaction facilitate the flip-flop of lipids accompanied by increasing levels of PS in the outer leaflet. This would sustain later steps in membrane fusion. Indeed, very recently a considerable flip-flop associated with PEG treatment and with PEG-induced fusion of RBC has been reported (Huang and Hui, 1990). Evidence for the relevance of the exposure of PS in the outer leaflet for fusion came from recent observations on osmotically-induced fusion of human erythrocytes in the presence of Ca^{2+} (Baldwin et al., 1990). It was shown that continuous swelling of erythrocytes in hypotonic media was accompanied by an influx of extracellular Ca^{2+} which facilitates a translocation of PS to the outer leaflet in agreement with recent findings (Verhoeven et al., 1990; Henseleit et al., 1990). Using an assay for procoagulant activity sensitive to the exposure of PS it was shown that the redistribution of PS precedes fusion of erythrocytes which could be induced in hypotonic medium. Interaction of Ca^{2+} with PS leads to a reduction of hydration of polar headgroups which allows a closer contact between adjacent membranes important for fusion (see above). It has been shown in model membranes that not only the acidic lipids are dehydrated by cations but their neutral, highly hydrated neighbors as PC may be dehydrated as well (Coorsen and Rand, 1988).

Endoplasmic fusion

   Further evidence for a regulatory role of phospholipid asymmetry in membrane fusion comes from investigations of endoplasmic fusion processes as exocytosis, the fusion of secretory vesicles with the plasma membrane   and

the release of vesicle content. Fusion events of the secretory pathway depend
on multiple components including membrane components, cytosolic proteins and
low molecular weight effector molecules such as Ca^{2+} and ATP (Stegmann et
al., 1989). It was proposed that the enrichment of the cytoplasmic leaflet
of synaptic vesicles in polyunsaturated PS and PE is relevant to the
mechanism of fusion, possibly involving nonbilayer structures (Deutsch and
Kelly, 1981). There is experimental support that the Ca^{2+}-triggered fusion
of synaptic vesicles is mediated by specific proteins interacting with Ca^{2+}
and the phospholipid bilayer. The cytoplasmic domain of the synaptic-vesicle
specific protein gp65 binds to acidic phospholipids as PS with high affinity.
Moreover, the possible interaction of this domain with the hydrophobic
interior of the bilayer suggests that it may insert into and structurally
modulate the bilayer. This implies a function in synaptic vesicle exocytosis
(Perin et al., 1990).

Exocytosis of chromaffin cells by fusion of chromaffin granules with
the plasma membrane is by far the best investigated fusion process of the
secretory pathway. Although the precise mechanism of granule exocytosis is
not known, conditions which facilitate membrane fusion are well documented.
In particular, the relevance of aminophospholipid has been shown. The
transversal phospholipid distribution of chromaffin granules is asymmetric
similar to plasma membranes with a preferential orientation of PS and PE to
the cytoplasmic leaflet while PC and SM are mainly on the exoplasmic
(luminal) leaflet (Buckland et al., 1978; Westhead, 1987). It has been shown
that an ATP-dependent translocase selectively transports PS from the luminal
to the cytoplasmic monolayer, provided that the medium contains ATP
suggesting that maintenance of PS orientation in granules posses a mechanism
comparable to the aminophospholipid translocase in human erythrocytes
(Zachowski et al., 1989). The relevance of the asymmetric distribution of
aminophospholipids in granules for exocytosis comes from the finding that
those proteins as synexin which seems to be involved in early fusion events
bind preferentially to membranes composed of acidic phospholipids and
aminophospholipids as phosphatidic acid, PS and PE, respectively, in the
presence of Ca^{2+} (Pollard et al., 1988). Synexin is a calcium binding protein
of 47 kDa which causes chromaffin granules as well as acidic liposomes to
aggregate and fuse (Hong et al., 1982; Pollard et al., 1988). In addition to
binding to granule membranes and liposomes, synexin binds also to the inner
leaflet of plasma membranes of chromaffin cells (Scott et al., 1985). From
investigations on liposomes it was concluded that synexin appears to catalyze
the aggregation of liposomes, but does not trigger fusion byitself. The
presence of PC in liposome membranes reduces drastically the binding of
synexin (Pollard et al., 1988) as well as any facilitation of fusion by
synexin (Papahadjopoulos et al., 1988). Liposomes containing only PC do not
aggregate in the presence of synexin and Ca^{2+}. Therefore, one might surmise
that the preferential orientation of PS and PE in the cytoplasmic leaflet of
granules and plasma membranes as well is a prerequisite for early fusion
events mediated by synexin. Recently, it was proposed that hydrophobic
synexin polymers enter both fusing membranes and form a bridge allowing
finally merging of membranes (Pollard et al., 1988). The lower hydration of
PS and PE, respectively, compared to PC and, moreover, their dehydration upon
binding of Ca^{2+} will presumably support this process.

ACKNOWLEDGMENT

A.H. was supported by the Deutsche Forschungsgemeinschaft (He 1928/1-1).

REFERENCES

Arnold, K., Herrmann, A., Gawrisch, K., and Pratsch, L., 1988, Water-mediated
    effects of PEG on membrane properties and fusion, in: "Molecular

mechanisms of membrane fusion", S. Ohki, D. Doyle, T.D. Flanagan, S.W. Hui, E. Mayhew, eds., Plenum Press, N.Y. and London.

Baldwin, J.M., O'Reilly, R., Whitney, M., and Lucy, J.A., 1990, Surface exposure of phosphatidylserine is associated with the swelling and osmotically-induced fusion of human erythrocytes in the presence of Ca^{2+}, Biochim. Biophys. Acta 1028: 14.

Bashford, C.L., 1988, Phospholipid flip-flop correlates with virus-membrane fusion rather with pore formation, studia biophysica 127: 155.

Ben-Bassat, I., Bensch, R.G., and Schrier, S.L., 1972, Drug induced erythrocyte membrane internalization, J. Clin. Invest. 51: 1833.

Bishop, W.P., and Bell, R.M., 1985, Assembly of the endoplasmic reticulum phospholipid bilayer: the phosphatidylcholine transporter, Cell 42: 51.

Bitbol, M., Fellmann, P., Zachowski, A., and Devaux, P.F., 1987, Ion regulation of phosphatidylserine and phosphatidylethanolamine outside-inside translocation in human erythrocytes, Biochim. Biophys. Acta 904: 268.

Bitbol, M., and Devaux, 1988, Measurement of outward translocation of phospholipids across human erythrocyte membrane, Proc. Natl. Acad. Sci. USA 85: 6783.

Blumenthal, R., 1987, Membrane Fusion, Curr. Top. Membr. Transp. 29: 203.

Blumenthal, R., 1988, Cooperativity in Viral Fusion, Cell Biophysics 12: 1.

Boland, R., Chyn, T., Roufa, Reyes, E., and Martonosi, A., 1977, The lipid composition of muscle cells during development, Biochim. Biophys. Acta 489: 349.

Brenner, R.B., 1984, Effect of unsaturated acids on membrane structure and enzyme kinetics, Prog. Lip. Res. 23: 69.

Buckland, R.M., Radda, G.K., and Shennan, C.D., 1978, Accessibility of phospholipids in the chromaffine granule membrane, Biochem. Biophys. Acta 513: 321.

Calvez, J.Y., Zachowski, A., Herrmann, A., Morrot, G., and Devaux, P.F., 1988, Asymmetric distribution of phospholipids in spectrin-poor erythrocyte-vesicles, Biochemistry 27: 5666.

Chejanovsky, N., Amselem, S., Zakai, N., Bahrenholz, Y., and Loyter, A., 1986, Membrane vesicles containing the Sendai virus binding glycoprotein, but not the viral fusion protein, fuse with phosphatidylserine liposomes at low pH, Biochemistry 25: 4810.

Chejanovsky, N., Nussbaum, O., Loyter, A., and Blumenthal, R., 1988, Fusion of enveloped viruses with biological membranes. Fluorescence dequenching studies, Subcellular Biochemistry 13: 415.

Chernomordik, L.V., Melikyan, G.B., and Chizmadzhev, Y.A., 1987, Biomembrane fusion: a new concept derived from model studies using two interacting planar lipid bilayers, Biochim. Biophys. Acta 906: 309.

Citovsky, V., Blumenthal, R., and Loyter, A., 1985, Fusion of Sendai virions with phosphatidylcholine-cholesterol liposomes reflects the viral activity required for fusion with biological membranes, FEBS Lett. 193: 135.

Coorssen, J., and Rand, R.P., 1988, Competitive forces between lipid membranes, studia biophysica 127: 53.

Cribier, S., Morrot, G., Neumann, J.-M., and Devaux, P.F., 1990, Lateral diffusion of erythrocyte phospholipids in model membranes comparison between inner and outer leaflet, Eur. Biophys. J. 18: 33.

De Kruijff, B., van Zoelen, E.J.J., and van Deenen, L.L.M., 1978, Glycophorin facilitates the transbilayer movement of phosphatidylcholine in vesicles, Biochim. Biophys. Acta 509: 537.

Deutsch, J.W.,. and Kelly, R.B., 1981, Lipids of synaptic vesicles: relevance to the mechanism of membrane fusion, Biochemistry 20: 378.

Devaux, P.F., Morrot, G., Herrmann, A., and Zachowski, A., 1988, Protein involvement in plasma membrane lipid asymmetry, studia biophysica 1-3: 183.

Devaux, P.F., 1990, The aminophospholipid translocase. A transmembrane lipid pump - Physiological significance, News Physiol. Sci. 5: 53.

Dimitrov, D.S., and Sowers, A.E., 1990, A delay in membrane fusion: Lag times observed by fluorescence microscopy of individual fusion events induced by an electric field pulse, <u>Biochemistry</u>.

Dressler, V., Schwister, K., Haest, C.W.M., and Deuticke, B., 1983, Dielectric breakdown of the erythrocyte membrane enhances transbilayer mobility of phospholipids, <u>Biochim. Biophys. Acta</u> 732: 304.

Düzgünes, N., Nir, S., Wilschut, J., Betntz, Newton, C., Portis, A., and Papahadjopoulos, D., 1981, Calcium- and magnesium-induced fusion of mixed phosphatidylserine/phosphatidylcholine vesicles: Effect of ion binding, <u>J. Membr. Biol.</u> 59: 115.

Düzgünes, N, 1985, Membrane fusion, <u>Subcell. Biochem.</u> 11: 195.

Ellens, H., Siegel, D.P., Alford, D., Yeagle, P.L., Boni, L., Lis, L.J., Qinn, P.J., and Bentz, J., 1989, Membrane fusion and inverted phases. <u>Biochemistry</u> 28:3692.

Evans, E., and Needham, D., 1988, Intrinsic colloidal attraction/repulsion between lipid bilayers and strong attraction induced by non-adsorbing polymers, <u>in</u>: "Molecular mechanisms of membrane fusion", S. Ohki, D. Doyle, T.D. Flanagan, S.W.Hui, E. Mayhew, eds., Plenum Press, N.Y. and London.

Grimaldi, S., Verna, R., Puri, A., Morris, S.J., and Blumenthal, R., 1988, Fusion of vesicular stomatitis virus with human red blood cell membranes: The role of phospholipid distribution, <u>in</u>: "Advances in Biotechnology of Membrane Ion Transport", P.L. Jorgensen and R. Verna, eds., Serono Symposia Publications from Raven Press 51, New York.

Hanahan, D.J., and Nelson, D.R., 1984, Phospholipids as dynamic participants in biological processes. <u>J. Lipid Res.</u> 25: 1528.

Haywood, A.M., and Boyer, B.P., 1984, Effect of lipid composition upon fusion of liposomes with Sendai virus membranes, <u>Biochemistry</u> 23: 4061.

Henseleit, U., Plasa, G., and Haest, C., 1990, Effects of divalent cations on lipid flip-flop in the human erythrocyte membrane, <u>Biochim. Biophys. Acta</u> 1029: 127.

Herbette, L., Blaisie J.K., Defoor, P., Fleischer, S., Bick, R.J., van Winkle, W.B., Tate, C.A., and Entman, M.L., 1984, Phospholipid asymmetry in the isolated sarcoplasmic reticulum membrane, <u>Archs. Biochem. Biophys.</u> 234: 235.

Herman, B.A., and Fernandez, S.M., 1978, Changes in membrane dynamics associated with myogenic cell fusion, <u>J. Cell. Physiol.</u> 94: 253.

Herman, B.A., and Fernandez, S.M., 1982, Dynamics and topographical distribution of surface glycoproteins during myoblast fusion. A resonance energy transfer study, <u>Biochemistry</u> 21: 3275.

Herrmann, A., and Müller, P., 1986, A model for the asymmetric lipid distribution in the human erythrocyte membrane, <u>Bioscience Rep.</u> 6: 185.

Herrmann, A., Zachowski, A., and Devaux, P.F., 1990a, Translocation and distribution of phospholipids across eukaryotic plasma membranes, <u>in</u>: "Biophysics of Cell Surface", Springer Series in Biophysics, R. Glaser and D. Gingell, eds, Springer-Verlag Berlin.

Herrmann, A., Zachowski, A., and Devaux, P.F., 1990b, The protein-mediated phospholipid translocation of endoplasmic reticulum has a low specifity, <u>Biochemistry</u> 29: 2023.

Herrmann, A., Clague, M.J., Puri, A., Morris, S.J., Blumenthal, R., and Grimaldi, S., 1990c, Effect of erythrocyte transbilayer phospholipid distribution on fusion with vesicular stomatitis virus, <u>Biochemistry</u> 29: 4054.

Herrmann, A., and Devaux, P.F., 1990, Alteration of the aminophospholipid translocase activity during in vivo and artificial aging of human erythrocytes, <u>Biochim. Biophys. Acta</u> 1027: 41.

Herrmann, A., Clague, M., and Blumenthal, R., 1991, The role of target membrane structure in fusion with influenza virus: effect of modulating erythrocyte transbilayer phospholipid distribution and cytoskeleton, in preparation.

Hoekstra, D., 1983, Topographical distribution if a membrane-inserted fluorescent phospholipid analogue during cell fusion, <u>Exp. Cell. Res.</u> 144: 482.

Hoekstra, D., de Boer, T., Klappe, K., and Wilschut, J., 1984, Fluorescence method for measuring the kinetics between biological membranes, <u>Biochemistry</u> 23: 5675.

Homan, R., and Pownall, H.J., 1987, Effect of pressure on phospholipid translocation, <u>J. Am. Chem. Soc.</u> 109: 4759.

Hong, K., Düzgünes, N., and Papahadjopoulos, D., 1982, Modulation of membrane fusion by calcium-binding proteins, <u>Biophys. J.</u> 37: 297.

Huang, S.K., and Hui, S.W., 1990, Fluorescence mesurements of fusion between human erythrocytes induced by poly(ethylene glycol), <u>Biophys. J.</u> 58: 1109.

Hui, S.W., Isac, T., Boni, L.T., and Sen, A., 1985, Action of polyethylene glycol on the fusion of human erythrocyte membranes, <u>J. Membr. Biol.</u> 84: 137.

Hullin, F., Bossant, M.-J., and Salem jr., N., 1991, Aminophospholipid molecular species asymmetry in the human erythrocyte plasma membrane, <u>Biochim. Biophys. Acta</u> 1061: 15.

Isrealeachvili, J.N., Marcelja, S., and Horn, R.G., 1980, Physical principles of membrane organization, <u>Quart. Rev. Biophys.</u> 13: 121.

Kent, C., Schimmel, S.D., and Vagelos, P.R., 1974, Lipid composition of plasma membranes from developing chick muscle cells in culture, <u>Biochim. Biophys. Acta</u> 360: 312.

Kielian, M., and White, J. 1984, Role of cholesterol in fusion of Semliki forest virus with membranes, <u>J. Virol.</u> 52: 281.

Klappe, K., Wilschut, J., Nir, S., and Hoekstra, D. , 1986, Parameters effecting fusion between Sendai virus and liposomes. Role of virus proteins, liposome composition and pH, <u>Biochemistry</u>, 25: 8252.

Knutton, S., 1979, Studies of membrane fusion: III. Fusion of erythrocytes with PEG, <u>J. Cell Sci.</u> 36: 61.

Lis, L.J., McAlister, M., Fuller, N., Rand, R.P., and Parsegian, V.A., 1982, Interactions between neutral phospholipid membranes, <u>Biophys. J.</u> 37: 657.

Maeda, T., Kawasaki, K., and Ohnishi, S.I., 1981, Interaction of influenza virus hemagglutinin with target membrane lipids is a key step in virus induced hemolysis and fusion at pH 5.2, <u>Proc. Natl. Acad. Sci. USA</u>, 87 4133.

Maksymiw, R., Sui, S., Gaub, H., and Sackmann, E., 1987, Electrostatic coupling of spectrin dimers to phosphatidylserine containing lipid lamellae, <u>Biochemistry</u> 26: 2983.

Mastromarino, P., Conti, C., Goldoni, P., Hauttecoeur, B., and Orsi, N, 1988, Characterization of membrane components of the erythrocyte involved in vesicular stomatitis virus attachment and fusion at acidic pH, <u>J. Gen. Virol.</u> 68: 2359.

McEvoy, L., Willimason, P., and Schlegel, R.A., 1986, Membrane phospholipid asymmetry as a determinant of erythrocyte recognition by macrophages, <u>Proc. Natl. Acad. Sci. USA</u> 83: 3311.

Michaelson, D.M. ,Barkai, G., Barenholz, Y., 1983, Asymmetry of lipid organization in cholinergic synaptic vesicle membranes, <u>Biochem. J.</u> 211: 155.

Middelkoop, E., 1989, Transmembrane phospholipid asymmetry in erythroid cells: mechanisms of maintenance, Ph.D. thesis, University of Utrecht (The Netherlands).

Morrot, G., Hervé, P., Zachowski, A., Fellmann, P., and Devaux, P.F., 1989, Aminophospholipid translocase of human erythrocytes: phospholipid substrate specifity and effect of cholesterol, <u>Biochemistry</u> 28: 3456.

Morrot, G., Zachowski, A., and Devaux, P.F., 1990, Partial purification and characterization of the human erythrocyte Mg^{2+}-ATPase. A candidate aminophospholipid translocase, <u>FEBS lett.</u> 266: 29.

Newton, A.C., and Koshland jr., D.E., 1990, Phosphatidylserine affects specifity of protein kinase C substrate phosphorylation and autophosphorylation, _Biochemistry_ 29: 6656.

Nir, S., Klappe, K., and Hoekstra, D., 1986, Mass action analysis of kinetics and extent of fusion between Sendai virus and phospholipid vesicles, _Biochemistry_, 25: 8261.

Ohki, S., 1982, A mechanism of divalent ion-induced phosphatidylserine membrane fusion, _Biochim. Biophys. Acta_ 689: 1.

Ohki, S., 1984, Effects of divalent cations, temperature, osmotic pressure gradient, and vesicle curvature on phosphatidylserine vesicle fusion, _J. Membr. Biol._ 77: 265.

Ohki, S., 1988, Surface tension, hydration energy and membrane fusion, _in_: "Molecular mechanisms of membrane fusion", S. Ohki, D. Doyle, T.D. Flanagan, S.W. Hui, and E. Mayhew, eds., Plenum Press, N.Y. and London.

Ohshima, H., and Ohki, S., 1985, Effects of divalent cations on the surface tension of a lipid monolayer-coated air/water interface, _J. Colloid Interface Sci._ 102: 85.

Op den Kamp, J.A.F. 1979, Lipid asymmetry in membranes, _A. Rev. Biochem._ 48: 47.

Quinn, P.J., Joo, F., and Vigh, L., 1989, The role of unsaturated lipids in structure and function, _Prog. Biophys. Molec. Biol._ 53: 71.

Papahadjopoulos, D., Meers, P.R., Hong, K., Ernst, J.D., Goldstein, I.M. and Düzgünes, N., 1988, Calicum-induced membrane fusion: From liposomes to cellular membranes, _in_: "Molecular mechanisms of membrane fusion", S. Ohki, D. Doyle, T.D. Flanagan, S.W. Hui, E. Mayhew, eds., Plenum Press, N.Y. and London.

Perin, M.S., Fried, V.A., Mignery, G.A., Jahn, R., and Südhof, T.C., 1990, Phospholipid binding by a synaptic vesicle protein homologous to the regulatory region of protein kinase C, _Nature_ 345: 1990.

Pollard, H.B., Rojas, E., Burns, L., and Parra, C., 1988, Synexin, calcium and the hydrophobic bridge hypothesisfor membrane fusion, _in_: "Molecular mechanisms of membrane fusion", S. Ohki, D. Doyle, T.D. Flanagan, S.W. Hui, E. Mayhew, eds., Plenum Press, N.Y. and London.

Poste, G., and Pasternak, C.A., 1978, Virus-induced cell fusion, _in_:"Membrane fusion (G. Poste and G.L. Nicholson, eds.),Elsevier/North-Holland Biomedical Press, Amsterdam.

Pratsch, L., and Donath, E., 1988, Poly-ethylene glycol depletion layers on human red blood cell surfaces measured by electrophoresis, _studia biophysica_ 123: 101.

Prives, J., and Shinitzky, M., 1977, Increased membrane fluidity precedes fusion of muscle cells, _Nature (London)_ 268: 761.

Puri, A., Winick, J., Lowy, J., Covell, D., Eidelman, O., Walter, A., and Blumenthal, R, 1988, Activation of vesicular stomatitis virus with cell by pretreatment at low pH, _J. Biol. Chem._ 263: 4749.

Rand, R.P., and Parsegian, V.A., 1989, Hydration forces between phospholipid bilayers, _Biochim. Biophys. Acta_ 988: 351.

Rando, R.R., 1988, Regulation of protein kinase C activity by lipids, _FASEB J._ 2:2348.

Reporter,M., and Norris, G., 1973, Reversible effects of lysolecithin of fusion of cultured rat muscle cells, _Differentiation_ 1: 83.

Ross, D.S., and Choppin, P.W., 1985, Biochemical studies on cell fusion. II. Control of fusion response by lipid alteration, _J. Cell Biol._ 101: 1591.

Sauro, V.S., Brown, G.A., Hamilton, M.R., Strickland, C.K., and Strickland, K.P., 1988, Changes in phospholipid metabolism dependent on calcium-regulated myoblast fusion, _Biochem. Cell Biol._ 66: 1110.

Schlegel, R., Tralka, T.S., Willingham, M.C., and Pastan, I., 1983, Inhibition of VSV binding and infectivity by phosphatidylserine: is phosphatidylserine a VSV-binding site?, _Cell_ 32: 639.

Schudt, C., Dahl, G., and Gratzl, M., 1976, Calcium-induced fusion of plasma membranes isolated from myoblasts grown in culture, <u>Cytobiology</u> 13: 211.

Schudt, C., and Pette, D, 1976, influence of monosaccharides, medium factors and enzymatic modification on fusion of myoblasts *in vitro*, <u>Cyto-biologie</u> 13: 74.

Scott, J.H., Creutz, C.E., and Pollard, H.B., 1985, Synexin binds in a calcium-dependent fashion to oriented chromaffin cell plasma membranes, <u>FEBS lett.</u> 180: 17.

Seddon, J.M., 1990, Structure of the inverted hexagonal (H_{II}) phase, and non-lamellar phase transitions of lipids, <u>Biochim. Biophys. Acta</u> 1031: 1.

Seigneuret, M., and Devaux, P.F., 1984, ATP-dependent asymmetric distribution of spin-labeled phospholipids in the erythrocyte membrane: relation to shape changes, <u>Proc. Natl. Acad. Sci. USA</u> 81: 3751.

Seigneuret, M., Zachowski, A., Herrmann, A., and Devaux, P.F., 1984, Asymmetric lipid fluidity in human erythrocyte membrane: New spin-label evidence, <u>Biochemistry</u> 23: 4271.

Session, A., and Horwitz, A.F., 1981, Myoblast aminophospholipid asymmetry differs from that of fibroblasts, <u>FEBS lett.</u> 134: 75.

Session, A., and Horwitz, A.F., 1983, Differentiation-related differences in the plasma membrane phospholipid asymmetry of myogenic and fibrogenic cell, <u>Biochim. Biophys. Acta</u> 728: 103.

Stegmann, T., Hoekstra, D., Scherphof, G., and Wilschut, J, 1986, Fusion activity of influenza virus, <u>J. Biol. Chem.</u> 261: 10966.

Stegmann, T., Doms, R., and Helenius, A., 1989, Protein-mediated membrane fusion, <u>Annu. Rev. Biophys. Biophys. Chem.</u> 18: 187.

Stubbs, C. D., and Smith, A., D., 1984, The modification of mammalian membrane polyunsaturated fatty acid composition in relation to membrane fluidity and function, <u>Biochim. Biophys. Acta</u> 779: 89.

Sune, A., Vidal, M., Morin, P., Saint-Marie, J., and Bienvenue, A., 1988, Evidence for bidirectional transverse diffusion of spin-labeled phospholipids in the plasma membrane of guinea pig blood cells, <u>Biochim. Biophys. Acta</u> 946: 315.

Tanaka, K.I., and Ohnishi, S.-I., 1976, Heterogeneity in the fluidity of intact erythrocyte membrane and its homogenization upon hemolysis, <u>Biochim. Biophys. Acta</u> 426: 218.

Tanaka, Y., and Schroit, A.J., 1983, Insertion of fluorescent phosphatidyl-serine into the plasma membrane of red blood cells. Recognition by autologous macrophages, <u>J. Biol. Chem.</u> 258: 11335.

Tullius, E.K., Williamson, P., and Schlegel, R.A., 1989, Effect of trans-bilayer phospholipid distribution on erythrocyte fusion, <u>Bioscience Rep.</u> 9: 623.

Van der Bosch, J., Schudt, C., and Pette, D., 1973, Influence of temperature, cholesterol, dipalmitoylecithin and Ca^{2+} on the rate of muscle cell fusion, <u>Exp. Cell Res.</u> 82: 433.

Van Meer, G., Davoust, J., and Simon, K., 1985, Parameters affecting low-pH mediated fusion of liposomes with the plasma membrane of cells infected with influenza virus, <u>Biochemistry</u> 24: 3599.

Van Oss, C.J., and Gilman, C.F., 1972, Phagocytosis as a surface phenomenon. I. Contact angles and phagocytosis of non-opsonized bacteria, <u>J. Reticuloendothel. Soc.</u> 3: 29.

Van Oss, C.J., 1986, Phagocytosis: An overview, <u>Meth. Enzymol.</u> 132: 3.

Verhoeven, B.M., Schlegel, R.A., and Williamson, P., 1990, Lipid asymmetry is rapidly lost and regained, 10th. Internat.Biophys.Congress 29.7.90-3.8.90, Vancouver, Abstract P.4.6.26, p. 353.

Verkleij, A.J., Leunissen-Bijvelt, J., de Kruijff, B., Hope, M., and Cullis, P.R., 1984,Non-bilayer structures in membrane fusion, <u>in</u>: Cell fusion, Ciba foundation Sympsoium 103: 45.

Wakelam, M.J.O., 1983, Inositol phospholipid metabolism and myoblast fusion, <u>Biochem. J.</u> 214: 77.

Wakelam, M.J.O., 1988, Myoblast fusion - A mechanistic analysis, <u>Curr. Top. Membr. Transp.</u> 32: 87.

Westhead, E.W., 1987, Lipid composition and orientation in secretory vesicles, <u>Ann. N.Y. Acad. Sci.</u> 493: 92.

White, J.M., Helenius, A., and Kartenbeck, J., 1982, Membrane fusion activity of influenza virus, <u>EMBO J.</u> 1: 217.

White, J., Kielian, M., and Helenius, A., 1983, Membrane fusion proteins of enveloped animal viruses, <u>Q. Rev. Biophys.</u> , 16:151.

White J, M., and Blobel, C.P., 1989, Cell-to-cell fusion, <u>Curr. Opin. Cell Biol.</u> 1: 934.

White, J., 1990, Viral and cellular membrane fusion, <u>Annu. Rev. Physiol.</u> 52: 675.

Wilschut, J., Düzgünes, N., Hoekstra, D., Papahadjopoulos, D., 1984, Modulation of membrane fusion by membrane fluidity. Temperature dependence of divalent cation induced fusion of phosphatidylserine vesicles, <u>Biochemistry</u> 24: 8.

Wilschut, J., Hoekstra, D., 1986, Membrane fusion in lipid vesicles as a model system, <u>Chem. Phys. Lip.</u> 40:145.

Wilschut, J. , 1989, Intracellular membrane fusion, <u>Curr. Opin. Cell Biol.</u>; 1:639.

Williamson, P., Algarin, L., Bateman, J., Choe, H.R., and Schlegel, R.A., 1985, Phospholipid asymmetry in human erythrocyte ghosts, <u>J. Cell. Physiol.</u> 123: 209.

Yamada, S., and Ohnishi, S., 1986, Vesicular stomatitis virus binds and fuses with phospholipid domain in target cell membranes, <u>Biochemistry</u> 25: 3703.

Yeagle, P.L., 1989, Lipid regulation of cell membrane structure and function, <u>FASEB J.</u> 3: 1833.

Zachowski, A., Fellmann, P., and Devaux, P.F., 1985, Absence of transbilayer diffusion of spin-labeled sphingomyelin in human erythrocytes. Comparison with the diffusion of several spin-labeled glycerophospholipids, <u>Biochim. Biophys. Acta</u> 815: 510.

Zachowski, A., Henry, J.P., and Devaux, P.F., 1989, Control of transmembrane lipid asymmetry in chromaffine granules by an ATP-dependent protein, <u>Nature</u> 340: 75.

Zachowski, A., and Devaux, P.F., 1990, Transmembrane movements of lipids, <u>Experientia</u> 46: 645.

Zwaal, R.F.A., 1988, Scrambling membrane phospholipids and focal control of blood clotting, <u>News Physiol. Sci.</u> 3: 57.

ANNEXIN-PHOSPHOLIPID INTERACTIONS IN MEMBRANE FUSION

Paul Meers*, Keelung Hong[‡] and Demetrios
Papaphadjopoulos[‡]

*Department of Pathology, William B. Castle
 Hematology Laboratory, Boston University School of
 Medicine, Boston , Massachusetts
[‡]Cancer Research Institute, University of California
 San Francisco, California

FUSION OF PHOSPHOLIPID VESICLES

Cellular secretion is a membrane fusion event of utmost
importance. The Ca^{2+} ion is critical in this process and its
intracellular concentration is highly regulated.   Therefore, Ca^{2+}
may play a direct or indirect role in these intracellular membrane
fusion processes. One early indication of a direct role in
membrane fusion was the observation that Ca^{2+} could induce
aggregation and leakage of contents from model membrane vesicles
composed of the most abundant naturally occuring acidic
phospholipid, phosphatidylserine (PS) (Papahadjopoulos and
Bangham, 1966) and that these phenomena were associated with
fusion of the vesicles (Papahadjopoulos et al., 1974).
Subsequently, the ability to produce large unilamellar vesicles
(LUV) (Szoka and Papahadjopoulos, 1978; Szoka et al., 1980;
Wilschut et al., 1980) and the development of fluorescent fusion
assays (Wilschut et al., 1980; Ellens et al., 1985) has allowed
detailed studies of the factors that mediate Ca^{2+}-dependent
membrane fusion, particularly fusion of PS vesicles (Wilschut et
al., 1980). A simple mass action model for the kinetics of
membrane fusion, that could describe or predict the results of
fusion assays under certain conditions, resulted from such studies
(Nir et al., 1980a; Bentz et al., 1983). A prominent feature of
this model was that the overall fusion process was treated as a
two step reaction involving first aggregation of vesicles with
rate constant C_{11} and dissociation constant D_{11} followed by
irreversible fusion with rate constant f_{11}. Using this rational
framework, the factors explicitly involved at each step can be
discussed separately.
Application of these methods to the fusion of PS vesicles has
helped delineate the ways in which Ca^{2+} acts in the overall fusion
process. For instance, Ca^{2+} can accelerate aggregation of PS
vesicles through charge neutralization (Nir et al., 1980b;
Mclaughlin, 1989), but a major factor that makes Ca^{2+} fusogenic is
the formation of an anhydrous "trans" intermembrane complex
(Portis et al., 1979; Rehfeld et al., 1981; Ekerdt and
Papahadjopoulos, 1982; Feigenson, 1986) where Ca^{2+} actually may
bind between the headgroups of PS molecules from separate apposed

Table I. Threshold concentrations of Ca^{2+} and Mg^{2+} (mM) for fusion of LUV composed of pure phospholipids.[a]

phospholipid	aggregation		fusion	
	Ca^{2+}	Mg^{2+}	Ca^{2+}	Mg^{2+}
phosphatidylinositol	3	6	-[b]	-
phosphatidylglycerol[c]	5	20	15	-
phosphatidylserine	2	5	2	-
phosphatidate	0.2	0.4	0.2	0.4

[a] All experiments were performed at 25° C at pH 7.4. Table adapted from Düzgünes et al., 1985.

[b] indicates no fusion

[c] data from Sundler(1984), Rosenberg et al.(1983) and N. Düzgünes, unpublished.

bilayers. The affinity for Ca^{2+} in this complex is much higher than in single bilayer ("cis") complexes (Portis et al., 1979; Ekerdt and Papahadjopoulos, 1982)· The dissociation constant for the Ca^{2+}-PS intermembrane complex has been measured and ranges from 0.1 to 3 μM depending on the species of PS (Feigenson, 1986), suggesting that this type of complex could participate in intracellular membrane fusion at physiologically relevant Ca^{2+} concentrations. The ability of Ca^{2+} to form an anhydrous intermembrane complex is crucial to fusion in this case, as dehydration of the phospholipid headgroups is a major energy barrier to close approach necessary for fusion and mixing of lipids from each interacting bilayer (Rand, 1981). Because Mg^{2+} does not form such a complex with PS (Portis et al., 1979), fusion of LUV is absolutely specific for Ca^{2+}, while both cations can cause aggregation of these vesicles. Fusion is also observed with some other phospholipid compositions but with different cation specificities and threshold concentrations (Table I). Much of the Ca^{2+} required for fusion of the acidic vesicles results from the necessity for charge neutralization. The resistance of PC toward Ca^{2+}-mediated fusion prevails even in liposomes in which the major bilayer components are fusion-susceptible (Hong et al., 1982b; Düzgünes et al., 1981). When PS is mixed with large percentages of phosphatidylethanolamine (PE), Mg^{2+} is also able to cause fusion. This is presumably due to the instability and low hydration of pure PE bilayers. Apposition of the PS/PE bilayers by any cation capable of charge neutralization of the acidic component is sufficient to allow fusion. The propensity of PE to form non-bilayer structures may be an important factor making such membranes fusogenic (Allen et al., 1990; Ellens et al., 1989). The mechanism of this fusion may be significantly different from fusion of PS vesicles induced by Ca^{2+}.

   Besides the dehydration of the headgroups which allows critically close apposition of interacting bilayers, rearrangements of the acyl chains of the phsopholipids are also important in the membrane fusion process, as increased molecular

motion induced by increased temperature enhances the rate of
membrane fusion (Düzgünes et al., 1984a; Bentz et al., 1985).
Conformational rearrangements of the acyl chains from bilayer
packing to some other phase such as the hexagonal II phase (Cullis
and Hope, 1978; Verkleij et al., 1979; Verkleij et al., 1980;
Siegel, 1984) or cubic phases (Ellens et al., 1989) may also be
important in the membrane fusion process, though these phases are
not generated in bulk during membrane fusion (Bearer et al.,
1982). Fusion may involve a small defect possessing some
characteristics of these non-bilayer phases or some other means of
exposing the hydrophobic acyl chains of interacting bilayers (Hui
et al., 1981; Horn, 1984; Marra and Israelachvilli, 1985).

These studies of Ca^{2+} dependent fusion have elucidated three
major steps in the fusion process. First long range electrostatic
repulsive forces must be overcome by double layer screening and
binding of cations to neutralize charge. Then water of hydration
at the surface of the bilayer must be stripped off to allow very
close approach of membrane surfaces. Finally a perturbation in
the bilayer packing of the phospholipids that may involve some
small transient intermediate structure may be necessary to promote
fusion of the bilayers. The importance of each step in any given
system may vary or they may even exist as one concerted step.

ROLE OF ANNEXINS IN INTRACELLULAR MEMBRANE FUSION

Despite the advances in elucidation of the mechanisms of
Ca^{2+}-dependent fusion of liposomes, the mechanisms of
intracellular membrane fusion remain unclear. Several facts
suggest that protein factors may help modulate Ca^{2+}-dependent
membrane fusion. One important fact is that the Ca^{2+} requirement
for liposome fusion (Table I) is several orders of magnitude
higher than the maximal Ca^{2+} concentration attained in the
cytoplasm in stimulated cells. For expample, in the chromaffin
cell system (Baker and Knight, 1978), an average Ca^{2+}
concentration of only approximately 1 µM is needed for secretion.
However a later study on depolarized chromaffin cells estimates
that the Ca^{2+} entry into the cytosolic space can reach up to 170-
340 µM in one second during a short depolarization (Artalejo et
al., 1987). The precise Ca^{2+} requirement for secretion in various
systems remains controversial. Part of the reason for the high
Ca^{2+} requirement for liposome fusion is probably the necessity to
screen the surface charge of acidic liposomes and the low
concentration of liposomes that must be used in fusion assays.
Factors that can overcome the charge repulsion and enhance the
rate of vesicle aggregation potentially decrease the Ca^{2+}
requirement. Indeed, acidic liposomes containing glycolipid can
fuse at significantly lower Ca^{2+} concentrations when pre-
aggregated by specific lectins (Sundler and Wijkander, 1983;
Düzgünes et al., 1984b). In this case, crosslinking of vesicles
probably allows the high affinity "trans" complex of Ca^{2+} and
phospholipids to form more easily due to more thermal collisions
per time between vesicle surfaces. A second, more compelling
argument for the existence of fusion-mediating proteins is the
problem of regulation and specificity within cells. Simply
regulating Ca^{2+} levels does not confer specificity on the site of
fusion and does not allow fine tuning of fusion through regulatory
cascades. It is also possible that movement of vesicles through
structurally complex cytoskeletal barriers could require protein
mediation. Although Ca^{2+}-phospholipid intermediates may
ultimately define the final step in intracellular fusion, it is

Table II. Effect of synexin on calcium-dependent aggregation and fusion of various phospholipid vesicles[a].

phospholipid composition	aggregation	fusion
PS, PS/PE, PA/PE[b]	enhanced	enhanced
PS/PC	enhanced	no effect
PI/PE	enhanced	inhibited

[a] From Hong et al., 1982a, Biophysical J. 37, 297-305

[b] PA – phosphatidate prepared from egg phosphatidylcholine, PC – egg phosphatidylcholine, PE – phosphatidylethanolamine transesterified from egg phosphatidylcholine, PI – bovine phosphatidylinositol, PS – bovine brain phosphatidylserine

likely that proteins exist to mediate and enhance specific interactions, thus sumperimposing temporal and spatial control.

The annexins are a group of Ca^{2+}-dependent membrane-binding proteins with a variety of putative intracellular and extracellular functions. These include Ca^{2+}-dependent membrane fusion (Creutz et al., 1978; Hong et al., 1981; Meers et al., 1987a; Meers et al., 1988a; Drust and Creutz, 1988; Ali et al., 1989) as well as membrane-cytoskeletal interactions (Gerke and Weber, 1984; Glenney, 1986; Ikebuchi and Waisman, 1990), intracellular signalling as kinase substrates (Fava, and Cohen, 1984; Glenney and Tack, 1985; Isacke et al., 1986; Creutz et al, 1987), control of cell growth (Fava and Cohen, 1984, Pepinsky and Sinclair, 1986), inhibition of blood coagulation (Funakoshi et al., 1987) and anchoring of cells to collagen (Pfäffle et al., 1988). Most of these putative functions depend on the common ability of annexins to bind to membrane phospholipids at Ca^{2+} concentrations in the range of 0.7 to several hundred µM. In addition, some annexins are able to cause the aggregation of model and biological vesicles in a Ca^{2+}-dependent manner. Structurally, there is a striking homology between the various members of the human family (40-50% identity) in the C-terminal 80-90% of the proteins (Pepinsky et al., 1988). This portion of the sequence consists of a repeated 60-80 amino acid segment that shows high intra-polypeptide homology. The annexin sequences do not contain a classical EF-hand Ca^{2+} binding site (Kretsinger and Creutz, 1986). A short N-terminal sequence is unique in composition and length in each annexin, suggesting that it may confer functional specificity. Modification of the N-terminus affects the Ca^{2+} sensitivity of several of the annexins (Schlaepfer and Haigler, 1987; Drust and Creutz, 1988; Ando et al., 1989). Though some proposals have been made for the structure of Ca^{2+} and phospholipid binding domains (Geisow et al., 1987), determination of the structure of membrane bound annexins has only recently begun with the first annexin crystal structure determination of annexin V (Huber et al., 1990a,b).

The first indication that annexins could mediate membrane
fusion came from one of the earliest isolated members of this
class, a protein called synexin (annexin VII) (Creutz et al.,
1978). Early investigations showed that synexin was capable of
aggregating chromaffin granules in a Ca^{2+}-dependent manner. Hong
et al.(1981, 1982a,b) showed that synexin could enhance fusion of
phospholipid vesicles composed of acidic phospholipids (Table II),
while other Ca^{2+} binding proteins had no effect. Synexin reduces
the Ca^{2+} concentration dependence for fusion of
phosphatidate/phosphatidylethanolamine vesicles by two orders of
magnitude from the millimolar range to approximately 10 μM.

MECHANISM OF SYNEXIN EFFECT ON MEMBRANE FUSION

We initiated studies to determine whether synexin enhances
aggregation of vesicles, actual fusion of vesicles or both. These
kinetic steps were isolated by making one or the other the rate
limiting step in the overall process. This was done by varying
phospholipid concentration, ionic compostion or phospholipid
composition (Meers et al., 1988a). For example, fusion of PS
liposomes is aggregation limited when induced by Ca^{2+} alone, but

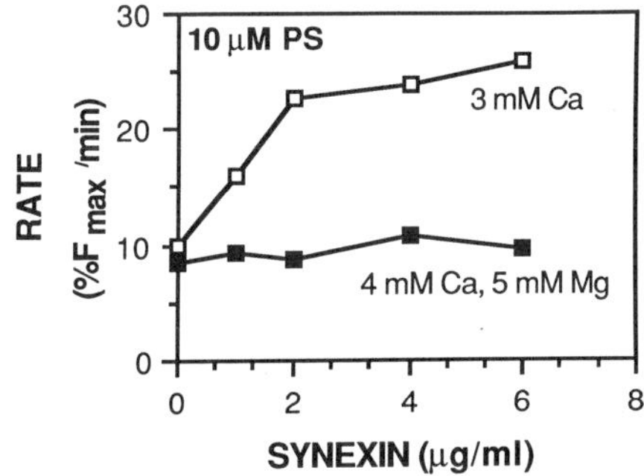

Figure 1. Overall rate of fusion of PS liposomes as initiated by
3 mM Ca^{2+} ( ) or 4 mM Ca^{2+} and 5 mM Mg^{2+} ( ). Ca^{2+} or
Ca^{2+} with Mg^{2+} was added at time zero to initiate fusion
of a 1 ml solution containing 10 μM PS and the indicated
amount of synexin, and the initial rate was measured. The
fusion was measured by the ANTS/DPX assay as described in
the text. The fluorescence of the fusion curves was
corrected for leakage as described in Bentz et al. (1983)
by setting the fluorescence intensity equal to the
original intensity plus one half of the intensity of the
leakage curve at the same time. All experiments were
performed at 25 °C in NaCl buffer.

when induced by Ca^{2+} plus Mg^{2+}, limited by fusion per se. Since
Mg^{2+} does not induce fusion itself and does not interfere with
synexin binding, it is clear from Figure 1 that synexin does not
increase the rate of fusion per se. In many other systems,
synexin also exerted positive effects on the overall rate of
fusion only under aggregation rate limiting kinetics (Meers et
al., 1988a). When the actual fusion step was rate limiting,
synexin was either inhibitory or had no effect (Meers et al.,
1987a; Meers et al., 1988a). Synexin-mediated fusion always
required a phospholipid composition itself susceptible to fusion

at some Ca^{2+} concentration. For example, at 50 mol% PC in PS, synexin can aggregate vesicles through its interaction with PS but fusion does not occur due to the inhibitory action of PC (Hong et al., 1981, 1982a,b). Thus fusion is still controlled by the phospholipid composition, consistent with the role of synexin as strictly an aggregator. In only one case has a marginal increase in the fusion rate constant, f_{11}, been reported. Synexin-mediated fusion at pH 6 of chromaffin granule ghosts that have been through several freeze-thaw cycles shows a very small (maximally 30%) increase in f_{11} at intermediate synexin concentration, while at high synexin concentration a 5-10 fold <u>decrease</u> in f_{11} was observed, (Nir et al., 1987) consistent with our results (Meers et al., 1987a, 1988a). The dominating effect of synexin in this study was a 10-fold increase in the aggregation rate. The very small increase in f_{11} could be related to denaturation of synexin which is inactivated by incubation at pH 3 for 30 seconds (P. Meers, unpublished results) or to the non-physiological effect of this low pH on chromaffin granules. Low pH enhancement of vesicular aggregation and fusion has been observed in a number of other systems such as chromaffin granule aggregation mediated by annexin II (Drust and Creutz, 1988), lactalbumin-mediated fusion of liposomes (Kim and Kim, 1986), clathrin-mediated fusion of liposomes (Maeza et al., 1989) and enhancement of liposome fusion with neutrophil plasma membranes (P. Meers, unpublished observation). We conclude that annexins probably assist in membrane fusion solely by enhancing the rate of aggregation while the rate of fusion itself is dependent on other membrane components.

Synexin, as a Ca^{2+}-dependent "aggregator", acts synergistically with promoters of the actual fusion step. Free fatty acids are generated upon stimulation of many secretory cells (e.g. Stenson and Parker, 1979; Waite et al., 1979; Walsh et al., 1981). They have also been shown to enhance fusion of small unilamellar vesicles composed of non-fusogenic phospholipids (Kantor and Prestegard, 1975, 1978). Fatty acids enhance the rate of the fusion step but not the aggregation step for a number of lipid vesicle systems (Meers et al., 1988b). Because of this kinetic specificity, they act synergistically with synexin. It is important to note that this synergy is a result of action at different kinetic steps and does not imply direct interaction between fatty acids and synexin. In fact all data thus far suggest that fatty acids (fusion promoters) and synexin (aggregation promoter) act at spatially separate sites. This distinction is quite important for determining the detailed mechanism by which a promoter of the overall fusion process acts in the cell.

In the presence of Ca^{2+} and the absence of vesicles, synexin self-associates or polymerizes to form extended rods (Creutz et al., 1979). It has been suggested that polymerization of synexin may be involved in vesicular aggregation. Some data suggests a limited role for polymerized synexin. The threshold concentration of Ca^{2+} required for polymerization (> 100 μM) (Creutz et al., 1979) is higher than that for binding to vesicles (6 μM) (Creutz et al., 1978). At 10-50 μM Ca^{2+} and sufficient protein and vesicle concentrations, most (> 70%) synexin is found associated with liposomes, presumably without polymerization to rods, under conditions where fusion occurs (Hong et al., 1982a,b). Polymerization of synexin prior to addition of vesicles for fusion inhibits the synexin effect (Meers et al., 1988a). These facts suggest that extended rods of synexin observed in the absence of phospholipids are probably not important in the fusion of

liposomes. However, there is some evidence for significant
activity of small polymers of synexin based on the observation of
vesicle-aggregating activity after partial polymerization and
involving a number of assumptions concerning the distribution of
polymerized products (Meers et al., 1988a). It should be noted
that a different kind of self-association could occur on the
vesicle surface, not related to the observed rod formation. At
this time, the relevant polymerization state of annexins for
promotion of membrane fusion remains unknown.

Synexin also promotes fusion in a system relevant to human
neutrophils. Evidence for a role in neutrophil degranulation was
obtained using specific granules and cytosolic proteins isolated
from this system (Meers et al., 1987a). Antibodies against bovine
liver synexin showed that human neutrophils contain this protein.
When cytosol from human neutrophils is exposed to a liposome
affinity column a group of three major proteins including the
putative synexin bind in a Ca^{2+}-dependent manner. Bovine synexin
and these neutrophil cytosol proteins mediate Ca^{2+}-dependent
aggregation of specific granules. Synexin also mediates Ca^{2+}-
dependent fusion of specific granules with liposomes. These data
suggest that synexin or a synexin-like protein may be involved in
human neutrophil degranulation. It is to be noted, however, that
fusion was not optimized in terms of Ca^{2+} concentration. Fusion
was not observed below approximately 100 μM Ca^{2+} under the
conditions of the experiments, a concentration which is much
higher than that observed after stimulation of the neutrophil.
One possible resolution is that transiently higher Ca^{2+}
concentrations occur locally in the cell, but quantitative
evidence for this phenomenon has been difficult to obtain in this
cell.

OTHER ANNEXINS AND MEMBRANE FUSION

Recently, fusion has been observed in the concentration range
of 1 μM Ca^{2+} using annexin II. This protein, referred to as
calpactin, exists as a heterotetramer with two annexin subunits
and two 10 kDa subunits. It allows aggregation of chromaffin
granules at a Ca^{2+} concentration as low as 0.7 μM (Drust and
Creutz, 1988) (if the pH is 6.0) and mediates exocytosis of
permeabilized chromaffin cells at 1-3 μM Ca^{2+} (Ali et al., 1989).
Thus annexins are implicated in membrane fusion at physiological
Ca^{2+} levels, and calpactin specifically in the chromaffin cell
system.

There have also been indications that lipocortin I (annexin
I) can interact with membranes at relatively low levels of Ca^{2+}
(Blackwood and Ernst, 1990).   The Ca^{2+} sensitivity of annexin I
can be modulated by alterations in its N-terminus. Some cells
contain endogenous proteases that cleave up to 29 amino acids from
the N-terminus (Chuah and Pallen, 1989). The truncated forms of
annexin I have a higher Ca^{2+} sensitivity with substantial membrane
binding in the micromolar range of Ca^{2+} concentration (Ando et
al., 1989). Des (1-9) annexin I has been shown to mediate fusion
of neutrophil-derived plasma membranes with phospholipid vesicles
in the 10 μM Ca^{2+} range (Oshry et al., 1991). Since annexin I is
an abundant protein in neutrophils, it may play an important role
in neutrophil degranulation and modification of the N-terminus may
be a regulatory mechanism.

Annexin IV (endonexin) has also been shown to mediate fusion
of liposomes (Meers et al. unpublished data) or chromaffin
granules (Zaks and Creutz, 1988). On the other hand annexins V

and VI inhibit fusion, despite very high homology between the
various annexins. The effect of annexin V (see below) can be
explained by binding in a monovalent manner such that only one
membrane surface is bound per monomer of protein.

MODEL FOR ANNEXIN-MEDIATED MEMBRANE FUSION

A conceptual model of annexin-mediated fusion consistent with
the experimental facts is presented in Figure 2. This model is
similar to that suggested by Hong et al.(1987) and Zaks and Creutz
(1988). Annexins V interacts with vesicles in a monovalent
fashion so that it cannot mediate aggregation and subsequent
fusion between vesicles. Other annexins interact with vesicles in
a polyvalent fashion so as to aggregate the vesicles. This could
involve protein-protein interactions. Annexins probably do not
normally accelerate the rate of fusion per se but only the
aggregation of vesicles. After aggregation, a number of
requirements must be satisfied so that fusion will occur. For

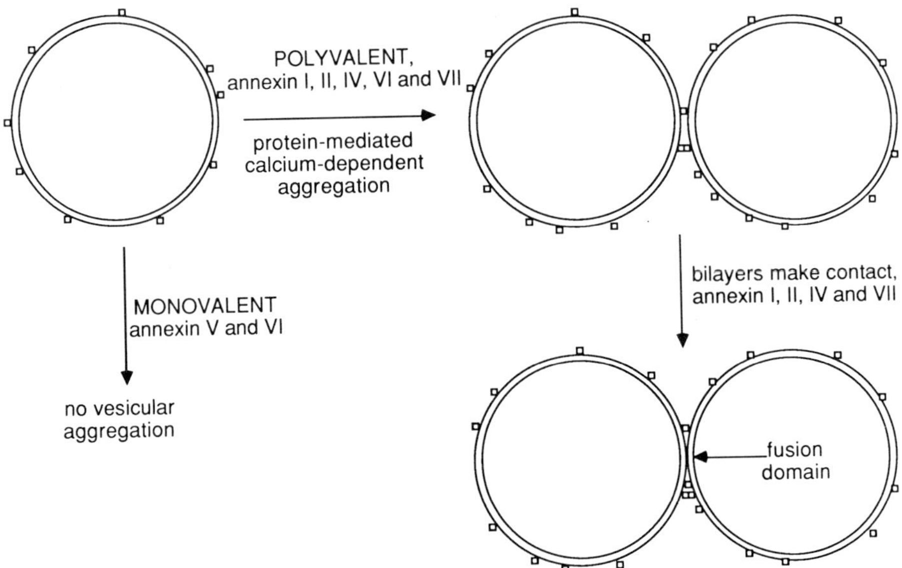

Figure 2. Conceptual model for annexin-mediated aggregation of
vesicles followed by fusion. Annexins are represented
as cubes approximately 4 nm on a side. A vesicle
approximately 150 nm in diameter is depicted with a
membrane thickness of approximately 4 nm. Binding by
mostly monomers is shown but self association of
annexins may also be important.

example, annexin VI mediates aggregation of vesicles under some
circumstances (Zaks and Creutz, 1990) but also can block fusion
(P. Meers, unpublished results). Fusion ultimately requires a
fusogenic phospholipid composition and is also inhibited by
phospholipids that inhibit fusion. Therefore fusion probably
occurs at a site involving solely phospholipids, remote from the
annexin binding site. In Figure 2, we have summarized these
results. It is likely that either an annexin monomer or polymer
mediates initial aggregation of vesicles. Subsequently, the
bilayers make contact at least transiently. This contact could be

enhanced by other annexin monomers or polymers, but the protein
molecules must not block intervesicle phospholipid-phospholipid
contact. In the case of PS, the avid "trans" complex for Ca^{2+} may
help to appose and dehydrate the surfaces of the vesicles leading
to fusion. Excess annexin binding could block intervesicle
contact in this model, consistent with the observed negative
effects of high concentrations of synexin (Meers et al., 1987a;
Nir et al., 1987; Meers et al., 1988a). The geometry and
stoichiometry of the prefusion complex and its odds of fusing will
depend on the angle and flexibility of annexin attachment, the
polymerization state of the annexin involved, the bilayer
curvature of the vesicle and its shape, the ability for the
bilayer to deform and appose to another bilayer, the relationship
of the relevant primary and secondary DLVO energy minima distances
to the interbilayer distance and the hydration forces for the
phospholipid at the apposition site.

MECHANISM OF ANNEXIN-LIPID BILAYER INTERACTION

Another important aspect of annexin mediated-membrane fusion
is the mechanism by which annexins interact with membrane
phospholipids. A simple sensitive assay is desirable to determine
the factors responsible for binding. We have found various
fluorescent phospholipid probes useful for this characterization
(Meers et al., 1987b, 1988c). Binding of annexins to phospholipid
vesicles causes dequenching of membrane-incorporated fluorophors
quenched in a concentration-dependent manner. For instance, when
synexin binds to vesicles containing 0.75 mol% each of N-(7-
nitrobenz-2-oxa-1,3-diazol-4-yl)dipalmitoyl-1-α-
phosphatidylethanolamine (NBD-PE) and N-(Lissamine rhodamine B
sulfonyl)dipalmitoyl-1-α-phosphatidylethanolamine (Rh-PE), an
increase in the NBD fluorescence is observed (Meers et al. 1991b).
NBD-PE is the donor of this resonance energy transfer pair and is
partially quenched by the presence of the acceptor Rh-PE. When
NBD-PE alone is incorporated into vesicles at 5 mol% of the total
phospholipid, it is partially self-quenched. Binding of annexins
to this type of vesicle also increases the NBD fluorescence. In
both cases the fluorescence change observed is dependent on the
simultaneous presence of both an annexin and Ca^{2+}. It is also
dependent on a fluorescent phospholipid derivative that is
quenched in a concentration-dependent manner.
Since the NBD group is interfacially localized and
potentially susceptible to artifactual effects (Silvius et al.,
1987, 1988; Düzgünes et al., 1988), a hydrophobic acyl chain probe
was also tested. 3-Palmitoyl-2-[1-pyrenedecanoyl]-L-α-
phosphatidylcholine (pyrene-PC) forms excited state complexes
called excimers at a sufficiently high proportion of the probe in
the membrane. When any annexins bind to PS vesicles containing
this probe, a decrease in the excimer-to-monomer ratio is
observed. The effect of binding of endonexin (annexin IV) is
shown in Figure 3a. Peaks from monomers at 377 and 396 nm
increase in intensity and the broad peak at 480 nm from excimers
decreases in intensity when endonexin binds in the presence of
Ca^{2+}. The increase in intensity of the 377 nm peak can be
followed as a function of time as in Figure 3b. From this figure
it is clear that the response is dependent on the protein
concentration within this range. Other non-annexin proteins do
not cause the observed effect. Based on the lack of probe
specificity and results indicating no change in fluorescence
lifetimes or direct annexin binding by this pyrene probe (Meers et

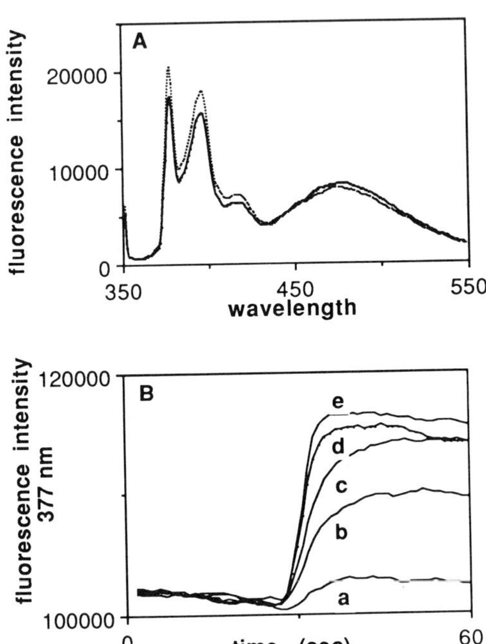

Figure 3. A. Emission spectrum of 5 mol% pyrene-PC incorporated
into phosphatidylderine vesicles at a total
phospholipid concentration of 0.25 μM. Each sample
contained 156 nM endonexin and was in a total volume of
2 ml. The solid curve is in buffer alone. The broken
curve contains 100 μM free Ca^{2+}. Endonexin was
purified by the method of Südhof et al.[74,75] with small
modifications. B. Dependence of probe response on
protein concentration. Endonexin at a final
concentration of 7.8 nM (**a**), 15.6 nM (**b**), 31.2 nM (**c**),
62.5 nM (**d**) or 125 nM (**e**) was added to vesicles
composed of 5 mol% pyrene-PC in PS at a total
phospholipid concentration of 1 μM.   At approximately
30 seconds a final concentration of 100 μM Ca^{2+} was
added to the sample and fluorescence at 377 nm was
monitored. The excitation wavelength was 344 nm. All
experiments were in 100 mM NaCl, 5 mM TES, 0.1 mM EDTA,
pH 7.4 at 25° C.

al., 1991a), the most likely cause of this effect is a decrease in
the effective bulk lateral mobility of membrane phospholipids upon
annexin binding. One speculation on the mechanism of this effect
is that each annexin molecule binds directly to several acidic
phospholipids, but does not deplete the acidic phospholipid in the
vesicle enough to cause an increase in excimer-to-monomer ratio
under the conditions of our experiments. Instead the excimer-to-
monomer ratio decreases because the annexin-phospholipid complexes
are large obstacles of reduced lateral mobility and therefore
restrict the lateral mobility of the probe molecules even though
they do not bind to the proteins (Eisinger et al., 1986; Saxton,
1987).

The reversibility of endonexin binding can be assessed using
the pyrene response. When EDTA is added to bound endonexin a
complete and rapid reversal of the fluorescence increase is
observed. The polycation spermine is also able to reverse the
annexin-mediated fluorescence increase (Figure 4). The presence
of spermine before Ca^{2+} addition also inhibits the fluorescence
increase. These results indicate that ionic interactions dominate
the binding of endonexin under the chosen conditions. Similar
results were obtained for annexin V (Meers et al., 1991a), I, VI
and VII (P. Meers, unpublished results).

Pyrene-PC fluorescence could also be used to measure
competition of unlabeled vesicles with labeled PS vesicles. The
effect of competing vesicles on the fluorescence increase induced
by binding of the 67 kDa calelectrin (annexin VI) is shown in
Figure 5. Clearly there is significant competition by unlabeled
PS vesicles. By binding the protein first to the labeled vesicles
and then adding the unlabeled vesicles the rate of exchange
between vesicles could also be monitored. In this case the
exchange is very slow on the time scale of minutes. Endonexin
binding to vesicles of varying percentages of PS in PC was also

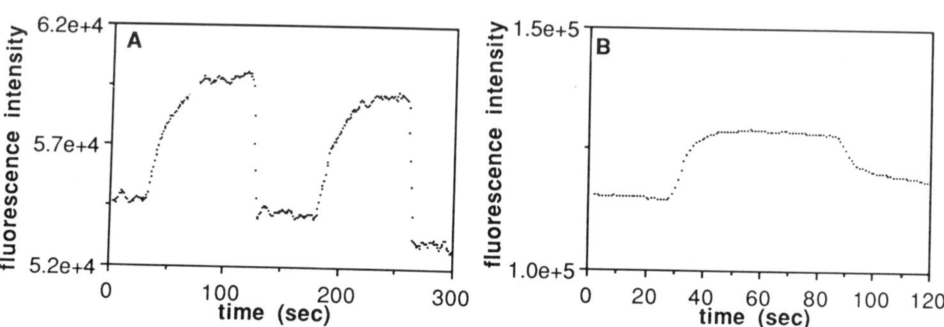

Figure 4. Reversibility of annexin binding. **A** - Sample contained
0.25 µM total phospholipid of 5 mol% pyrene-PC in PS
and 15 nM endonexin. Additions were 100 µM Ca^{2+} at 30
sec, 110 µM EDTA at 120 sec, 110 µM Ca^{2+} at 180 sec
and 110 µM EDTA at 250 sec. Experimental conditions
were as in Figure 3. **B** - 31 nM endonexin and 1 µM
total phospholipid composed of 5 mol% pyrene-PC in PS.
100 µM Ca^{2+} was added at the first arrow and 100 µM
spermine at the second arrow. Spermine alone had no
effect, but inhibited the Ca^{2+} response.

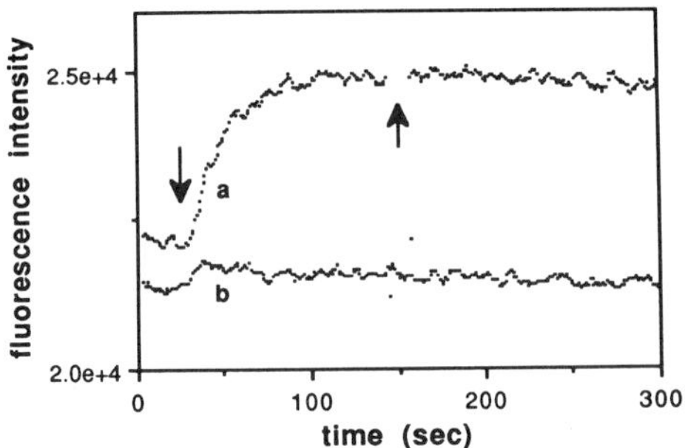

Figure 5. Competition of vesicles for annexin binding. Initial preparations contained 7.5 nM of the 67 KDa calelectrin and either 0.25 μM of 5 mol% pyrene-PC in PS (**a**) or 0.25 μM of 5 mol% pPC in PS along with 5 μM PS (**b**). In **a** and **b** 100 μM Ca^{2+} was added at the first arrow. In **a** 5 μM PS was added at the second arrow. Other experimental conditions were as in Figure 4. The 67 kDa calelectrin was prepared by the method of Südhof et al. (1984, 1987) with modifications.

measured by competion and is shown in Table III. The effect of phospholipid composition was observed both in directly labeled vesicles and by competition of unlabeled vesicles with labeled vesicles. There is an apparent cut-off in the binding of endonexin when the PS composition drops below approximately 25-50%. This is probably related to the necessity for the protein to access a minimum number of PS molecules within the surface area covered by that protein. Binding to phosphatidate is also observed using this probe. Competition experiments with annexin V showed almost no specificity among the acidic phospholipids PS, PA and PG, but no binding to PI (Meers et al., 1991a).

These results taken together indicate that annexins appear to bind to phospholipids by mainly reversible and largely non-specific ionic interactions and probably do not normally insert deeply into the membrane. This interaction does, however, perturb the lateral mobility of a number of the phospholipids, possibly by creating obstacles for the diffusion of probe molecules. The studies leading to these conclusions were designed to focus on binding alone in the absence of the effects of aggregation and fusion of vesicles. Under conditions where annexins mediate vesicle aggregation, other modes of binding could exist.

STRUCTURE OF MEMBRANE-BOUND ANNEXINS

A number of structural models for annexin binding to phospholipids have been proposed despite a paucity of structural data. In the case of synexin (annexin VII), several disparate models for interaction with membranes have been proposed. In one recent model (Lelkes and Pollard, 1991), polymers of synexin span phospholipid bilayers, crosslink vesicles and subsequently "split" in such a way as to allow phospholipids to leave the bilayer

Table III. Effect of phospholipid composition on the change in pyrene-PC fluorescence due to endonexin (annexin IV) binding.

phospholipid composition	% fluorescence increase[a]	competition index[b]
PS	13.0	0.95
PS/PC (3/1)	–	0.95
PS/PC (1/1)	3.2	0.84
PS/PC (1/3)	0.4	0.01
PC	0	–
PA	8.5	–

[a] Percentage increase in pyrene-PC fluorescence at 377 nm upon addition of 100 μM Ca^{2+} to 1 μM total phospholipid with no competing unlabeled vesicles. Vesicles were of the noted composition and contained 5 mol% pyrene-PC. Final endonexin concentration was 31 nM. Experiments were performed in 100 mM NaCl, 5 mM TES, 0.1 mM EDTA, pH 7.4 at 25° C. Endonexin was purified by the method of Südhof et al.[74,75] with small modifications.

[b] Samples contained 47 nM endonexin and 2 μM total phospholipid of vesicles composed of PS with 5 mol% pyrene-PC. Also included was 2 μM total phospholipid of competing unlabeled vesicles of composition indicated in the first column. The competition index was defined as $\frac{I_{max} - I_{obs}}{I_{max} - I_{1/2}}$ where I_{max} is the maximal % increase without competing vesicles, I_{obs} is the observed % increase in fluorescence and $I_{1/2}$ is the % increase for half of the original amount of protein (10 nM), i.e. $I_{1/2}$ is the % increase expected if endonexin bound equally well to the competing vesicles. An index of 1.0 denotes equal binding to the labeled and competing vesicles, while 0 indicates no binding to the competing vesicles. Experiments were performed in 100 mM NaCl, 5 mM TES, 0.1 mM EDTA, pH 7.4 at 25° C.

organization and move along the surface of the synexin forming a "fusion pore". The evidence cited for this model includes electrophysiological measurements demonstrating that synexin binding can increase the capacitance of a bilayer and that some of the annexins appear to mediate the formation of Ca^{2+} channels (Burns et al., 1989; Rojas et al., 1991), or are partially present in biological membranes in a form that requires detergent for removal (Sheets et al., 1987). The Ca^{2+} channels appear to require extreme conditions such as pH 6 and very high Ca^{2+} concentrations (30 mM) on one side of the membrane. Ca^{2+} channel formation may not be connected with fusion as annexin V apparently forms channels (Rojas et al., 1991; Huber et al., 1990a) but is inhibitory to fusion (Oshry et al., 1991). Transmembrane insertion of annexins would be expected to involve long hydrophobic stretches of amino acids and to be irreversible. Annexins do not appear to contain long enough hydrophobic stretches for a transbilayer helix. In phospholipid bilayers there is as yet no evidence other than electrophysiological for an irreversibly bound form of annexin V (Meers, 1990; Meers et al., 1991a) or annexin I, IV, VI or VII (Meers et al., 1991b, this publication) at physiological pH. Data indicating that annexins do not significantly increase the rate of fusion per se in most circumstances suggest that a membrane-inserted form of annexins is a minor or fusion-irrelevant species. A more recent model (Guy et al., 1991) seems to no longer deal with the notion of a polymeric synexin fusion pore in favor of a Ca^{2+} channel generated by an individual synexin molecule with a "TIM barrel" structural motif. This model seems inconsistent with the reported crystal structure of annexin V in the absence of phospholipids (Huber et al., 1990a), a protein highly homologous to most of the synexin sequence. The resolution of these inconsistencies will be of great interest.

Our approach to study the structure of the membrane bound annexins has been to try to identify and characterize important aspects of protein-lipid interactions using spectroscopic methods. Because of the difficulty of crystallization of a membrane-bound protein in the presence of a phospholipid bilayer for detailed X-ray diffraction analysis, it is desirable to use other approaches to study this form as well. One method has been to use the intrinsic tryptophan fluorescence of the annexins (Meers, 1990). Most human annexins contain a single tryptophan, providing an ideal site specific probe. For example, human lipocortin V (annexin V) contains a single tryptophan in the third consensus sequence. When this protein binds to vesicles composed of 50% phosphatidylserine and 50% phosphatidylcholine in the presence of 100 μM Ca^{2+}, a marked increase in fluorescence intensity is observed, accompanied by a red shift indicating a conformational change upon binding (Meers, 1990). The tryptophan fluorescence of the bound lipocortin V is quenched strongly when PC is replaced by a derivative with a nitroxide moiety at the 5 postion of the acyl chain (Figure 6). These results predicted that the third consensus sequence probably makes close contact with the phospholipids of the membrane. Quenching by derivatives with the nitroxide more deeply localized in the membrane is weaker, suggesting an interfacial location for the tryptophan (Figure 6). The single tryptophan in lipocortin I (annexin I), located in the N-terminal region, shows little interaction with the phospholipid quenchers suggesting that this tryptophan does not make close

contact with the phospholipids. Subsequently, it was shown that
the tryptophan in the third consensus sequence of calpactin
(annnexin II) is in close proximity to a Tb^{3+} binding
site (Marriott et al., 1990). Taken together, these results
suggest the possibility that the third consensus sequence in some
annexins is involved in phospholipid binding and could even
participate in a ternary annexin-Ca^{2+}-phospholipid complex, where
the Ca^{2+} ion has both protein and phospholipid ligands.

The recent crystalization and X-ray diffraction analysis of
annexin V in the absence of phospholipids (Huber et al., 1990a,
1990b) along with our previous predictions concerning the
orientation with respect to the bilayer of tryptophan-containing
portions of annexins have led to a plausible model for the

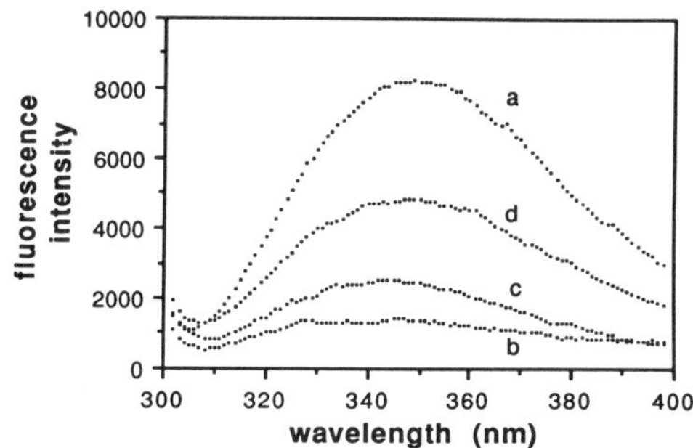

Figure 6. Effect of phospholipid quenchers on the fluorescence
of tryptophan in annexin V. Samples consisted of 10 µg/ml
lipocortin V and 50 µM total phospolipid in 100 µM free
Ca^{2+}, 100 µM EDTA, 100 mM NaCl, 5 mM TES, pH 7.4 at 25° C.
The phospholipid compositions were **a**) PS/PC (1/1), **b**)
PS/5-PC (1/1), **c**) PS/12-PC (1/1) and **d**) PS/16-PC (1/1).

interaction of this annexin with phopholipid bilayers. In Figure
7 is shown the α carbon backbone of annexin V with Ca^{2+} binding
sites deduced from the diffraction analysis. The protein is
nearly toroidal in shape with the view in the figure of the side
of the torus. All the Ca^{2+} binding sites are located on one side
of the protein and are incompletely liganded, containing six
instead of the preferred seven ligands. The approximate location
of tryptophan 186 and the N-terminal region are shown in the
figure. Our data place this tryptophan near the interface of the
membrane while the N-terminal region is expected to be oriented
away from the membrane surface by inference. This alignment
places the partially liganded Ca^{2+} ions along the surface of the
membrane where they can utilize phospholipid headgroups as
ligands. Hence the properties of annexin V are explained.
Binding is monovalent in terms of vesicles and ionic and
reversible in nature. We expect that certain aspects of this
structure will also apply to other annexins, but the identities of

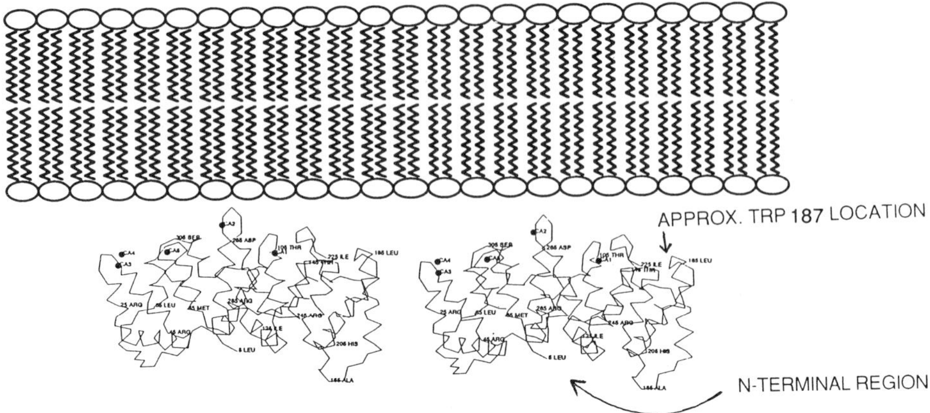

Figure 7. Model for binding of annexin V to membranes based on tryptophan localization in the presence of phospholipids and crystal structure in the absence of phospholipids. The annexin C_α-chain tracing is taken from Huber et al. (1990b). A bilayer approximately 4 nm across is depicted. The protein and bilayer are approximately to scale. Ca^{2+} binding sites determined in the crystal are shown as dark dots. Location of tryptophan 187 and the N-terminal region of the molecule are depicted.

the additional annexin or phospholipid binding sites that confer a polyvalent, vesicle-aggregating nature on certain annexins remain unknown. Further structural characterization will help define the mechanism of annexin binding and its effects on fusion of membranes.

CONCLUDING REMARKS

Membrane fusion, as we have defined it, may be a critical step in exocytosis, and therefore it is very important to understand the mechanism at the molecular level. Our approach has been the establishment of a relatively simple experimental model system amenable to a detailed study of the role of individual components of this complex reaction. This system involves vesicle fusion, starting with phospholipid vesicles (liposomes) and proceeding to isolated secretory granules and plasma membrane vesicles. The relative advantages of this cell-free system are as follows: i) Simplicity of starting material, with incremental complexity. ii) Accessibility of detailed kinetic studies by the use of sensitive fluorescence assays. iii) Independent assessment of the specificity of various lipids, proteins, metal ions and other "effectors". iv) Possible study of structure-function relationships of the individual components.

The kinetic analysis of fusion has proved to be extremely important in defining the role of specific components. This is because the overall fusion reaction involves two in-series reactions, one the aggregation of the vesicles, and the other,

the actual fusion of their membranes. Thus, it is essential to
define whether a particular reactant is active in the step
promoting the aggregation (recognition, close contact) or fusion
(mixing of membranes and vesicle contents). The overall fusion
rate could be affected by participation in either step, depending
on which is the rate-limiting reaction. Our results with synexin
indicate that its ability to enhance the overall vesicle fusion
reaction is due to increasing the aggregation rate and not the
rate of fusion per se. Polyamines such as spermine and spermidine
seem to have a similar effect by the same criteria. On the
contrary, long chain fatty acids, such as arachidonate, seem to
have an effect in enhancing the rate of fusion itself, rather than
the rate of aggregation.

Studies of structural aspects of annexin binding to membranes
using spectroscopic methods has allowed us to help define several
essential features in annexin-membrane interactions that probably
relate to their actions in the membrane fusion process. In
particular, annexins appear to interact with membranes by mainly
reversible relatively non-specific ionic forces. In the case of
annexin V, binding probably involves bridging by Ca^{2+} ions between
one surface of the protein molecule and the bilayer. This leaves
open the question of how some annexins link two vesicles
simultaneously.

Taking into account the specificity and careful regulation of
intracellular membrane fusion, annexins are likely to constitute
only one component of a much more complex system. Further
detailed studies of the role of various cytoplasmic and membrane
proteins is needed for a thorough understanding of the mechanism
of membrane fusion during exocytosis. The vesicle fusion system
we have reviewed here has proven to be a valuable tool in that
pursuit.

[†] This investigation was supported by research grant 1570 C-1 from
the Massachusetts Chapter of the American Cancer Society (P.M.), a
postdoctoral fellowship (P.M.) from the Arthritis Foundation and
partially by grant GM-28117 (K.H. and D.P) and grant GM-41790-01A1
(P.M.) from the National Institutes of Health.

REFEENCES

Ali, S.M., Geisow, M.J., and Burgoyne, R.D. , 1989, *Nature* 340,
313-315.
Allen-T-M. Hong-K. Papahadjopoulos-D., 1990, *Biochemistry* 29,
2976-2985.
Ando, Y., Imamura, S., Hong, Y.-M., Owada, M. K., Kakunaga, T. and
Kannag, R., 1989, *J. Biol. Chem.* 264, 6948-6955.
Artalejo, C. R., Garcia, A. G. and Aunis, D., 1987, *J. Biol.
Chem.* 262, 915-926.
Baker, P. F. and Knight, D. E., 1978, *Nature* 276, 620-622.
Bearer, E. L., Düzgünes, N., Friend, D. S. and Papahadjopoulos,
D., 1982, *Biochim. Biophys. Acta* 693, 93-98.
Bentz J., Nir, S. and Wilschut, J., 1983, *Colloids and Surfaces*
6, 333-363.
Bentz, J., Düzgünes, N. and Nir, S., 1985, *Biochemistry* 24, 1064-
1072.
Blackwood, R. A. and Ernst, J. D., 1990, *Biochem. J.* 266, 195-
200.
Burns, A. L., Magendzo, K., Shirvan, A., Srivastava, M., Rojas,
E., Alijani, M. R. and Pollard, H. B., 1989, *Proc. Nat. Acad.
Sci.* 86, 3798-3802.
Chuah, S. Y. and Pallen, C. J., 1989, *J. Biol. Chem.* 264, 21160-
21166.

Creutz, C. E., Pazoles, C. J. and Pollard, H. B., 1978, J. Biol. Chem. 253, 2858-2866.

Creutz, C. E., Pazoles, C. J. and Pollard, H. B., 1979, J. Biol Chem. 254, 553-558.

Creutz, C. E., Zaks, W. J., Hamman, H. C. Crane, S., Martin, W. H., Gould, K. L., Oddie, K. M. and Parsons, S. J., 1987, J. Biol. Chem. 262, 1860-1868.

Cullis, P. R. and Hope, M. J., 1978, Nature 271, 672-674.

Drust, D. S. and Creutz, C. E., 1988, Nature 331, 88-91.

Düzgünes, N., Wilschut, J., Fraley, R. and Papahadjopoulos, D., 1981, Biochim. Biophys. Acta 642, 182-195.

Düzgünes, N., Paiement, J., Freeman, K., Lopez, N. G., Wilschut, J. and Papahadjopoulos, D., 1984a, Biochemistry 23, 3486-3494.

Düzgünes, N., Hoekstra, D., Hong, K. and Papahadjopoulos, D., 1984b, FEBS Lett. 173, 80-84.

Düzgünes, N., Allen, T. M., Fedor, J. and Papahadjopoulos, D., 1988, Why fusion assays disagree. in: "Molecular Mechanisms of Membrane Fusion", S. Ohki, D. Doyle, T. D. Flanagan, S. W. Hui and E. Mayhew, eds., Plenum Press, New York, NY.

Eisinger, J., Flores, J. and Petersen, W. P., 1986, Biophys. J. 49, 987-1001.

Ekerdt, R. and Papahadjopoulos, D., 1982, Proc. Nat. Acad. Sci. U.S.A. 79, 2273-2277.

Ellens, H., Bentz, J. and Szoka, F. C., 1985, Biochemistry 24, 3099-3106.

Ellens, H., Siegel, D. P., Alford, D., Yeagle, P. L., Boni, L., Lis, L. J., Quinn, P. J. and Bentz, J., 1989, Biochemistry 28, 3692-3703.

Fava, R. A. and Cohen, S., 1984, J. Biol. Chem. 259, 2636-2645.

Feigenson, G. W., 1986, Biochemistry 25, 5819-5825.

Funakoshi, T., Heimark, R. L., Hendrickson, L. E., Mc Mullen, B. A. and Fujikawa, K., 1987, Biochemistry 26, 5572-5578.

Geisow, M. J., Walker, J. H., Boustead, C. and Taylor, W., 1987, Biosci. Rep. 7, 289-298.

Gerke, V. and Weber, K., 1984, EMBO J. 3, 227-233.

Glenney, J., 1986, Proc. Nat. Acad. Sci. U.S.A. 83, 4258-4262.

Glenney, J. R. and Tack, B. F., 1985, Proc. Nat. Acad. Sci. USA 82, 7884-7888.

Guy, H. R., Rojas, E. M., Burns, A. L. and Pollard, H. B., 1991, Biophys. J. 59, 372a.

Hong, K., Düzgünes, N. and Papahadjopoulos, D., 1981, J. Biol. Chem. 256, 3641-3644.

Hong, K., Düzgünes, N. and Papahadjopoulos, D., 1982a, Biophys. J. 37, 297-305.

Hong, K., Düzgünes, N., Ekerdt, R. and Papahadjopoulos, D., 1982b, Proc. Nat. Acad. Sci. USA 79, 4642-4644.

Hong, K., Meers, P. and Papahadjopoulos, D., 1987, Protein modulation of liposome fusion , in: "Cell Fusion", A. E. Sowers, ed., Plenum Press, New York, New York.

Horn, R. G., 1984, Biochim. Biophys. Acta 778, 224-228.

Huber, R., Römisch, J. and Paques, E.-P., 1990a, EMBO J. 9, 3867-3874.

Huber, R., Schneider, M., Mayr, I., Römisch, J. and Paques, E.-P., 1990b, FEBS Lett. 275, 15-21.

Hui, S. W., Stewart, T. P., Boni, L. T. and Yeagle, P. L., 1981, Science 212, 921-923.

Ikebuchi, N. W. and Waisman, D. M., 1990, J. Biol. Chem. 265, 3392-3400.

Isacke, C. M., Trowbridge, I. S. and Hunter, T., 1986, Mol. Cell Biol. 6, 2745-2751.

Kantor, H. L. and Prestegard, J. H., 1975, Biochemistry 14, 1790-1795.

Kantor, H. L. and Prestegard, J. H., 1978, <u>Biochemistry</u> 17, 3592-3597.

Kim, J. and Kim, H., 1986, <u>Biochemistry</u> 25, 7867-7874.

Kretsinger, H. and Creutz, C. E., 1986, <u>Nature</u>, London, 320, 573.

Lelkes, P. I. and Pollard, H. B., 1991, Cytoplasmic determinants of exocytotic membrane fusion. in: "Membrane Fusion", J. Wilschut and D. Hoekstra, eds., Marcel Dekker, Inc. New York, N.Y.

Maezawa, S., Yoshimura, T., Hong, K., Düzgünes, N. and Papahadjopoulos, D., 1989, <u>Biochemistry</u> 28, 1422-1428.

Marra, J. and Israelachvili, J., 1985, <u>Biochemistry</u> 24, 4608-4618.

Marriott, G., Kirk, W. R., Johnsson, N. and Weber, K., 1990, <u>Biochemistry</u> 29, 7004-7011.

McLaughlin, S., 1989, <u>Ann. Rev. Biophys. Chem.</u> 18, 113-136.

Meers, P., Ernst, J. D., Hong, K., Düzgünes, N., Fedor, J., Goldstein, I. M. and Papahadjopoulos, D., 1987a, <u>J. Biol. Chem.</u> 262, 7850-7858.

Meers, P., Papahadjopoulos, D. and Hong K., 1987b, <u>Biophys. J.</u> 51, 169a.

Meers, P., Bentz, J., Alford, D., Nir, S., Papahadjopoulos, D. and Hong, K., 1988a, <u>Biochemistry</u> 27, 4430-4439.

Meers, P., Hong, K. and Papahdjopoulos, D., 1988b, <u>Biochemistry</u> 27, 6784-6794.

Meers, P., Papahadjopoulos, D. and Hong, K., 1988c, <u>Biophys. J.</u> 53, 319a.

Meers, P., 1990, <u>Biochemistry</u> 29, 3325-3330.

Meers, P., Daleke, D., Hong, K. and Papahadjopoulos, D., 1991a, <u>Biochemistry</u> 30, 2903-2908.

Meers, P. Hong, K. and Papahadjopoulos, D., 1991b, submitted to <u>Biochem. Biophys. Res. Comm.</u>

Nir, S., Bentz, J. and Wilschut, J., 1980a, <u>Biochemistry</u> 19, 6030-6036.

Nir, S., Bentz, J. and Portis, A. R. Jr., 1980b, <u>Adv. Chem. Ser.</u> 188, 75-106.

Nir, S., Stutzin, A. and Pollard, H. B., 1987, <u>Biochim. Biophys. Acta</u> 903, 309-318,

Oshry, L., Meers, P., Mealy, T. and Tauber A. I., 1991, <u>Biochim. Biophys. Acta</u>, in press.

Papahadjopoulos, D. and Bangham, A. D., 1966, <u>Biochim. Biophys. Acta</u> 126, 185-188.

Papahadjopoulos, D., Poste, G., Schaeffer, B. E. and Vail, W. J., 1974, <u>Biochim. Biophys. Acta</u> 352, 10-28.

Pepinsky, R. B. and Sinclair, L. K., 1986, <u>Nature</u> 321, 81-84.

Pepinsky, R. B., Tizard, R., Mattaliano, R. J., Sinclair, L. K., Miller, G. T., Browning, J. L., Chow, E. P., Burne, C., Huang, K.-S., Pratt, D., Wachter, L., Hession, C., Frey, A. Z. and Wallner, B. P., 1988, <u>J. Biol. Chem.</u> 263, 10799-10811.

Pfäffle, M., Ruggiero, F., Hofmann, H., Fernández, M. P., Selmin, O., Yamada, Y., Garrone, R. and von der Mark, K., 1988, <u>EMBO J.</u> 7, 2335-2342.

Portis, A., Newton, C., Pangborn, W. and Papahadjopoulos, D., 1979, <u>Biochemistry</u> 18, 780-790.

Rand, R. P., 1981, <u>Ann. Rev. Biophys. Bioeng.</u> 10, 277-314.

Rehfeld, S. J., Düzgünes, N., Newton, C., Papahadjopoulos, D. and Eatough, D. J., 1981, <u>FEBS Lett. 123</u>, 249-251.

Rojas, E., Pollard, H. B., Haigler, H. T., Parra, C. and Burns, A. L., 1991, <u>J . Biol. Chem.</u> 265, 21207-21215.

Rosenberg, J., Düzgünes, N. and Kayalar, C., 1983, <u>Biochim. Biophys. Acta</u> 735, 173-180.

Saxton, M. J., 1987, <u>Biophys. J.</u> 52, 989-997.

Schlaepfer, D. and Haigler, H. T., 1987, <u>J. Biol. Chem.</u> 262, 6931-6937.

Sheets, E. E., Giugni, T. D., Coates, G. G., Schlaepfer, D. D. and Haigler, H. T., 1987, <u>Biochemistry</u> 26, 1164-1172.

Siegel, D., 1984, <u>Biophys. J.</u> 399-420.

Silvius, J. R., Leventis, R., Brown, P. M. and Zuckermann, M., 1987, <u>Biochremistry</u> 26, 4279-4287.

Silvius, J. R., Leventis, R. and Brown, P. M., 1988, 'Slow artifacts' in assays of lipid mixing between membranes' in: "Molecular Mechanisms of Membrane Fusion", S. Ohki, D. Doyle, T. D. Flanagan, S. W. Hui and E. Mayhew, eds., Plenum Press, New York, NY.

Stenson, W. F. and Parker, C. W., 1979, <u>J. Clin. Invest.</u> 64, 1457-1465.

Südhof, T. C. and Stone, D. K., 1987, <u>Methods Enzymol.</u> 139, 30-35.

Südhof, T. C., Ebbecke, M., Walker, J. H., Fritsche, U. and Boustead, C., 1984, <u>Biochemistry</u> 23, 1103-1109.

Sundler, R. and Wijkander, J., 1983, <u>Biochim. Biophys. Acta</u> 730, 391-394.

Sundler, R., 1984, Role of phospholipid head group structure and polarity in the control of membrane fusion. in: "Biomembranes" vol. 12, M. Kates and L. A. Manson eds., Plenum Press, New York, N.Y.

Szoka, F. and Papahadjopoulos, D., 1978, <u>Proc. Nat. Acad. Sci. USA</u> 75, 4194-4198.

Szoka, F., Olson, F., Heath, T., Vail, W., Mayhew, E. and Papahadjopoulos, D., 1980, <u>Biochim. Biophys. Acta</u> 601, 559-571.

Verkleij, A. J., Mombers, C., Gerritsen, W. J., Leunissen-Bijvelt, L. and Cullis, P. R., 1979, <u>Biochim. Biophys. Acta</u> 555, 358-361.

Verkleij, A. J., van Echteld, C. J. A., Gerritsen, W. J., Cullis, P. R. and de Kruiff, B., 1980, <u>Biochim. Biophys. Acta</u> 600, 620-624.

Waite, M., DeChatelet, L. R., King, L. and Shirley, P. S., 1979, <u>Biochem. Biophys. Res. Commun.</u> 90, 984-992.

Walsh, C. E., Waite, B. M., Thomas, M. J. and DeChatelet, L. R., 1981, <u>J. Biol. Chem.</u> 256, 7228-7234.

Wilschut, J., Düzgünes, N., Fraley, R. and Papahadjopoulos, D., 1980, <u>Biochemistry</u> 19, 6011-6021.

Zaks, W. J. and Creutz, C. E., 1988, Membrane fusion in model systems for exocytosis. in: "Molecular Mechanisms of Membrane Fusion", S. Ohki, D. Doyle, T. D. Flanagan, S. W. Hui, and E. Mayhew, eds., Plenum Press, New York, N.Y.

Zaks, W. J. and Creutz, C. E., 1990, <u>Biochim. Biophys. Acta</u> 1029, 149-160.

BIOLOGICAL CONSEQUENCES OF ALTERATIONS IN THE PHYSICAL PROPERTIES OF MEMBRANES

Richard M. Epand

Department of Biochemistry, McMaster University Health
Sciences Centre, Hamilton, Ontario, Canada, L8N 3Z5

INTRODUCTION

Membranes serve important functions for cell organization and for
the transduction of signals from the external environment to the cell
interior. These functions are generally determined by specific molecules
in the membrane such as receptors and membrane bound enzymes. The
functioning of these receptors and enzymes are modulated by the nature
of their physical environment, i.e. the efficiency of signal transduction
and the activity of membrane-bound enzymes will be affected by the nature
of the membrane surrounding the specific functional sites. In addition,
some properties of membranes such as permeability or membrane fusion may
not be absolutely dependent on the presence of specific proteins but may
also occur by non-specific mechanisms. In this review, we will focus on
the modulation of viral fusion, protein kinase C activity and insulin
signalling in adipocytes as examples of membrane functions that are
modulated by the physical properties of the membrane. An earlier review
of our work in this area has recently appeared (Epand, 1990a).

Physical Properties of Membranes

There have been many efforts to evaluate the physical properties of
membranes and to attempt to correlate alterations in these properties
with changes in function. Many of these studies have used either
fluorescent probes, such as 1,6-diphenyl-1,3,5-hexatriene, to measure
fluorescence depolarization or nitroxide-labelled lipids to measure
hyperfine interactions with electron spin resonance. These methods have
the advantage that they can be applied to complex biological membranes
so that a parameter of the physical state of the membrane can be compared
with a functional property of the same membrane. However, there are
certain limitations inherent to the use of probe molecules. One is that
the probe itself may perturb, at least locally, the properties of the
membrane so that the parameter being measured is not directly related to
an intrinsic property of the membrane. Furthermore, there is evidence
that biological membranes are composed of distinct domains (Karnovsky
et al., 1982) so that a further complication arises from the non-uniform
distribution of the probe among domains in the membrane. The behaviour
of these probes in membranes is dependent on both orientational order and

rates of motion which are generally anisotropic. The measured behaviour
of the probe is often used to simply define a single parameter of
"fluidity" or an order parameter to describe the state of the membrane.
It would be expected that properties such as the rate of lateral
diffusion in the membrane would be altered by changes in "fluidity" or
in the order parameter. However, there is recent evidence that the acyl
chain order parameters of predeuterated dimyristoylphosphatidylcholine
is independent of the presence of a non-bilayer forming lipid, dioleoylphosphatidylethanolamine, despite the fact that the non-bilayer forming
lipid would be expected to alter the physical properties of the membrane
(Fenske *et al.*, 1990). Another property of the membrane is the nature
of the membrane surface including hydration, headgroup packing electrostatic charge and hydrogen bonding. These probes would be expected to
be less sensitive to changes in the nature of the membrane surface.
However certain functional properties, such as membrane fusion or the
ability of membrane proteins to enter the membrane environment, may be
highly dependent on the nature of the membrane surface. Increased
hydrophobicity of the membrane surface or the presence of packing
defects would be expected to lead to increased rates of membrane fusion
or greater protein penetration. There have been some efforts to measure
membrane surface hydrophobicity using fluorescent probes. Probes which
partition into the hydrophobic-hydrophilic membrane interface (Weber and
Farris, 1979; Chong, 1988) as well as phospholipids containing a
covalently linked fluorescent probe (Kimura and Ikegami, 1985) have been
used. The changes in the fluorescence properties of dansylphosphatidylethanolamine in membranes which are prone to fuse suggest that membrane
fusion is promoted by increased surface hydrophobicity (Ohki and Arnold,
1990). The fluorescence emission wavelength of probes in membranes,
however, is also dependent on the rate of dipolar solvent relaxation
(Sommer *et al.*, 1990). Small changes in the membrane surface, which
would be difficult to measure experimentally, may produce large changes
in certain membrane functions.

Our studies were primarily concerned with determining whether the
modulation of biological activity by substances which partition into
membranes can be explained by the resulting alteration of membrane
physical properties. As an initial effort to evaluate whether such a
relationship might exist, we compared the effects of these membrane
additives on the phase behaviour of model membranes with their effects
on biological function. The model system we selected was the bilayer
(L_α) to hexagonal (H_{II}) phase transition temperature of synthetic
phosphatidylethanolamines.

Lipid Polymorphism

It has long been known that phospholipid bilayers can undergo gross
morphological rearrangements to micellar, cubic or hexagonal phases
(Seddon, 1990). It has been suggested that the formation of non-bilayer
phases may have a functional role in biological membranes (Cullis *et
al.*, 1986). Localized regions of nonbilayer structure, such as interlamellar associations (ILAs) (Siegel *et al.*, 1989), or an alteration of
the properties of the bilayer may be responsible for the apparent
correlation between the propensity to form the H_{II} phase and changes in
membrane properties. Examples of membrane properties that may be
correlated with both H_{II}-phase propensity and membrane function, but do
not represent the formation of specific structures such as ILA, are the
stability of the bilayer, the number of point and line defects of no
specific structure, the hydration of the membrane surface and the
density of headgroup packing at the surface. We recently demonstrated

that membrane properties can be altered concomitant with a change in the propensity to form the H_{II} phase but while the membrane was still in the bilayer phase (Epand and Leon, 1991). The apparent solvent environment of ϵ-dansyl-L-Lysine markedly changes as the L_α-H_{II} transition temperature is approached but while the membrane is still in a bilayer arrangement. This change is completely suppressed by agents that raise the L_α-H_{II} transition temperature. The changes detected by ϵ-dansyl-L-Lysine are likely to reflect changes in membrane surface properties since this probe would not penetrate deeply into the membrane.

We have evaluated the relationship between the shift of the L_α-H_{II} transition temperature caused by incorporation of a substance into a model membrane and the effect of this substance on membrane functional properties. Our studies did not specifically distinguish between changes in functional properties caused by alterations in the bulk physical properties of the bilayer and altered function resulting from changes in the amount of non-bilayer structures present in the membrane. The relative stability of the L_α and H_{II} phases result, at least in part, from changes at the membrane surface. Even the subtle change of solvent from H_2O to ^2H_2O causes a shift of several degrees in the L_α-H_{II} transition temperature (Epand, 1990b). Dehydration and increased interlipid hydrogen bonding favour H_{II} phase formation. Of course, other factors also affect this equilibrium. The forces affecting the L_α-H_{II} equilibrium have been divided into those that affect monolayer curvature and those that affect hydrocarbon packing in the H_{II} phase (Gruner, 1985). However, both of these properties are affected by any substance that partitions into the membrane. Thus, the L_α-H_{II} transition temperature will be affected by several factors in addition to the nature of the membrane surface. Nevertheless, in general substances which, when added to a membrane, raise the L_α-H_{II} transition temperature will increase hydration and headgroup volume. Although the correlation may not be quantitative, one would expect additives to the membrane that raise the L_α-H_{II} transition to affect certain membrane properties in an opposite fashion from those additives that lower this transition temperature (Yeagle, 1989; Hui and Sen, 1989).

We have assessed the effect of substances on the L_α-H_{II} transition temperature using differential scanning calorimetry by measuring a series of samples of fully hydrated synthetic phosphatidylethanolamines containing various mol fractions of the additive. At least at low mol fractions, the shift in the L_α-H_{II} transition temperature is usually proportional to the mol fraction of additive. The proportionality constant will be referred to as the slope (it is the slope of a plot of transition temperature vs mol fraction of additive).

Viral Fusion

A number of substances that raise the L_α-H_{II} transition temperature in model membranes have been shown to have antiviral activity (Table I). Certainly all substances which raise the L_α-H_{II} transition temperature do not exhibit antiviral activity and even those bilayer stabilizers which do show antiviral activity are not effective against infection by a large range of viruses. The group of agents listed in Table I represent compounds of diverse chemical structures which are all sufficiently hydrophobic to spontaneously partition from the aqueous to the membrane phase. None of the agents are particular potent inhibitors of viral replication, except perhaps for the apolipoprotein A-I, with micromolar concentrations generally required for antiviral activity. This suggests that the action of these compounds is on the bulk biophysical properties of the membrane and that these effects may be non-specific. A possible mechanism by which agents that raise the L_α-H_{II} transition temperature exhibit antiviral activity is by inhibiting fusion of the membrane of encapsulated viruses with the target membrane. Increased tendency to form the H_{II} phase has often been associated with increased rates of membrane fusion. In some cases, the inhibition of membrane fusion by anti-viral agents that suppress H_{II} phase formation has been demonstrated directly. Cyclosporin A (McKenzie *et al.*, 1987), the apolipoprotein A-I (Owens *et al.*, 1990), cholesterol sulfate (Watarai *et al.*, 1990) and tromantadine (Ickes *et al.*, 1990) all inhibit viral-induced cell-cell fusion. Although the antiviral action of carbobenzoxy-L-Ser-L-Leu-NH$_2$ has not been shown to be caused by the inhibition of viral fusion, this peptide derivative has been shown to inhibit exocytosis of intracellular vesicles (Mundy and Strittmatter, 1985), a process requiring membrane fusion, as well as the fusion of myoblasts to myotubes (Couch and Strittmatter, 1983). In all of the above examples inhibition of membrane fusion was studied in cell systems. These systems are complex and the inhibition of membrane fusion may not be a direct consequence of the agent altering the physical properties of the membrane but rather may result from a change in the activity of an enzyme or a change in ion flux across the membrane. A more detailed mechanistic study was done with the inhibition of the fusion of Sendai virus by cholesterol sulphate (Cheetham *et al.*, 1990a). It was shown that this amphiphile inhibits viral-induced hemolysis as well as the transfer of octadecylrhodamine from the virus to red blood cell ghosts or to liposomes composed of egg phosphatidylethanolamine containing 5% of the viral receptor, the ganglioside G_{D1A}. The IC_{50} of 3 μM was similar for the three systems. Neither adamantanol sulphate, a sulphated amphiphile which is not a bilayer stabilizer, nor cholesterol inhibited viral fusion. Cationic and zwitterionic as well as other anionic sterol-based amphiphiles which are bilayer stabilizers also inhibit viral-induced hemolysis. The peptide carbobenzoxy-D-Phe-L-Phe-Gly is another antiviral agent that has been shown to inhibit both vesicle-vesicle as well as virus-vesicle fusion in model systems (Kelsey *et al.*, 1990).

Thus several of the anti-viral agents listed in Table I in addition to raising the L_α-H_{II} transition temperature also inhibit membrane fusion either in biological or in model systems. It is thus likely that their effect on membrane physical properties results in the inhibition of membrane fusion which contributes to their antiviral activity. However, this still doesn't define the molecular mechanism by which these agents act. Bilayer stabilzers may decrease the binding of the virus to the target membrane, insertion of the viral fusion protein into the target membrane, the formation of fusion intermediates or disassembly of the virus subsequent to initial fusion. The rate determining step in membrane fusion and perhaps even the fusion mechanisms are not identical for all viruses which may explain some of the selectivity of bilayer stabilizers for different viruses. A number of excellent recent reviews on the mechanisms of viral fusion by Hoekstra, Helenius, Haywood, Loyter, Blumenthal and others have recently appeared in Wilschut and Hoekstra (1991).

Protein Kinase C

A brief review discussing the role of membrane biophysical properties in the regulation of protein kinase C (PKC) activity has recently appeared (Epand and Lester, 1990). Modulation of the activity of membrane-bound enzymes by hydrophobic and amphiphilic compounds can occur by specific mechanisms involving binding to sites on the enzymes as well as by non-specific mechanisms which would include an alteration of the physical properties of the membrane surrounding the enzyme.

Table I.   Some Antiviral Agents Which Raise the L_α-H_{II} Transition Temperature

Agent	Virus[a]	Ref.	Slope[b]	Ref.
Cholesterol sulfate	Sendai	Cheetham *et al.* 1990a	173±8	Cheetham *et al.* 1990a
	Bovine immuno-deficiency virus	Watarai *et al.* 1990		
Tromantidine	Herpes Simplex	Rosenthal *et al.* 1982	17±1	Cheetham *et al.* 1987
Cyclosporin A	Cytomegalovirus	Gui *et al.* 1982	Positive[c]	Epand *et al.* 1987a
	Herpes Simplex	McKenzie *et al.* 1987		
Apolipo-protein A-I	Herpes Simplex	Srinivas *et al.* 1990	Large positive values[c]	Epand *et al.* 1991a
	HIV	Owens *et al.* 1990		
Carboben-zoxy-D-Phe-L-Phe-Gly	Measles	Richardson *et al.* 1980	84±5	Epand, 1986
Carboben-zoxy-D-Phe-L-Phe-Gly-D-Phe-D-Leu-D-Leu	Measles	Lobl *et al.* 1988	98±9	Lobl *et al.* 1988
Carboben-zoxy-L-Ser-L-Leu-NH$_2$	Measles	Epand *et al.* 1987	27±3	Epand *et al.* 1987b

[a] Virus against which antiviral activity was shown.
[b] Slope of a plot of a L_α-H_{II} tansition temperature of dielaidoylphos-phatidylethanolamine vs mol fraction additive in units of K/mol fraction additive.
[c] Non-linear effect. Potent bilayer stabilizer at low mol fractions.

Often, however, modulators may act by a combination of mechanisms. In the case of PKC, for example, there is a requirement for phosphatidylserine for activity (Nishizuka, 1986) but this requirement is not absolute since other anionic lipids (Hannun *et al.*, 1986) and other stereoisomers of phosphatidylserine (Lee and Bell, 1989) can substitute. In addition, there are a large number of uncharged and zwitterionic compounds which affect PKC activity. We have shown that there is an empirical correlation between the effect of this class of compounds on the L_α-H_{II} phase transition temperature of model membranes and their effect on PKC activity. Compounds that lower the L_α-H_{II} transition temperature are activators of PKC, while those that raise this transition temperature are inhibitors. There is great structural diversity in both inhibitors and activators (Epand and Lester, 1990) with inhibitors including acyl carnitines, phosphatidylcholine, peptides and sterol-based amphiphiles; while activators include diacylglycerols, hydrocarbons, organic solvents and hydrophobic steroid derivatives. This lack of structural specificity and the observed correlation with the behaviour of these substances in model systems suggests that the effects are caused by a change in the membrane physical properties rather than by binding to a site on PKC. However, there certainly are hydrophobic and amphipathic substances that bind to specific sites on PKC. For example, phorbol 12-myristate 13-acetate lowers the L_α-H_{II} transition temperature (unpublished results) and as expected is an activator of PKC. However, phorbol esters activate PKC with a high potency and specificity which almost certainly results from binding to a specific site on the enzyme (Nakamura *et al.*, 1989) and not from alteration of the physical properties of the surrounding membrane. An intermediate case is the diacylglycerols. These uncharged H_{II} phase-promoters have a potency at least 1000-fold lower than the phorbol esters but a high degree of structural specificity is required (Rando, 1988; Molleyres and Rando, 1988). The low potency suggests some non-specific effects may be involved but the structural specificity indicates some binding to the enzyme or at least a specificity for the ability of this class of compounds to enter the region of the membrane surrounding PKC. A more clear-cut example suggesting a non-specific effect is the comparison of the 8-methyl and the 8-n-butyl derivatives of distearoylphosphatidylcholine. The former compound raises the L_α-H_{II} transition temperature and is an inhibitor of PKC while the latter compound lowers this transition temperature and is an activator of PKC (Epand *et al.*, 1991b). It is unlikely that two such structurally similar molecules can bind to a site on PKC and in one case activate but in the other case inhibit the enzyme. However, if one considers that these zwitterionic amphiphiles are altering the physical properties of the membrane environment of PKC, then their opposite effects on PKC activity are readily explicable in terms of their opposing actions on lipid phase propensity. The mechanism by which neutral and zwitterionic substances alter the membrane environment to affect PKC activity is not known. It is not likely to be a result of inhibiting the formation of non-bilayer phase intermediates since similar effects of activators and inhibitors are observed in assays done both in phospholipid vesicles as well as in Triton micelles where lipid polymorphism cannot be manifested. It is possible that amphiphiles and hydrophobic substances affect the hydration and surface packing density of the membrane, which in turn alters the L_α-H_{II} transition temperature as well as the activity of PKC. How this putative change in surface properties alters enzyme activity can only be conjectured but may involve the ability of PKC to penetrate into the membrane or it may alter the conformation that the enzyme adapts in the membrane.

Insulin Receptor Signalling in Adipocytes

One of the antiviral agents listed in Table I is carbobenzoxy-L-Ser-L-Leu-NH$_2$. It was indicated that this peptide can also inhibit exocytosis and myoblast fusion. In addition, carbobenzoxy-L-Ser-L-Leu-NH$_2$ has been shown to inhibit insulin-stimulated glucose transport in adipocytes (Aiello *et al.*, 1986). A number of carbobenzoxy-dipeptide-amides were found to inhibit the insulin effect and we showed that their inhibitory potency correlated with the extent of their effect on the L$_\alpha$-H$_{II}$ transition temperature in model membranes (Epand *et al.*, 1987b). These uncharged peptides are sparsely soluble in water and would be expected to spontaneously partition into a membrane environment. They exhibit a number of effects on biological membranes including the inhibition of cytochalasin B binding to red cells, alteration of erythrocyte shape (Aiello *et al.*, 1986) as well as inhibition of myoblast fusion (Couch and Strittmatter, 1983), intracellular vesicular transport (Strous *et al.*, 1988) and exocytosis (Baxter *et al.*, 1983; Mundy and Strittmatter, 1985). It is therefore possible that the inhibition of insulin-promoted glucose transport in adipocytes is a consequence of their effect on the bulk biophysical properties of the membrane. In accord with this suggestion is our finding that several different small peptides inhibit insulin effects in adipocytes. The inhibitory action of these peptides is independent of their amino acid composition, sequence or charge but is related to the extent to which these peptides raise the L$_\alpha$-H$_{II}$ transition temperature in model membranes (Epand *et al.*, 1991). In addition, hexane and DL-threo-dihydrosphingosine which lower the L$_\alpha$-H$_{II}$ transition temperature in model membranes, increase the basal rate of glucose uptake in adipocytes. Another effect of insulin on adipocytes, the stimulation of protein synthesis, is inhibited by a bilayer-stabilizing peptide and is promoted by DL-threo-dihydrosphingosine. Thus, there is an empirical relationship between the effects of membrane additives on insulin-dependent functions in adipocytes and their effects on lipid phase behaviour in model membranes. The action of the bilayer stabilizing peptide may be explained, at least in part, by its inhibition of insulin-dependent protein phosphorylation (Epand *et al.*, 1991c). It is possible that the tyrosine kinase activity of the insulin receptor is affected by the physical properties of its surrounding membrane in a manner analogous to what we have described for PKC.

CONCLUSIONS

In this review, we have emphasized the correlation between effects of certain substances on lipid polymorphism in model membranes and their effects on membrane functional properties. This certainly is not the only factor determining such diverse functions as viral fusion, PKC activity and insulin signalling. All of these are protein-mediated phenomenon and include membrane fusion promoted by viral proteins, the catalytic activity of a membrane-bound enzyme protein and the functioning of a receptor protein for the peptide hormone insulin. However, all of these protein-mediated phenomenon are modulated by the nature of the membrane environment. Measurement of the effect of membrane additives on the L$_\alpha$-H$_{II}$ transition temperature in model systems can be of predictive value for distinguishing between inhibitors and activators of certain membrane functions. Several limitations of this generalization must be recognized. Inhibitors generally raise the L$_\alpha$-H$_{II}$ transition temperature but many compounds which do this also have detergent-like action and disrupt bilayers to form micelles. Thus, those amphiphiles that promote micelle formation may have a dual action on bilayer membranes. On the one hand, they make the membrane surface more polar and tightly packed, but on the other hand they disrupt the bilayer structure by solubilizing membranes in the form of micelles. The former "bilayer stabilizing" effect would be inhibitory to the membrane functions we have discussed while the latter effect might be stimulatory. At low concentrations, where their ability to destabilize the bilayer is less, micelle-forming amphiphiles such as lysophosphatidylcholine (Nayendova et $al.$, 1990) or the apolipoprotein A-I (Table I) can inhibit membrane fusion without greatly disrupting the bilayer structure. Also micelle-forming amphiphiles such as short chain phosphatidylcholines activate PKC in a manner independent of phosphatidylserine. These zwitterionic amphiphiles raise the L_α-H_{II} transition temperature and would therefore be expected to inhibit PKC. We suggest that it is difficult to predict the effect of detergents on PKC activity. Just as with fusion, there may be dual effects caused by bilayer disruption as well as by changes in the physical properties of the membrane caused by an amphiphile that can raise the L_α-H_{II} transition temperature. These two alterations in membrane properties may produce opposite effects on PKC activity. Of course, there are many compounds, such as those listed in Table I (except for the apolipoprotein A-I), which raise the L_α-H_{II} phase transition but do not have significant micelle-forming tendencies. The effects of such substances on membrane fusion or PKC activity as inhibitors can be predicted more reliably. Another consideration is that the effect of a substance on lipid polymorphism may depend on the conditions used such as pH or ion concentrations. For example, the good bilayer stabilizer and inhibitor of fusion, cholesterol sulfate, loses its stabilizing effect on bilayers and promotes fusion in the presence of Ca^{2+} (Cheetham et $al.$, 1990b). Furthermore, for certain biological functions, other membrane properties may predominate in importance over effects on lipid phase propensity. Even for the case of PKC where there is a good correlation for neutral and zwitterionic compounds between effects on lipid polymorphism in model membranes and effects on enzyme activity, this correlation does not hold for cationic amphiphiles (Epand, 1987). Cationic amphiphiles are inhibitors of protein kinase C, regardless of their effects on lipid polymorphism. Because of these other factors, but mainly because of differences in the ability of membrane additives to interact with specific functional membrane components, the relationship between effects on the L_α-H_{II} equilibrium and membrane function is not quantitative. Nevertheless, keeping in mind these complexities, this simple correlation is generally remarkably useful in predicting which substances will be activators and which will be inhibitors of a membrane function.

Having established this correlation doesn't directly indicate anything about mechanism. In the case of viral fusion, many steps are required between the state where virus and target are separated to the final fusion product. These steps include binding of the virus to the target, establishment of a fusion competent site and final membrane fusion. The intermediate steps can include processes such as rearrangement of viral proteins and/or target receptors to attain the required cooperativity, deformation of the target membrane, penetration of the viral fusion protein and alteration of the physical properties of the target and/or viral membranes. By determining which of these processes is sensitive to the presence of substances which alter the physical state of the membrane one can gain a better understanding of the natural mechanism of viral fusion and can direct experiments to further evaluate the role of membrane physical properties in a

particular step in viral fusion. For example, it has been suggested
that the hydrophobic amino terminal segments of many viral fusion
proteins insert into membranes as an α-helix oriented at an angle to the
bilayer normal (Brasseur *et al.*, 1990). Such an orientation would
destabilize bilayers and would promote increased hexagonal phase
propensity. This may be a mechanism for inducing viral fusion which is
inhibited by bilayer stabilizing amphiphiles. Similarly for PKC,
several steps may be affected by membrane physical properties including
the depth of burial of PKC in the membrane, the conformation attained
by PKC, the association of PKC with phosphatidylserine and Ca^{2+}, etc.
Analysis of the particular step which is affected by zwitterionic and
uncharged amphiphiles and hydrocarbons will indicate the nature of the
physical change in the membrane which gives rise to changes in enzyme
activity. Finally, the case for insulin signalling is less well
developed. The range of compounds tested is more limited because of the
requirement to maintain cell viability and the number of potential
targets of these agents is greater because of the complexity of the
system of intact live adipocytes. We wish to test individual potential
targets, such as the tyrosine kinase activity of the insulin receptor
in an isolated model system and determine if it is sensitive to the same
agents that affect insulin function in intact adipocytes.

SUMMARY

Viral fusion, PKC activity and insulin signalling in adipocytes
have been shown to be sensitive to the presence of substances in the
membrane which alter its propensity for forming non-bilayer phases.
Thus, the behaviour of simple model phospholipid systems provides a good
criterion for predicting certain membrane functions. The identification
of these correlations provides a new criterion for designing drugs which
will affect these membrane functions. It also may yield new information
about the mechanism of these membrane-dependent processes.

ACKNOWLEDGEMENTS

We are grateful to the Medical Research Council of Canada for
financial support of our research. I wish to thank the members of my
laboratory including Mr. Remo Bottega, James Cheetham, Alan Stafford,
Ms. Tina McCallum, as well as Drs. Richard Callaghan, Raquel Epand,
Guillermo Seinisterra, Nie Song-Qing, and Leon van Gorkom for their
contributions to this work and for critically reading this manuscript.

REFERENCES

Aiello, L. P., Wessling-Resnick, M., and Pilch, P. F., 1986, Dipeptide
    metalloendoprotease substrates are glucose transport inhibitors and
    membrane structure perturbants, <u>Biochemistry</u>, 25:3944.

Baxter, D. A., Johnston, D., and Strittmatter, W. J., 1983, Protease
    inhibitors implicate metalloendoprotease in synaptic transmission
    at the mammalian neuromuscular junction. <u>Proc. Natl. Acad. Sci.
    USA</u>, 80:4174.

Brasseur, R., Vandenbranden, M., Cornet, B., Burny, A., Ruysschaert,
    J.-M., 1990, Orientation into the lipid bilayer of an asymmetric
    amphipathic helical peptide located at the N-terminus of viral
    fusion proteins, <u>Biochim. Biophys. Acta</u>, 1029:267.

Cheetham, J. J., and Epand, R. M., 1987, Comparison of the interaction of the antiviral chemotherapeutic agents amantadine and tromantadine with model phospholipid membranes, <u>Bioscience Reports</u>, 7:225.

Cheetham, J. J., Epand, R. M., Andrews, M., and Flanagan, T. D., 1990a, Cholesterol sulfate inhibits the fusion of Sendai Virus to biological and model membranes, <u>J. Biol. Chem.</u>, 265:12404.

Cheetham, J. J., Chen, R. J. B., and Epand, R. M., 1990b, Interaction of calcium and cholesterol sulphate induces membrane destabilization and fusion: Implications for the acrosome reaction, <u>Biochim. Biophys. Acta</u>, 1024:367.

Chong, P. L.-G., 1988, Effects of hydrostatic pressure on the location of Prodan in lipid bilayers and cellular membranes, <u>Biochemistry</u>, 27:399.

Couch, C. B., and Strittmatter, W. J., 1983, Rat myoblast fusion requires metalloendoprotease activity, <u>Cell</u>, 32:257.

Cullis, P. R., Hope, M. L., and Tilcock, C. P., 1986, Lipid polymorphism and the roles of lipids in membranes, <u>Chem. Phys. Lipids</u>, 40:127.

Epand, R. M., 1986, Virus replication inhibitory peptide inhibits the conversion of phospholipid bilayers to the hexagonal phase, <u>Bioscience Reports</u>, 6:647.

Epand, R. M., 1987, Properties determining whether substances will be activators or inhibitors of protein kinase C, <u>Chem. Biol. Interac.</u>, 63:239.

Epand, R. M., 1990a, Relationship of phospholipid hexagonal phases to biological phenomena, <u>Biochem. Cell Biol.</u>, 68:17.

Epand, R. M., 1990b, Hydrogen bonding and the thermotropic transitions of phosphatidylethanolamines, <u>Chem. Phys. Lipids</u>, 52:227.

Epand, R. M., and Lester, D. S., 1990, The role of membrane biophysical properties in the regulation of protein kinase C activity, <u>Trends in Pharm. Sci.</u>, 11:317.

Epand, R. M., and Leon, B. T.-C., 1991, Hexagonal phase forming propensity detected in phospholipid bilayers with a fluorescence probe, <u>Biochemistry</u>, submitted for publication.

Epand, R. M., Epand, R. F., and McKenzie, R. C., 1987a, Effects of viral chemotherapeutic agents on membrane properties, <u>J. Biol. Chem.</u>, 262:1526.

Epand, R. M., Lobl, T. J., and Renis, H. E., 1987b, Bilayer stabilizing peptides and the inhibition of viral infection: Antimeasles activity of carbobenzoxy-Ser-Leu-amide, <u>Bioscience Reports</u>, 7:745.

Epand, R. M., Epand, R. F., Anantharamaiah, G. M., and Segrest, J. P., 1991a, Apolipoprotein A-I is a potent inhibitor of hexagonal phase formation, <u>Biochim. Biophys. Acta</u>, submitted for publication.

Epand, R. M., Epand, R. F., Leon, B. T.-C., Menger, F. M., and Kuo, J. F., 1991b, Evidence for the regulation of the activity of protein kinase C through changes in membrane properties, <u>Bioscience Reports</u>, submitted for publication.

Epand, R. M., Stafford, A. R., and Debanne, M. T., 1991c, The action of insulin in rat adipocytes and membrane properties, <u>Biochemistry</u>, in press.

Fenske, D. B., Jarrell, H. C., Guo, Y., and Hui, S. W., 1990, Effect of unsaturated phosphatidylethanolamine on the chain order profile of bilayers at the onset of the hexagonal phase transition. A ^{2}H NMR study, <u>Biochemistry</u>, 29:11222.

Gui, X. E., Ho, M., and Camp, P. E., 1982, Effect of cyclosporin A on murine natural killer cells, <u>Infect. Immun.</u>, 36:1123.

Gruner, S. M., 1985, Intrinsic curvature hypothesis for biomembrane lipid composition: a role for nonbilayer lipids, <u>Proc. Natl. Acad. Sci. USA</u>, 82:3665.

Hannun, Y. A., Loomis, C. R., and Bell, R. M., 1986, Protein kinase C activation in mixed micelles: Mechanistic implications of phospholipid, diacylglycerol and calcium dependencies, <u>J. Biol. Chem.</u>, 261:7184.

Hui, S.-W., and Sen, A., 1989, Effects of lipid packing on polymorphic phase behaviour and membrane properties, <u>Proc. Natl. Acad. Sci. USA</u>, 86:5825.

Ickes, D. E., Venetta, T. M., Phonphok, Y., and Rosenthal, K. S., 1990, Tromantadine inhibits a late step in herpes simplex virus type 1 replication and syncytium formation, <u>Antiviral Research</u>, 14:75.

Karnovsky, M. J., Kleinfield, A. M., Hoover, R. L., and Klausner, R. D., 1982, The concept of lipid domains in membranes, <u>J. Cell Biol.</u>, 94:1.

Kelsey, D. R., Flanagan, T. D., Young, J., and Yeagle, P. L., 1990, Peptide inhibitors of enveloped virus infection inhibit phospholipid vesicle fusion and Sendai virus fusion with phospholipid vesicles, <u>J. Biol. Chem.</u>, 265:12178.

Kimura, Y., and Ikegami, A., 1985, Local dielectric properties around polar region of lipid bilayer membranes, <u>J. Membrane Biol</u>, 85:225.

Lee, M.-H., and Bell, R. M., 1989, Phospholipid functional groups involved in protein kinase C activation, phorbol ester binding, and binding to mixed micelles, <u>J. Biol. Chem.</u>, 264:14797.

Lobl, T. L., Renis, H. E., Epand, R. M., Maggiora, L. L., and Wathen, M. W., 1988, Peptides as potential virus inhibitors: Synthesis and bioassay of five respiratory syncytial virus peptide analogs with antimeasles activity, <u>Int. J. Peptide Protein Res.</u>, 32:326.

McKenzie, R. C., Epand, R. M., and Johnson, D. C., 1987, Cyclosporin A inhibits herpes simplex virus-induced cell fusion but not virus penetration into cells, <u>Virology</u>, 159:1.

Molleyres, L. P., and Rando, R. R., 1988, Structural studies on the diglyceride-mediated activation of protein kinase C, <u>J. Biol. Chem.</u>, 263:14832.

Mundy, D. I., and Strittmatter, W. J., 1985, Requirement for metalloendoprotease in exocytosis: Evidence in mast cells and adrenal chromaffin cells, <u>Cell</u>, 40:645.

Nakamura, H., Kishi, Y., Pajares, M. A., and Rando, R. R., 1989, Structural basis of protein kinase C activation by tumor promoters, <u>Proc. Natl. Acad. Sci. USA</u>, 86:9672.

Naydenova, S., Lalchev, Z., Petrov, A. G., and Exerowa, D., 1990, Pure and mixed lipid black foam films as models of membrane fusion, <u>Eur. Biophys. J.</u>, 17:343.

Nishizuka, Y., 1986, Studies and perspectives of protein kinase C, <u>Science</u>, 233:305.

Ohki, S., and Arnold, K., 1990, Surface dielectric constant, surface hydrophobicity and membrane fusion, <u>J. Membrane Biol.</u>, 114:195.

Owens, R. J., Anantharamiah, G. M., Kahlon, J. B., Srinivas, R. V., Compans, R. W., and Segrest, J. P., 1990, Apolipoprotein A-I and its amphipathic helix peptide analogues inhibit human immunodeficiency virus-induced syncytium formation, <u>J. Clin. Invest.</u>, 86:1142.

Rando, R. R., 1988, Regulation of protein kinase C activity by lipids, <u>FASEB J.</u>, 2:2348.

Richardson, C. D., Scheid, A., Choppin, P. W., 1980, Specific inhibition of paramyxovirus and myxovirus replication by oligopeptides with amino acids similar to those at the N-termini of the F1 or HA2 viral polypeptides, <u>Virology</u>, 105:204.

Rosenthal, K. S., Sokol, M. S., Ingram, R. L., Subramanian, R., and Fort, R. C., 1982, Tromantadine: Inhibitor of early and late events in Herpes Simplex Virus replication. <u>Antimicrobial Agents and Chemotherapy</u>, 22:1031.

Seddon, J. M., 1990, Structure of the inverted hexagonal (H_{II}) phase, and non-lamellar phase transitions of lipids, <u>Biochim. Biophys. Acta</u>, 1031:1.

Siegel, D. P., Burns, J. L., Chestnut, M. H., and Talmon, Y., 1989, Intermediates in membrane fusion and bilayer/nonbilayer phase transitions imaged by time-resolved cryo-transmission electron microscopy, <u>Biophys. J.</u>, 56:161.

Sommer, A., Paltauf, F., and Hermetter, A., 1990, Dipolar solvent relaxation on a nonosecond time scale in ether phospholipid membranes as determined by multifrequency phase and modulation fluorometry, <u>Biochemistry</u>, 29:11134.

Srinivas, R. V., Birkedal, B., Owens, R. J., Anantharamaiah, G. M., Segrest, J. P., and Compans, R. W., 1990, Antiviral effects of apolipoprotein A-I and its synthetic amphipathic analogs, <u>Virology</u>, 176:48.

Strous, G. J., van Kerkhof, P., Dekker, J., and Schwartz, A. L., 1988, Metalloendoprotease inhibitors block protein synthesis, intracellular transport, and endocytosis in hepatoma cells, J. Biol. Chem., 263:18197.

Watarai, S., Onuma, M., Yamamoto, S. and Yasuda, T., 1990, Inhibitory effect of liposomes containing sulfatide or cholesterol sulfate on syncytium formation induced by bovine immunodeficiency virus-infected cells, J. Biochem., 108:507.

Weber, G., and Farris, F. J., 1979, Synthesis and spectral properties of a hydrophobic fluorescent probe: 6-propionyl-2-dimethylamino-naphthalene, Biochemistry, 18:3075.

Wilschut, J., and Hoekstra, D., 1991, "Membrane Fusion", Marcel Dekker, New York.

Yeagle, P. L., 1989, Lipid regulation of cell membrane structure and function, FASEB J. 3:1833.

EVIDENCE FOR MULTIPLE STEPS IN ENVELOPED VIRUS BINDING

Anne M. Haywood

Departments of Pediatrics and Microbiology
Box 777, University of Rochester Medical Center
Rochester, NY 14642, U. S. A.

SUMMARY

This paper reviews the data that indicate that the binding of viruses to cell is far more complex than simple binding of a ligand to a receptor and involves several steps. Viral surface glycoproteins have several domains, so these proteins have the potential of binding more than one component of the cell surface. Several studies have indicated viral binding is followed by rearrangements in the viral proteins or the viral surface.

Paramyxoviruses are known to bind to sialoglycoconjugates, and this binding appears to involve several steps. At 0-4°C this binding is weak and easily reversible. A stronger adhesion develops in one or more steps when the temperature is raised. The data suggest the main effect of the temperature change is upon the viral proteins. Kinetic studies with Sendai virus have shown that "stabilization of binding" is the rate-limiting step for fusion. Engulfment of Sendai virus by liposomes or cells occurs as the temperature is raised and presumably reflects enhanced binding.

Polymers, e.g., ficoll and dextran, increase the strength of adhesion between virus and receptor-containing liposomes even at low temperatures. One possible mechanism is that polymers cause changes in the organization or conformation of the viral HN proteins or of the ganglioside receptors. A second possible mechanism is that the polymers are excluded from the region between bound virus and liposome. As a result the osmolarity in the fluid between the bound regions is lower than in the bulk fluid, and this difference should cause the virus and liposome to be forced together. The increased strength of adhesion in the presence of polymers may facilitate "stabilization of binding" and engulfment of the virus. The presence of polymers also accelerates viral membrane fusion. Polymers in serum and extracellular fluid might have similar effects.

Understanding of the different steps in viral binding should open the possibility for developing new classes of antiviral agents.

INTRODUCTION

Viruses infect only certain organs in animals and infect only certain cell types both in the animal and in cell culture. The presence of specific receptors on the susceptible tissues is one factor that determines these viral tropisms. Implicit in this concept has been the idea that there is *one* specific molecule on the cell surface to which the virus binds and that the initial binding to the virus completes the role of the receptor. This is likely to be a considerable oversimplification of the virus and cell surface interactions. This paper reviews the data that are compatible with the hypothesis that binding involves several steps.

Such considerations are not unique to enveloped viruses. In the early bacteriophage studies, it was postulated that bacteriophage adsorbed to bacteria according to the von Smoluchowski coagulation equation. This equation states that the adsorption constant should be proportional to the diffusion constant of the particle (virus), to the radius of the sphere (host cell) to which the particle is adsorbing and to the fraction of particles that are irreversibly bound. However, it was then recognized that the data were not consistent with this, and bacteriophage adsorption was postulated to be a two-step process with the second step being temperature-dependent [Stent and Wollman, 1952]. Further, the cell binding that is necessary for phagocytosis to occur [Wright and Silverstein, 1986], for cell-cell recognition [Brandley and Schnaar, 1986] and for cell adhesion to extracellular matrix proteins [Lotz et al., 1989] may involve similar sequential steps. Therefore, the binding of enveloped viruses may be representative of much of biologic binding. The binding of paramyxoviruses and orthomyxoviruses has been extensively studied, so they will be used as the principle examples of enveloped virus binding.

THE STRUCTURE OF THE ENVELOPED VIRUSES IS COMPATIBLE WITH MULTIPLE INTERACTIONS

Paramyxovirus membrane glycoproteins have multiple domains, and the HN protein is a tetramer

Sendai virus is a mouse paramyxovirus type 1, which is very similar to the human paramyxovirus type 1. The virion (virus particle) has a diameter of about 2000 Å and an isoelectric point of 4.0 [Haywood, unpublished data]. In PBS, pH 7.4, it has a ζ potential of -17 mV [Haywood, 1974]. The virus particle contains six established virus-coded proteins. These include the three membrane proteins, which are the HN (hemagglutinin-neuraminidase), F (fusion) and M (matrix) proteins, and the three internal proteins, which are the NP (nucleocapsid), L (large) and P (phosphoprotein) proteins. The virion also contains host-derived actin. With so few proteins, it is logical that the viral proteins should have multiple functions.

The membrane glycoproteins, the HN and F proteins, both have a long glycosylated domain external to the viral membrane, a hydrophobic domain spanning the membrane and a short hydrophilic domain inside the membrane. The external portions are visible by electron microscopy as spikes on the cell surface. The HN protein binds sialic acid-containing receptors, which causes red cells to hemagglutinate, and has neuraminidase (sialidase) activity. It is responsible for much or all of the virus binding to the host cell. The HN protein is present in the virus as dimers and tetramers [Markwell and Fox, 1980], and microscopy of solubilized tetramers shows a box-shaped head that has four identical subunits and measures about 10 by 10 nm [Thompson et al., 1988]. The strength of the viral binding should be affected by how many of the HN spike tetramers are bound and how many monomers in each tetramer are bound at any time. Mapping of the Sendai virus HN protein with monoclonal antibodies has shown at least four domains [Örvell and Grandien, 1982; Portner et al., 1987a]. Iorio et al. [1989] pointed out that receptor recognition by another paramyxovirus, Newcastle disease virus, might involve two domains of the HN protein. New binding sites on the HN protein could be made available either by a conformational change of the individual monomers or by alterations in the monomer interactions.

The F protein is the second glycoprotein on the surface of the virus and is also visible as a spike. It is synthesized as the F_0 protein and then cleaved by proteases into the active form, which contains the disulfide-linked F_1 and F_2 products of the cleavage. The virus requires the active form of the protein to enter the cell or to cause membrane fusion. Studies with monoclonal antibodies also indicate at least four sites on the F protein [Portner et al., 1987b]. In addition to having a major role in the fusion process the F protein has been postulated also to have binding activity [Peterhans et al., 1983].

The viral M or membrane protein forms a matrix on the inner side of the membrane. Its exact form is unclear, but it is thought to form a layer on the inside of the virion membrane and associate with the membrane lipid. The M protein interacts with the glycoproteins, with actin [Giuffre et al., 1982], and with the ribonucleoprotein (RNP). In some ways it functions as does the cellular cytoskeleton and has been dubbed a viroskeleton [Kim et al., 1979]. Although it is loosely organized, in some ways it may also function as does the viral nucleocapsid present in some other enveloped viruses. Inside the viral membrane is a negative-strand, 15-kilobase-long RNA associated with nucleoprotein and the L and P proteins, which have RNA polymerase activity. The virion also contains other enzymatic activities such as one or more protein kinase and protease activities. A kinase activity is associated with the L protein [Einberger et al., 1990]. Therefore changes in the viral glycoproteins as a result of binding could be transmitted to the interior of the virus and cause changes in the M protein viroskeleton and/or activate some of the viral enzymes such as the protein kinase. Such changes could lead to alterations of the entire virion and thereby alter its binding, its fusion capacity and/or its ability to disassemble.

Binding at 37°C causes changes over the entire virion

Knutton [1976, 1978] investigated the changes in the Sendai viral envelope that follow viral binding to cells at 37°C and precede infection. After binding at 37°C the virus particles are no longer spherical but develop a convoluted profile. Such virus particles are characterized in freeze-fracture replicas by the appearance of smooth linear ridges on E faces and by a complementary arrangement of linear grooves on P faces with a rearrangement and change of the intramembranous particles. These changes in the structural organization of the viral envelope only take place when virus particles are bound to cells. Free virus particles incubated at 37°C remain spherical and display the same morphology as virus particles at 4°C. This rearrangement requires an active F protein. Only virus particles having this altered morphology appear to be capable of fusion with the erythrocyte membrane. The smooth ridged regions of the viral envelope seem to be the initial site of fusion. It is likely that the M protein participates in this reorganization. Thus the binding of virus initiates changes in the virus necessary for subsequent interactions between the virus and host.

Sindbis virus, an alphavirus, also undergoes changes upon interacting with the cell surface [Flynn et al., 1990]. As demonstrated by monoclonal antibody studies, its E1/E2 glycoprotein spike undergoes a structural rearrangement that seems necessary for viral entry. The spikes that bind the antibody are probably not those that bind the cell, which suggests binding to cells initiates propagation of the change over the entire virus. As with the changes in Sendai virus upon binding, the changes in Sindbis virus not only require cell contact but also are temperature-dependent. CD4 is a receptor for human immunodeficiency virus type-1 (HIV-1) and addition of soluble CD4 causes the HIV surface glycoprotein gp120 to dissociate from its complex with the HIV transmembrane protein gp41. This dissociation is also temperature-dependent and only occurs after more than 50% of the gp120 molecules on a virion are occupied [Moore et al., 1990].

ORTHOMYXOVIRUSES AND PARAMYXOVIRUSES BIND TO SIALOGLYCO-CONJUGATES

Until the late 1980s the only viruses for which receptors were known were the orthomyxoviruses (e.g., influenza viruses) and paramyxoviruses (e.g., Sendai virus, Newcastle disease virus, mumps, and the human parainfluenza viruses). Research on receptors for these viruses began in the 1940s, so much more is known about their receptors than about the receptors for other viruses. Further, there have been extensive studies on the influenza virus hemagglutinin protein, which binds the influenza receptors.

In 1941 Hirst first showed that influenza viruses cause hemagglutination [Hirst, 1941]. This observation allowed quantification of influenza viruses by red cells instead of by measuring lethality in mice, which at that time was a great advance. Hirst [1942] then showed that influenza and paramyxoviruses destroy their receptors after continued incubation. Thus if red cells are incubated with a particular strain of virus, the red cells lose their receptors for that particular strain. The suggestion was made that viruses "browse"-namely roll over the surface of the cell by binding and then digesting receptors until there are no more receptors left. Burnet et al. [1946] showed that *Vibrio cholerae* also has the virus "receptor destroying enzyme". In addition they described what they called the virus "receptor gradient", namely, paramyxoviruses and subtypes of influenza could be ordered so that cells rendered resistant to agglutination with a given virus failed to be agglutinated by virus strains earlier in the series but were still agglutinated by those in the gradient succeeding it. In 1958 the influenza "receptor destroying enzyme" was demonstrated to be a sialidase [Gottschalk and Thomas, 1958]. This made it evident that the viral receptors contained sialic acid (substituted neuraminic acid). Because some sialomucoproteins inhibit viral binding, it was assumed the viruses bound to sialoglycoprotein receptors.

In 1974 it was shown that gangliosides in liposomes can serve as Sendai virus receptors, and the receptor activity requires specific gangliosides [Haywood, 1974, 1975a]. Markwell et al. [1981] showed that cells can no longer bind Sendai virus after their sialic acid residues are removed by sialidase. Only gangliosides with the terminal sequence NeuAcα2,3Gal or NeuAcα2,8NeuAcα2,3Gal have receptor activity. When these gangliosides are added to sialidase-treated cells, they restore the ability of the cells to be infected by Sendai virus. Different influenza virus isolates recognize different sialyloligosaccharide sequences [Rogers and Paulson, 1983; Rogers and D'Souza, 1989]. The differences in receptor sialyl-oligosaccharide sequences plus differences in the specificity of different viral sialidases explain the receptor gradient.

In 1981 the structure of the influenza virus hemagglutinin was determined in detail by X-ray crystallography, and it was found to be a trimer [Wilson et al., 1981]. In 1988 X-ray crystallography of influenza hemagglutinin complexed to sialic acid showed that a pocket of conserved amino acids on the influenza hemagglutinin is filled by sialic acid [Weis et al., 1988]. This study showed multiple hydrogen bonds and van der Waals contacts. The charge on the sialic acid carboxyl group is buried. This kind of bonding seems to be representative of the protein-sugar interactions of carbohydrate-binding proteins [Quiocho, 1986].

Thus studies between 1941 and 1989 seemed to have resulted in a fairly detailed description of the initial binding of the orthomyxoviruses and paramyxoviruses. However, as will be described below, the initial binding to sialoglycoconjugates is probably only a part of the total binding process.

Since 1985 the receptors for some other viruses have been identified. They were identified mainly by monoclonal antibodies or by transfecting fragments of DNA from cells with receptors into cells without receptors and studying the DNA fragment responsible for making a previously resistant line permissive for infection. These methods are mostly limited to the analysis of protein receptors. They demonstrated that members of the immunoglobulin superfamily are receptors for several viruses, i.e., ICAM-1 for picornaviruses [Greve et al., 1989; Staunton et al., 1990; Tomassini et al., 1989] and CD4 for human immunodeficiency virus when the host is a hematopoietic cell [Weiss et al., 1989]. The CR2 glycoprotein is a receptor for Epstein-Barr virus [Cooper et al., 1988].

It is possible that most enveloped viruses have specific host membrane proteins as receptors while the orthomyxoviruses and paramyxoviruses have sialic acid groups as receptors. However, it seems more likely that all enveloped viruses bind to several molecules and that these include both carbohydrate groups that are specific for the individual viruses plus some specific cell membrane protein. Thus it would be interesting to ask if the ortho-myxoviruses and paramyxoviruses also have protein receptors. A suggestion of such dual binding comes from the literature on herpes simplex virus (HSV), which has on its surface multiple glycoproteins that participate in binding and entry. WuDunn and Spear [1989] showed that HSV binds to heparan sulfate and suggested this was the first step in a cascade of interactions. Johnson et al. [1990] discussed that HSV entry involves an initial binding that involves sites on the cell that are numerous but before entry the gD protein must bind to a limited set of cell surface receptors that are different from those required for the initial viral adsorption. Thus it is possible that the initial binding is to a carbohydrate followed by binding to a protein and/or a carbohydrate-protein complex.

PRIMARY BINDING OF SENDAI VIRUS TO SIALOGLYCOCONJUGATES

The initial binding of Sendai virus to a cell or liposome includes recognition of sialic acid by the HN protein. It seems probable that the binding of the Sendai virus HN protein to sialic acid is similar to the binding of the influenza hemagglutinin to sialic acid and involves hydrogen bonds and van der Waals contacts. Influenza virus and paramyxovirus binding studies conventionally have been performed at low temperatures (0-4°C) to avoid the viral sialidase activity present at higher temperatures. The initial binding of Sendai virus to gangliosides occurs at these low temperatures and is very weak. This binding causes the virus to adhere to ganglioside-containing liposomes when they are centrifuged at 9,000 x g [Haywood, 1974], but the viruses are released from the liposomes when centrifuged at high speeds, e.g., 300,000 x g [Haywood & Boyer, 1982].

The initial studies of influenza virus binding in the early 1940s led to the suggestion that viruses could "browse" over the cell surface. When image processing and Nomarski optics were used to view Sendai virus, the virus indeed seemed to migrate randomly over the surface of ganglioside-containing liposomes [Haywood, unpublished data]. Since the microscope stage was not temperature-controlled, the temperature at which this migration occurred is unknown. This migration or "browsing" could be due to binding of the virus to sialic acid groups followed by digestion of those groups as initially suggested. It also could be due to the rapid making and breaking of multiple weak bonds, which would allow the virus to roll over the cell surface. This would constitute a two-dimensional walk over the cell surface. As pointed out by Adam and Delbrück [1968], if a particle goes from free diffusion in three dimensions to movement in two dimensions over a surface, this can be of considerable advantage in reducing the time of diffusion unless the diffusion coefficient is higher in two dimensions. If the virus migrates over the surface by making and breaking attachments to sialoglycoconjugates, the rate would depend upon the speed of these processes. Adam and Delbrück also pointed out that being caught on a surface could also improve the acquisition of molecules from a stream of air or liquid. Thus initial weak binding of virus to sialoglycoconjugates would allow the virus to attach wherever on a cell surface it happened to collide and then to wander over the surface until it found a favorable locus for penetration. This would allow the virus to utilize the cell glycocalyx instead of being hindered by it. A gradient of gangliosides on the cell surface, such as might occur if gangliosides are associated with a glycoprotein receptor, could also direct the virus. Alternatively, rapid browsing due to sialidase activity might make a local region of membrane with a temporarily reduced surface charge that therefore could approach the virus more closely.

SECONDARY BINDING (STABILIZATION OF BINDING) OF SENDAI VIRUS TO SIALOGLYCOCONJUGATES

Changes in the nature of the binding of Sendai virus to ganglioside-containing liposomes or to cells upon raising the temperature are indicated because engulfment occurs only at higher temperatures and kinetic studies reveal a "stabilization of binding" at higher temperatures. This secondary binding causes stronger adhesion of virus to liposomes.

Engulfment of virus is temperature-dependent and results from binding of the viral HN proteins to receptors

When liposomes contain receptor gangliosides, raising the temperature results in engulfment ("phagocytosis") of virus by liposomes [Haywood, 1975b]. The liposome moves around the virus to cover much or all of it, but it never fuses with itself to form a closed vesicle as occurs in phagocytosis in cells. The engulfment of the virus is due to binding of the receptors in the liposomes to the viral HN proteins, which are present over the entire surface of the virus. For the many viral HN proteins to remain bound, the bonds must not be easily reversible. Envelopment, like fusion, does not occur at 0-4°C and does occur at 37°C. This is true for liposomes with lipid compositions that are fluid at 0-4°C, so the temperature requirement for engulfment is determined by factors other than lipid phase. The exact temperature threshold for engulfment of viruses has not been determined, but it seems likely that, as occurs with viral membrane fusion and phagocytosis in macrophages, engulfment requires temperatures that are greater than 18-20°C. The temperature requirement appears to relate to some viral function.

The changes occurring in virus binding at higher temperatures are likely to be determined by the viral proteins. Membranes of egg-grown viruses contain lipids that are very unsaturated and so are in the fluid phase in the temperature range being studied. On the other hand, there is a marked increase in the mobility of Sendai virus membrane proteins above 18-22°C [Lee et al., 1983; Hoekstra et al., 1989], and it has been suggested the proteins form aggregates at lower temperatures. The viral enzymes, such as the neuraminidase and the protein kinase, become active at higher temperatures and may cause changes that would be reflected in the binding. Interactions between the F protein and HN protein or between the M protein and the F and HN proteins could cause changes in the glycoproteins and in the binding. Additional domains on the viral glycoproteins might be able to bind only at higher temperatures. Sendai virus proteins might also change conformation when bound at higher temperatures, and this might either alter the binding affinity or expose new binding sites.

Initial binding at 4°C with increased adhesion at higher temperatures is a pattern that is also evident in other examples of biologic adhesion, such as macrophage phagocytosis, cell-cell adhesion, and cell adhesion to extracellular matrix proteins. Macrophage phagocytosis [Wright & Silverstein 1986] bears many similarities to liposomal engulfment of virus. Macrophage phagocytosis requires the binding of macrophage receptors to ligands over the entire particle surface. Binding occurs at 4°C, but engulfment only occurs above 18-20°C. Wright and Silverstein emphasized the macrophage receptors must be mobile for phagocytosis to occur. Binding to the macrophage receptors does not always result in engulfment. Macrophages have to be stimulated, e.g., by lymphokines or phorbol esters, before they engulf particles. Among the suggestions they make for this requirement for activation is that the macrophage receptors are regulated by a reversible reaction, such as phosphorylation. At higher temperatures the cytoskeleton may change the distribution of receptors. If activated macrophages are put on a surface containing IgM, which does not promote phagocytosis, proteins of molecular weights up to 200,000 can be found in the interstices between the macrophage and the surface. However if activated macrophages are put on a surface containing IgG, which binds macrophage receptors and promotes phagocytosis, they adhere so tightly that proteins are excluded from the interface [Wright & Silverstein, 1984]. Exclusion of proteins could also contribute to tight binding, since it would decrease the osmotic pressure in the region between the macrophage and its bound surface (see below). Cell-cell adhesion also involves recognition of cell surface carbohydrates that can occur at 4°C followed by a second step that requires higher temperatures [Brandley and Schnaar, 1986]. The second step results in increased adhesion as measured by resistance of the adhesion to centrifugal fields. This adhesion can be modeled with polymers derivatized with carbohydrates, and at 37°C the strength of adhesion increases for 10-20 minutes or longer [Guarnaccia and Schnaar, 1982]. Cell adhesion to extracellular matrix proteins involves an initial binding at 4°C and after incubation at 37° is more than tenfold stronger [Lotz et al., 1989]. This strengthening of adhesion requires cytoskeletal involvement. Lotz et al. discuss the increase in strength of adhesion at higher temperatures and the cytoskeletal involvement in terms of a peeling model. In this model, if a bound surface is flexible (the example given is scotch tape), the strength of the adhesion is determined by the bonds along the length of the border of the contact. If, however, the binding surface is rigid so it cannot be peeled away but must be removed in its entirety, the bonds in the entire area of contact contribute to the strength of the adhesion. Changes in the cytoskeleton both can increase the rigidity of the membrane and can cause clustering of receptors. Clustered receptors should cause greater adhesion than widely separated receptors that can be peeled away one by one.

"Stabilization of binding" is a prerequisite for viral membrane fusion

In studying the kinetics of Sendai virus-membrane interactions Tsao and Huang [1986] described a step they called stabilization of binding, which corresponds to a second step in the binding process. In their experiments Tsao and Huang added virus in large excess to liposomes that contained ganglioside G_{D1a} and phosphatidylethanolamine (PE). They found the association between Sendai virus and liposomes follows pseudo-first-order, consecutive irreversible, three-step kinetics under these conditions of virus in large excess. The first step is fast binding, which is almost completed at zero time. They suggested that this binding represents weakly associated transient complexes between virus and liposomes, and only a small fraction of liposomes remain associated with virus after centrifugation. This would seem to correspond to the weak binding of virus observed at low temperatures. The next step follows pseudo-first-order kinetics and was called stabilization of binding. After this step, virus stays bound during centrifugation at 165,000 x g unless dithiothreitol (DTT), which alters the HN protein, is added. The last step is the actual membrane fusion, which follows zero-order kinetics. Stabilization of binding is the rate-limiting step in the steps leading to fusion. The binding stabilization and the fusion step each require the active F protein. Since Tsao and Huang used liposomes that had a gel-liquid-crystalline phase transition between 15°C and 25°C, it is impossible to to know if the increase in binding stabilization around 20°C in their experiments is due to changes in the liposomal lipids or in the viral proteins.

Engulfment of virus ("phagocytosis") precedes membrane fusion. This is shown by the fact that Sendai virus fuses with the leading edge of the region of a receptor-containing liposome that is engulfing the virus [Haywood and Boyer, 1981; Haywood, 1988, 1991]. The radius of curvature of the liposome at this leading edge is markedly reduced, which potentiates the close approach of viral and liposomal membranes and tends to destabilize the liposomal membrane. At present it is impossible to tell if binding stabilization causes, results from, or is the same as engulfment of the virus.

POLYMERS MODULATE SENDAI VIRUS BINDING TO RECEPTORS

Not only the properties of the virus particle but also the environment affects the strength of Sendai virus adhesion to receptors. Cells, of course, have a glycocalyx and are usually grown in media that contain serum and therefore macromolecules. In contrast, the interaction between purified viruses and receptor-containing liposomes is devoid of any free macromolecules. The adhesion of Sendai virus to ganglioside-containing liposomes in the cold can be strengthened by the presence of carbohydrate polymers, specifically ficoll and dextran [Haywood and Boyer, 1986]. In the presence of these polymers Sendai virus remains bound to liposomes containing receptor gangliosides even when centrifuged at high speeds (300,000 x g).

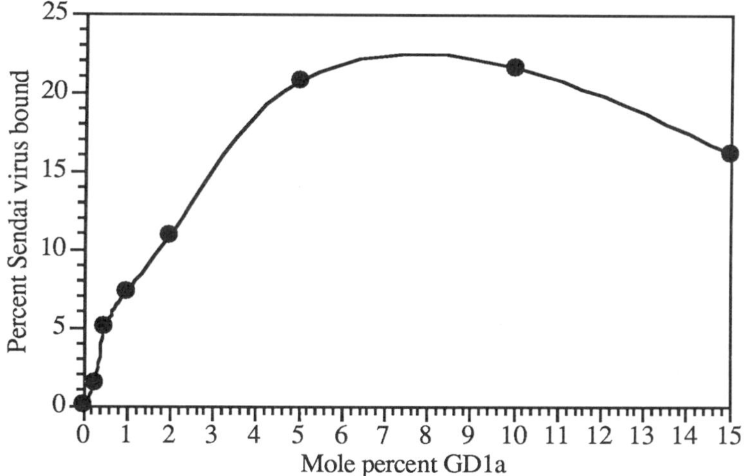

Figure 1. Dependence of viral adhesion upon receptor (ganglioside G_{D1a}) concentration. Sendai virus (5 μg of protein) in PBS (phosphate buffered saline without divalent cations) was added to 1 μmol of liposomes composed of egg PC and ganglioside G_{D1a} in the mole percents indicated. After 1 hr at 0°C, ficoll was added to yield a final concentration of 14% (w/v). The sample was layered over 14% ficoll and under 12% ficoll, which was under PBS. It was then centrifuged at 54,000 rpm (about 300,000 x g) for 40 minutes in an SW60 rotor at 4°C. The viruses bound to liposomes went to the 12% ficoll-PBS interface, and free virus sedimented through the 14% ficoll to a 60% sucrose pad. A small amount (3.5%) of the viruses always goes to the 12% ficoll-PBS interface with all kinds of liposomes and is thought to represent trapping. This was subtracted from the total amounts bound. The bound virus is expressed as percentage of total virus. Data are taken from Haywood and Boyer [1986].

Ficoll and dextran enhance Sendai virus adhesion to receptors

Enhancement of viral adhesion in the presence of ficoll is only noted when receptor-containing liposomes are used and not when other liposomes are used [Haywood and Boyer, 1986]. Thus ficoll enhances adhesion to liposomes containing ganglioside G_{D1a}, which has receptor activity, but does not enhance adhesion to liposomes containing ganglioside G_{M1}, which does not have receptor activity.

The increased adhesion in the presence of ficoll is dependent upon both the concentration of receptor gangliosides and the concentration of ficoll. In the presence of 14% ficoll the fraction of virus adhering to liposomes after centrifugation at 300,000 x g increases in proportion to the liposomal G_{D1a} concentration up to between 2 and 5 mole percent G_{D1a} (Figure 1). The decrease in adhesion when the liposomes contain more than 10 mole percent G_{D1a} is probably because of the increased net negative charge of the liposomes. Previous work has shown that the amount of Sendai virus fusion with liposomes containing ganglioside G_{D1a} and phosphatidylcholine (PC) increases with the concentration of G_{D1a} up to about 2 mole percent and then levels off [Haywood & Boyer, 1982]. Therefore, the G_{D1a} concentrations required for enhanced binding and for fusion are fairly similar.

The effect of ficoll concentration upon adhesion of Sendai virus to liposomes containing 5 mole percent G_{D1a} is shown in Figure 2. The enhancement of viral adhesion begins to be apparent at ficoll concentrations between 2 and 4% (w/v) and increases as the ficoll concentration increases.

The effect of ficoll upon viral adhesion to receptor-containing liposomes is reversible. Virus and liposomes that had been together at 0°C for 1 hr in the presence of 14% ficoll were diluted 7-fold with PBS before centrifugation. After dilution, the viruses no longer remained bound to the liposomes upon high speed centrifugation [Haywood & Boyer, 1986]. This

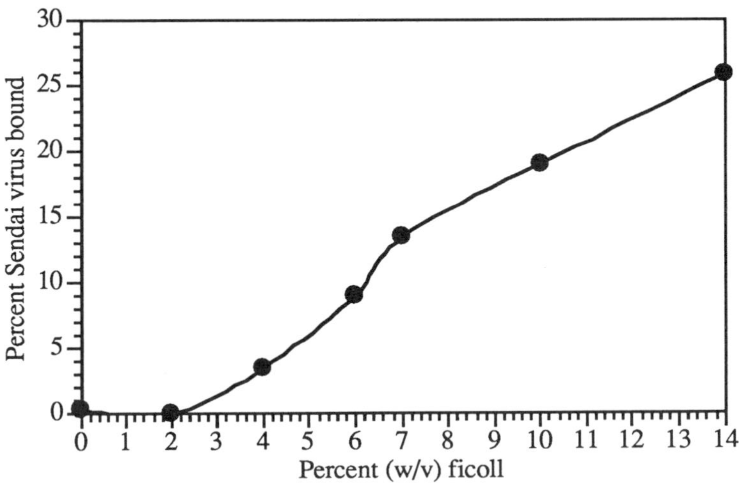

Figure 2. Dependence of viral adhesion upon ficoll concentration. Sendai virus (27 μg of protein) was added to 1 μmol of liposome composed of egg PC and 5 mole percent G_{D1a} in PBS. Ficoll M_w 400,000 was added at different concentrations and left for 1 hr at 0°C. The samples were layered over 20% sucrose and under PBS and were centrifuged at 54,000 rpm (210,000-290,000 x g) for 40 minutes in an SW60 rotor at 4°C. The viruses bound to liposomes went to the sample-PBS interface and free virus sedimented through the 20% sucrose to a 60% sucrose pad. A small amount (4.5%) of the viruses goes to the sample-PBS interface with all kinds of liposomes and is thought to represent trapping. This amount was subtracted from the total amounts bound. The bound virus is expressed as percentage of total virus. Data are taken from Haywood and Boyer [1986].

also indicates that only binding and not fusion had occurred, which is consistent with the requirement for higher temperatures for membrane fusion to occur.

The order of addition of virus, liposomes and ficoll appeared to make no difference. Thus two of the three components, (i.e., viruses and liposomes, viruses and ficoll, and liposomes and ficoll) were each left together for 15 minutes at 0°C before addition of the third component. After an additional 45 minutes at 0°C the mixtures were centrifuged. There were no differences in the amounts of binding [Haywood & Boyer, 1986]. Hoekstra et al. [1989] showed that preincubation of Sendai virus with polyethylene glycol (PEG) did make an effect on subsequent behavior of the virus and that the effect of preincubation was very slow, so that it was not pronounced after 15 minutes but was after an hour. If PEG and ficoll are acting by similar mechanisms, it is possible that premixing of pairs of components containing ficoll might have had a more marked effect with a longer time of premixing.

In one experiment Sendai virus (92 µg) and 0.5 µmol of receptor-containing liposomes (composed of egg PC, egg PE, cholesterol and ganglioside G_{D1a} in the mole ratios of 7, 3, 6, and 0.5) in PBS were kept at 0-4°C for 1 hr with and without 10% ficoll. They were then examined with Nomarski optics and image processing. Because the stage was not temperature-controlled, the temperature during the observation period was unknown and the microscope light undoubtedly warmed the sample. The sample containing ficoll appeared to contain more liposomal aggregates than the sample without ficoll. Increased formation of aggregates by virus is consistent with increased strength of adhesion. The "browsing" seemed more marked in the ficoll-containing sample. It is possible that the ficoll helps the virus keep its hold on the liposome surface while it browses. Interestingly, Hoekstra et al. [1989] noted that PEG stabilizes the binding of Sendai virus particles because in the presence of PEG there is less release of virus when the temperature is raised from 4° to 37°C (which allows the neuraminidase to become active).

The mechanism by which ficoll modulates adhesion of virus to receptor-containing liposomes is not known. Because the enhancement of viral adhesion to receptor-containing liposomes occurs in the cold, an enzymatic mechanism is unlikely. One possible mechanism is the polymers cause changes in the organization or conformation of the viral HN proteins or of the ganglioside receptors. This is similar to the mechanism suggested for the effects of dextran and BSA upon the binding of lectins to their receptors in liposomes [Ketis and Grant, 1982, 1983; Grant and Peters, 1984]. The binding of wheat germ agglutinin to glycophorin in liposomes [Ketis and Grant, 1980] or of concanavalin A to band 3 protein in liposomes [Ketis and Grant, 1982] is of low affinity. However, when bovine serum albumin (BSA) or dextran (molecular weight 500,000) is present, the binding mimics the positive cooperativity observed when lectins bind to receptors in cells. Ketis and Grant point out that the lectins have four binding sites and multiple binding should increase the affinity. However, the oligosaccharide chains of these receptor glycoproteins have considerable freedom of motion, which may make multiple binding difficult. Grant and Ketis suggest that an adsorbed layer of dextran or BSA in some way organizes the receptors or lectins so that there is more opportunity for stable polydentate binding. Similarly the viral HN spike is a tetramer, so changes in the proteins and gangliosides could result in the binding of more monomers in each tetramer. Ficoll and dextran might disperse aggregation of gangliosides or of the viral glycoproteins or change the orientation or conformation of either the gangliosides or the proteins.

A second possible mechanism by which ficoll might modulate adhesion of virus to receptor-containing liposomes is exclusion of the ficoll from the gap between the bound virus and liposome. This would result in a lower osmolarity in the gap than in the bulk fluid, and the resulting osmotic pressure should drive the virus and liposome together. Several hydrophilic polymers have been demonstrated to be completely or partially excluded from the aqueous layer next to membranes and this layer is referred to as the exclusion layer [Arnold et al., 1988; de Gennes, 1988; Yamazaki et al., 1989]. An example is polyethylene glycol (PEG), a polymer that promotes membrane aggregation and fusion and that is very soluble in water and tends to restructure water. The difference in the osmolarity between the exclusion layer and the bulk fluid causes vesicles or cells to aggregate in the presence of PEG. Evans and Needham [1988] found that the adhesion energies of vesicles in the presence of different dextrans are consistent with the hypothesis that the attraction stress caused by the polymers is due to the osmotic pressure reduction in the region between the vesicles. Ficoll might similarly cause viruses and liposomes to aggregate, and this aggregation could give the viruses the opportunity to recruit gangliosides or rearrange their HN proteins to produce either a greater number or an increased affinity of the HN-ganglioside bonds. Polymers might also be excluded from the region between virus and bound liposomes/cells because of steric interference. The network formed by the HN proteins and their bound receptors should form a molecular sieve, which would exclude polymers according to their dimensions. This would account well for why the effects of ficoll are seen with receptor-containing liposomes and not with other liposomes. Exclusion of polymers from the regions of receptor binding should also cause a difference between the osmolarity of the bulk fluid and the osmolarity of the fluid between the bound virus and liposome. This difference in osmolarity should cause water to be removed from the gap between the virus and liposome and should cause the virus and liposome to be forced together. This in turn might also cause alterations in the membrane components that would strengthen the HN protein-receptor binding and that would assist fusion. Not only synthetic polymers but also biologic polymers, such as proteins, are likely to be excluded from the bound regions.

If polymer exclusion is responsible for the increased viral adhesion, the adhesion should be affected by the molecular weight of the polymers. In 14% Ficoll M_w 70,000 twenty-six percent of viruses remains bound to liposomes after centrifugation and in 14% Ficoll M_w 400,000 thirty percent of viruses remains bound. Therefore, the increase in the molecular weight of ficoll does not cause a large difference in binding. However, ficoll is a highly branched polymer of sucrose and so tends to be globular. Dextran, on the other hand, is a flexible linear polymer of glucose with sparse short branches. Therefore the radius of gyration of dextran should be more dependent than that of ficoll upon molecular weight [Luby-Phelps et al., 1988]. Dextrans with an average molecular weight of 9,000, 40,600, and 249,000 were obtained from Sigma Chemical Co.. Dextran M_w 9,000 behaves like ficoll in that it reproducibly enhances adhesion of Sendai virus only to liposomes containing gangliosides with receptor activity. Thus after one hr at 0°C with 14% dextran 9,000, 23.6% of the viruses remain bound to liposomes containing 5 mole percent G_{D1a} during centrifugation at 300,000 x g; whereas, only the usual background level of 4.2% of the viruses remain bound to liposomes containing 5 mole percent G_{M1} [Haywood and Boyer, 1986]. When dextran of molecular weight 40,000 or 249,000 were used, for reasons that are not clear there were some problems in reproducibility of results. However, there was a pattern of increase of viral adhesion as the molecular weight of the dextran increased.

Ficoll accelerates viral membrane fusion

Preliminary data indicate that Sendai virus membrane fusion occurs more rapidly in the presence of ficoll. Thus after 10 minutes at 40°C, five percent of the viruses have fused in the absence of ficoll but thirty percent have fused in the presence of 14% ficoll. This indicates that the enhancement of adhesion by polymers helps the intermediate binding steps that precede membrane fusion and is compatible with the idea that polymers may facilitate stabilization of binding and engulfment of virus. Sendai virus with the F_0 protein, the uncleaved inactive form of the F protein, does not fuse in the presence of ficoll, so ficoll does not bypass the requirement for an active F protein for fusion.

Hoekstra et al. [1989] have shown that in the presence of small amounts (4% w/v) of PEG 8000 there is a 1.5-fold increase in the initial binding rate of Sendai virus with red cell ghosts but that the rate of virus fusion increases by approximately 5-fold. They attributed this to the dehydrating effect of PEG, but it is possible that PEG at these low concentrations is mainly enhancing stabilization of binding. Hoekstra et al. performed their kinetics at much lower viral concentrations than did Tsao and Huang [1986], so what they term binding (measured by centrifuging at 10,000 x g) is the initial attachment of virus to cells and a second-order reaction. Their rate constant for membrane fusion is a first-order reaction and probably includes both the stabilization of binding described by Tsao and Huang and fusion. Yamazaki and Ito [1990] have suggested that PEG-induced membrane fusion is due to the deformation of the membranes that follows the osmotic stress resulting from the local imbalance of osmolarity between the exclusion layer and the bulk phase. This is somewhat similar to the observation that virus binding forces a liposome to form a tightly curved region and that fusion occurs at this stressed region [Haywood and Boyer, 1981; Haywood, 1988; 1991].

At present the biologic significance of the effects of polymers such as ficoll upon virus adsorption is open to question. Serum and extracellular fluid contain polymers that might similarly assist in tight binding as could mobile components of the glycocalyx. If further work shows that the enhanced fusion in the presence of polymers is related to the stabilization of binding, this would contribute substantially to understanding the mechanism of membrane fusion. As described below polyanionic polymers adsorb to virus and inhibit their binding to cells, so that if nonadsorbing polymers do play a biologic role in viral binding and fusion, adsorbed polymers may make impossible the action of nonadsorbing polymers.

THE STEPS IN BINDING ARE POTENTIAL TARGETS FOR ANTIVIRAL AGENTS

The different stages in viral binding are potential targets for antiviral agents. The ideal antiviral agent should inhibit a viral but not a host function. The viral-coded proteins determine most viral functions. For those viruses that code for only a few proteins, the majority of the proteins are in the virus particle and many are involved in viral entry and exit. These include the proteins involved in the different stages of binding. The most obvious approach is to inhibit binding of the viral ligand with its cellular receptor. As reviewed by Mitsuya et al. [1990] HIV infection can be blocked by a soluble form of its receptor, the protein CD4, and this approach is being tested clinically. Possible problems with this approach are the fact that the protein receptors are likely to play an important host function as well as a viral function, the fact the inhibitors are protein and likely to induce antibody formation, and the fact that viruses can mutate to become resistant to specific inhibitors, such as soluble CD4. The presence of several stages in binding makes possible antiviral strategies other than blocking binding to a protein receptor. Thus it may be possible to inhibit the initial viral binding or to inhibit the rearrangements that allow secondary binding to occur.

An example of possibly clinically usable antiviral agents that inhibit binding are polyanionic polymers, such as dextran sulfate. Polyanionic polymers have long been recognized to inhibit viral binding [De Somer et al., 1968; Mitsuya et al., 1988], but little is known about the mechanism. Schols et al. [1990] have ascribed the inhibitory effect of polyanions on HIV binding to direct interaction with the HIV glycoprotein gp120, and dextran sulfate binds to Sendai virus [Ohki et al., 1991]. If viruses bind first to components of the glycocalyx such as heparan sulfate and sialic acid residues, polyanionic polymers may compete for this binding. If polymers in serum or extracellular fluid act *in vivo* as ficoll and dextran act with viruses and receptor-containing liposomes, polyanionic polymers might interfere with this action. For instance, the polyanionic polymers might adsorb to virus tightly enough that they can not be excluded and thereby prevent stabilization of binding. Many polyanionic polysaccharides have some anticoagulant activity but otherwise seem to be relatively nontoxic. Further they are a poor stimulus for antibody production. If they bind to virus nonspecifically on the basis of charge [Ohki et al., 1991], they are not likely to induce viral resistance and so could be candidates for use in long term therapy of persistent viral infections. Since a variety of polyanionic polymers are active against viruses, it should be possible to design agents with favorable pharmacological properties. Thus further work on the stages of binding and the agents that inhibit these stages have the potential to lead to the development of classes of antiviral agents that are relatively nontoxic and would be applicable to many virus groups.

ACKNOWLEDGMENTS

Parts of the investigations from the author's laboratory were supported by grants PCM 78-08931 and PCM 82-05896 from the National Science Foundation and by grant AI-15540 from the National Institutes of Health. Some of the unpublished results were obtained at the Institut für Immunologie und Virologie der Universität Zürich while the author was supported by Senior International Fellowship TWO1147 from the Fogarty International Center of the National Institutes of Health.

REFERENCES

Adam, G., and Delbrück, M., 1968, Reduction of dimensionality in biological diffusion processes, in: "Structural Chemistry and Molecular Biology," A. Rich, and N. Davidson, eds., W. H. Freeman Press, pp. 198-215.

Arnold, K., Herrmann, A., Gawrisch, K., and Pratsch, L., 1988, Water-mediated effects of PEG on membrane properties and fusion, in: "Molecular Mechanisms of Membrane Fusion," S. Ohki, D. Doyle, T. D. Flanagan, S. W. Hui, and E. Mayhew, eds., Plenum Publ. Corp., New York, pp. 255-272.

Brandley, B. K., and Schnaar, R. L., 1985, Phosphorylation of extracellular carbohydrates by intact cells: Chicken hepatocytes specifically adhere to and phosphorylate immobilzed N-acetylglucosamine, J. Biol. Chem., 260: 12474-12483.

Brandley, B. K., and Schnaar, R. L., 1986, Cell-surface carbohydrates in cell recognition and response, J. Leukocyte Biol., 40:97-111.

Burnet, F. M., McCrea, J. F., and Stone, J. D., 1946, Modification of human red cells by virus action. I. The receptor gradient for virus action in human red cells, Brit. J. Exp. Path., 27:228-236.

Cooper, N. R., Moore, M. D., and Nemerow, G. R., 1988, Immunobiology of CR2, the B lymphocyte receptor for Epstein-Barr virus and the C3d complement fragment, Ann. Rev. Immunol., 6:85-113.

de Gennes, P. G., 1988, Model polymers at interfaces, in: "Physical Basis of Cell-Cell Adhesion," P. Bongrand, ed., CRC Press, Inc, Boca Raton, pp. 39-60.

De Somer, P., De Clercq, E., Billiau, A., Schonne, E., and Claesen, M., 1968, Antiviral activity of polyacrylic and polymethacrylic acids. I. Mode of action in vitro, J. Virol., 2:878-885.

Einberger, H., Mertz, R., Hofschneider, P. H., and Neubert, W. J., 1990, Purification, renaturation, and reconstituted protein kinase activity of the Sendai virus large (L) protein: L protein phosphorylates the NP and P proteins in vitro, J. Virol., 64:4274-4280.

Evans, E., and Needham, D., 1988, Attraction between lipid bilayer membranes in concentrated solutions of nonadsorbing polymers: Comparison of mean-field theory with measurements of adhesion energy, Macromolecules, 21:1822-1831.

Flynn, D. C., Meyer, W. J., MacKenzie, J. M. Jr., and Johnston, R. E., 1990, A conformational change in Sindbis virus glycoproteins E1 and E2 is detected at the plasma membrane as a consequence of early virus-cell interaction, J. Virol., 64:3643-3653.

Giuffre, R. M., Tovell, D. R., Kay, C. M., and Tyrrell, D. L. J., 1982, Evidence for an interaction between the membrane protein of a paramyxovirus and actin, J. Virol., 42:963-968.

Gottschalk, A., 1958, The influenza virus neuraminidase, Nature, 181:377-378.

Grant, C. W. M., and Peters, M. W., 1984, Lectin-membrane interactions: Information from model systems, Biochim. Biophys. Acta, 779:403-422.

Greve, J. M., Davis, G., Meyer, A. M., Forte, C. P., Yost, S. C., Marlor, C. W., Kamarck, M. E., and McClelland, A., 1989, The major human rhinovirus receptor is ICAM-1, Cell, 56:839-847.

Guarnaccia, S. P., and Schnaar, R. L., 1982, Hepatocyte adhesion to immobilized carbohydrates. I. Sugar recognition is followed by energy-dependent strengthening, J. Biol. Chem., 257:14288-14292.

Haywood, A. M., 1974, Characteristics of Sendai virus receptors in a model membrane, J. Mol. Biol., 83:427-436.

Haywood, A. M., 1975a, Model membranes and Sendai virus: Surface-surface interactions, in: "Negative Strand Viruses," R. D. Barry, and B. W. J. Mahy, Eds., Academic Press, London, 2, pp. 923-928.

Haywood, A. M., 1975b, 'Phagocytosis' of Sendai virus by model membranes, J. Gen. Virol., 29:63-68.

Haywood, A. M., and Boyer, B. P., 1981, Initiation of fusion and disassembly of Sendai virus membranes into liposomes, Biochim. Biophys. Acta, 646:31-35.

Haywood, A. M., and Boyer, B. P., 1982, Sendai virus membrane fusion: Time course and effect of temperature, pH, calcium, and receptor concentration, Biochemistry, 21:6041-6046.

Haywood, A. M., and Boyer, B. P., 1986, Ficoll and dextran enhance adhesion of Sendai virus to liposomes containing receptor (ganglioside G_{D1a}), Biochemistry, 25:3925-3929.

160

Haywood, A. M., 1988, 'Entry' of enveloped viruses into liposomes, in: "Molecular Mechanisms of Membrane Fusion," S. Ohki, D. Doyle, T. D. Flanigan, S. W. Hui, and E. Mayhew, Eds., Plenum Publishing Corp, New York, pp. 427-440.

Haywood, A. M., 1991, Fusion of virus membranes with phospholipid vesicles at neutral pH, in: "Membrane Fusion," J. Wilschut, and D. Hoekstra, eds., Marcel Dekker, New York, Ch 16, pp. 337-374.

Hirst, G. K., 1941, The agglutination of red cells by allantoic fluid of chick embryos infected with influenza virus, Science, 94:22-23.

Hirst, G. K., 1942, Adsorption of influenza hemagglutinins and virus by red blood cells, J. Exp. Med., 76:195-209.

Hoekstra, D., Klappe, K., Hoff, H., and Nir, S., 1989, Mechanism of fusion of Sendai virus: role of hydrophobic interactions and mobility constraints of viral membrane proteins. Effects of polyethylene glycol, J. Biol. Chem., 264:6786-6792.

Iorio, R. M., Glickman, R. L., Riel, A. M., Sheehan, J. P., and Bratt, M. A., 1989, Functional and neutralization profile of seven overlapping antigenic sites on the HN glycoprotein of Newcastle disease virus: monoclonal antibodies to some sites prevent viral attachment, Virus Research, 13: 245-262.

Johnson, D. C., Burke, R. L., and Gregory, T., 1990, Soluble forms of herpes simplex virus glycoprotein D bind to a limited number of cell surface receptors and inhibit viral entry into cells, Virology, 64:2569-2576.

Ketis, N. V., Girdlestone, J., and Grant, C. W. M., 1980, Positive cooperativity in a (dissected) lectin-membrane glycoprotein binding event, Proc. Natl. Acad. Sci. USA, 77:3788-3790.

Ketis, N. V., and Grant, C. W. M., 1982, Co-operative binding of concanavalin A to a glycoprotein in lipid bilayers, Biochim. Biophys. Acta, 689:194-202.

Ketis, N. V., and Grant, C. W. M., 1983, Time-dependent lectin binding to isolated receptors in model membranes, Biochim. Biophys. Acta, 730:359-368.

Kim, J., Hama, K., Miyake, Y., and Okada, Y., 1979, Transformation of intramembrane particles of HVJ (Sendai virus) envelopes from an invisible to visible form on aging of virions, Virology, 95:523-535.

Knutton, S., 1976, Changes in viral envelope structure preceding infection, Nature, 264:672-673.

Knutton, S., 1978, The mechanism of virus-induced cell fusion, Micron, 9: 133-154.

Lee, P. M., Cherry, R. J., and Bächi, T., 1983, Correlation of rotational mobility and flexibility of Sendai virus spike glycoproteins with fusion activity, Virology, 128:65-76.

Lotz, M. M., Burdsal, C. A., Erickson, H. P., and McClay, D. R., 1989, Cell adhesion to fibronectin and tenascin: Quantitive measurements of initial binding and subsequent strengthening response, J. Cell Biol., 109:1795-1805.

Luby-Phelps, K., Lanni, F., and Taylor, D. L., 1988, The submicroscopic properties of cytoplasm as a determinant of cellular function, Ann. Rev. Biophys. Biophys. Chem., 17:369-396.

Markwell, M. A. K., and Fox, C. F., 1980, Protein-protein interactions within paramyxoviruses identified by native disulfide bonding or reversible chemical cross-linking, J. Virol., 33:152-166.

Markwell, M. A. K., Svennerholm, L., and Paulson, J. C., 1981, Specific gangliosides function as host cell receptors for Sendai virus, Proc. Natl. Acad. Sci. USA, 78:5406-5410.

Mitsuya, H., Looney, D. J., Kuno, S., Ueno, R., Wong-Staal, F., and Broder, S., 1988, Dextran sulfate suppression of viruses in the HIV family: Inhibition of virion binding to CD4+ cells, Science, 240:646-649.

Mitsuya, H., Yarchoan, R., and Broder, S., 1990, Molecular targets for AIDS therapy, Science, 249:1533-1544.

Moore, J. P., McKeating, J. A., Weiss, R. A., and Sattentau, Q. J., 1990, Dissociation of gp120 from HIV-1 virions induced by soluble CD4, Science, 250:1139-1142.

Ohki, S., Arnold, K., Srinivasakumar, N., and Flanagan, T. D., 1991, Effect of dextran sulfate on fusion of Sendai virus with human erythrocyte ghosts, Biomed. Biochim. Acta, 50: 199-206.

Örvell, C., and Grandien, M., 1982, The effects of monoclonal antibodies on biologic activities of structural proteins of Sendai virus, J. Immunol., 129: 2779-2787.

Peterhans, E., Baechi, T., and Yewdell, J., 1983, Evidence for different receptor sites in mouse spleen cells for the Sendai virus hemagglutinin-neuraminidase (HN) and fusion (F) glycoproteins, <u>Virology</u>, 128: 366-376.

Portner, A., Scroggs, R. A., and Metzger, D. W., 1987a, Distinct functions of antigenic sites of the HN glycoprotein of Sendai virus, <u>Virology</u>, 158:61-68.

Portner, A., Scroggs, R. A., and Naeve, C. W., 1987b, The fusion glycoprotein of Sendai virus: sequence analysis of an epitope involved in fusion and virus neutralization, <u>Virology</u>, 157:556-559.

Quiocho, F. A., 1986, Carbohydrate-binding proteins: Tertiary structures and protein-sugar interactions, <u>Ann. Rev. Biochem.</u>, 55:287-315.

Rogers, G. N., and Paulson, J. C., 1983, Receptor determinants of human and animal influenza virus isolates: Differences in receptor specificity of the H3 hemagglutinin based on species of origin, <u>Virology</u>, 127:361-373.

Rogers, G. N., and D'Souza, B. L., 1989, Receptor binding properties of human and animal H1 influenza virus isolates, <u>Virology</u>, 173:317-322.

Schols, D., Pauwels, R., Desmyter, J., and De Clercq, E., 1990, Dextran sulfate and other polyanionic anti-HIV compounds specifically interact with the viral gp120 glycoprotein expressed by T-cells persistently infected with HIV-1, <u>Virology</u>, 175:556-561.

Staunton, D. E., Dustin, M. L., Erickson, H. P., and Springer, T. A., 1990, The arrangement of the immunoglobulin-like domains of ICAM-1 and the binding sites for LFA-1 and rhinovirus, <u>Cell</u>, 61:243-254.

Stent, G. S., and Wollman, E. L., 1952, On the two-step nature of bacteriophage adsorption, <u>Biochim. Biophys. Acta</u>, 8:260-269.

Thompson, S. D., Laver, W. G., Murti, K. G., and Portner, A., 1988, Isolation of a biologically active soluble form of the hemagglutinin-neuraminidase protein of Sendai virus, <u>J. Virol.</u>, 62:4653-4660.

Tomassini, J. E., Graham, D., DeWitt, C. M., Lineberger, D. W., Rodkey, J. A., and Colonno, R. J., 1989, cDNA cloning reveals that the major rhinovirus receptor on HeLa cells is intercellular adhesion molecule 1, <u>Proc. Natl. Acad. Sci. USA</u>, 86:4907-4911.

Tsao, Y.-s., and Huang, L., 1986, Kinetic studies of Sendai virus-target membrane interactions: Independent analysis of binding and fusion, <u>Biochemistry</u>, 25:3971-3976.

Weis, W., Brown, J. H., Cusack, S., Paulson, J. C., Skehel, J. J., and Wiley, D. C., 1988, Structure of the influenza virus haemagglutinin complexed with its receptor, sialic acid, <u>Nature</u>, 333:426-431.

Weiss, R. A., Clapham, P. R., McClure, M., and Marsh, M., 1989, The CD4 receptor for the AIDS virus, <u>Biochem. Soc. Trans.</u>, 17:644-647.

Wilson, I. A., Skehel, J. J., and Wiley, D. C., 1981, Structure of the haemagglutinin membrane glycoprotein of influenza virus at 3 Å resolution, <u>Nature</u>, 289:366-373.

Wright, S. D., and Silverstein, S. C., 1984, Phagocytosing macrophages exclude proteins from the zones of contact with opsonized targets, <u>Nature</u>, 309: 359-361.

Wright, S. D., and Silverstein, S. C., 1986, Overview: the function of receptors in phagocytosis, <u>in</u>: "Handbook of Experimental Immunology," D. M. Weir, L. A. Herzenberg, C. Blackwell, and L. A. Herzenberg, eds., Blackwell Scientific Publications, Oxford, Vol. 2, pp. 41.1-41.14.

WuDunn, D., and Spear, P. G., 1989, Initial interaction of herpes simplex virus with cells is binding to heparan sulfate, <u>J. Virol.</u>, 63:52-58.

Yamazaki, M., Ohnishi, S.-i., and Ito, T., 1989, Osmoelastic coupling in biological structures: Decrease in membrane fluidity and osmophobic association of phospholipid vesicles in response to osmotic stress, <u>Biochemistry</u>, 28:3710-3715.

Yamazaki, M., and Ito, T., 1990, Deformation and instability in membrane structure of phospholipid vesicles caused by osmophobic association: Mechanical stress model for the mechanism of poly(ethylene glycol)-induced membrane fusion, <u>Biochemistry</u>, 29:1309-1314.

INHIBITION OF SENDAI VIRUS FUSION AND PHOSPHOLIPID VESICLE FUSION: IMPLICATIONS FOR THE PATHWAY OF MEMBRANE FUSION

Philip L. Yeagle[†], Daniel R. Kelsey[†], Thomas D. Flanagan[§], Joyce Young[†]

[†]Department of Biochemistry and [§]Department of Microbiology
School of Medicine and Biomedical Sciences
State University of New York at Buffalo
Buffalo, New York 14214

INTRODUCTION

Membrane fusion is an essential step in the infection cycle of enveloped viruses. Fusion may occur with the plasma membrane or may occur by endocytosis of the virion followed by acidification and subsequent pH-induced fusion. Enveloped viruses possess an outer limiting membrane containing glycoproteins responsible for recognition of the target cell and mediation of the fusion event. Fusion may be facilitated by a dedicated protein, such as the F protein of Sendai, or the ability to facilitate fusion and the binding of the virion to the target membrane may involve a single protein species. Fusion of the viral envelope with the target membrane allows the entry of the viral genome and initiation of replication. At present an adequate understanding of the mechanism of the viral membrane fusion process is lacking.

The events leading up to the fusion of two membranes have been suggested to include the following events (Bentz and Ellens 1988):

1. aggregation or adhesion of the membranes that will fuse;
2. close approach of the lipid bilayers of the membranes, leading to removal of some of the water separating the membranes (partial dehydration);
3. destabilization of the bilayer at the point of fusion (two bilayers closely opposed will not spontaneously fuse);
4. mixing of the bilayers and ultimate separation from the point of fusion into the new membrane structure(s).

Many enveloped viruses are effective at promoting this fusion event. Among the most studied are Sendai which fuses at the plasma membrane of the cell (Haywood 1988), and influenza, which may fuse in endocytic vesicles after pH reduction by the vesicle H^+ ATPases (Skehel et al. 1982). These viruses possess an outer limiting membrane containing glycoproteins responsible for recognition of the target cell and mediation of the fusion event between the viral envelope and the target cell plasma membrane. Following the fusion event in a successful infection, the virus then undergoes disassembly (uncoating) whereupon viral replication can begin.

Current hypotheses for the mechanism of enveloped virus fusion postulate a role for the envelope glycoproteins. Among these glycoproteins is a class of "fusion" proteins which appear to be required in the fusion process. In Sendai virus, this protein has been identified as the F or fusion protein (Hsu, Scheid, and

Cell and Model Membrane Interactions
Edited by S. Ohki, Plenum Press, New York, 1991

Choppin 1979, Nakanishi, Uchida, and Okada 1982, Scheid and Choppin 1974) and in RSV, the fusion activity is associated with another F protein (Elango et al. 1985). Similar glycoproteins for HSV I fusion have been identified (Cai, Gu, and Person 1988, Manservigi, Spear, and Buchan 1977, Roizman and Sears 1991, Sarmiento, Haffey, and Spear 1979). Many investigators have suggested that these fusion proteins cause membrane bilayer destabilization (Daniels et al. 1985, Doms, Helenius, and White 1985, Skehel et al. 1982), which is promoted by a "fusogenic" sequence common to many enveloped viruses (Ohnishi and Murata 1988).

Exposure of the "fusogenic sequence" to the target membrane is apparently required for enveloped virus fusion (Asano and Asano 1984, Daniels et al. 1985, Doms, Helenius, and White 1985, Novick and Hoekstra 1988, Ruigrok et al. 1988, Skehel et al. 1982). This is supported by the fact that peptides encompassing the fusogenic regior of influenza virus are themselves fusogenic, suggesting that the membrane fusion activity lies in that "fusogenic" sequence (Ohnishi and Murata 1988). Recent data suggest that the analogous peptides from HIV are also fusogenic (Retalsk et al. 1989). Recently, a model has been proposed which describes the interaction of these fusion proteins with the membrane bilayer (Brasseur et al. 1990). This model states that the fusion proteins enter the membrane, as a helix whose long axis sits obliquely in the bilayer at an angle of about 60° from the membrane/water interface.

It was the sequence of these "fusion peptides" that led others to investigate the ability of apparent homologues of these fusion peptides to inhibit viral infection. Richardson, et al. (1980) have shown that some oligopeptides, with sequences reminiscent of the hydrophobic N-termini of viral fusion proteins, were capable of inhibiting measles and Sendai virus fusion with target cells. The mechanism of this inhibition was not clear. Richardson and Choppin (1983) presented evidence suggesting that the inhibitory peptides, in particular Z-D-Phe-L-PheGly, bind to receptor sites on the cell membrane and do not bind to the envelope of measles virus. Much earlier work indicated that similar peptides were capable of inhibiting Herpes Simplex virus infections (Nicolaides et al. 1968). These peptide inhibitors have proven useful in the present work because they have now been shown to be inhibitors of membrane fusion facilitated by some enveloped viruses

The details of the pathways of fusion for enveloped viruses are not known. However, it is appropriate to now briefly examine what is known about the more simple membrane fusion mechanisms. Two kinds of fusion pathways have been described in model membrane fusion studies. One utilizes a mechanism involving calcium and phosphatidylserine (or cardiolipin) which may involve a gel (dehydrated) phase formation by the calcium - phosphatidylserine complex (Hoekstra 1982a, Hoekstra 1982b, Leventis et al. 1986, Papahadjopoulos et al. 1977, Silvius and Gagne 1984). Other investigators have ruled out this pathway for virus fusion (Stegmann et al. 1985, White, Kartenbeck, and Helenius 1982).

The second pathway has been suggested to involve non-lamellar "isotropic" structures, I_S, identified in ^{31}P NMR spectra of lipid dispersions as intermediates in membrane fusion (Ellens, Bentz, and Szoka 1986, Ellens et al. 1989, Gagne et al. 1985, Siegel et al. 1989a). This pathway is of major interest here in that the I_S could result from viral fusion protein activity and provide step 3 for fusion in the above scheme. Recently, a relationship between the presence of the isotropic resonance in the ^{31}P NMR spectrum of phospholipid vesicles and membrane fusion was reported (Ellens, Bentz, and Szoka 1986) for LUV of N-monomethyl-dioleoylphosphatidylethanolamine They showed that membrane fusion in this system began at the first appearance of I_S (Gagne et al. 1985) and increased with increase in the prevalence of I_S, below the L_α to H_{II} phase transition temperature. These and other data suggested that the rate of vesicle membrane fusion, measured by a contents-mixing fluorescence assay, was directly proportional to the percent of the total membrane lipid found in I_S. On the basis of studies such as these, the hypothesis was advanced that I_S was involved in the mechanism of membrane fusion for these lipid systems as an intermediate on the fusion pathway (Ellens et al. 1989).

It is not known whether intermediates resembling this I_S are involved in viral fusion. The data from our laboratories suggests that there may be a step in common on the pathway of fusion of the LUV of N-methyl DOPE and the fusion of Sendai virus with target membranes. The experiments described in this chapter indicate that Sendai virus will fuse with N-methyl DOPE LUV in a receptor-independent mechanism. Hydrophobic peptide inhibitors of enveloped viral infection inhibit the fusion not only of Sendai virions with cells but also the fusion of the N-methyl DOPE LUV. The target of these inhibitors in the LUV fusion appears to be I_S. These experiments shed light on the pathway of fusion of those enveloped viruses that fuse at neutral pH.

MATERIALS AND METHODS

N-methyl dioleoylphosphatidylethanolamine (N-methyl DOPE[1]) was obtained from Avanti Polar Lipids, Birmingham, AL. Octadecylrhodamine B chloride (R_{18}), 1-Aminonaphthalene-3,6,8-trisulfonic acid (disodium salt) (ANTS) and p-xylene bis(pyridinium bromide) (DPX) were from Molecular Probes, Inc., Junction City, OR. Carbobenzoxy-D-Phe-L-PheGly (Z-D-Phe-L-PheGly), carbobenzoxy-D-Phe (Z-D-Phe), carbobenzoxy-L-Phe (Z-L-Phe) and carbobenzoxy-Gly-L-Phe (Z-Gly-L-Phe) were purchased from Sigma. Carbobenzoxy-L-Phe-L-Tyr (Z-L-Phe-L-Tyr) was obtained from Chemical Dynamics Corporation, South Plainfield, New Jersey.

Peptide synthesis and derivatization
L-Phe-L-Phe and Gly-L-Phe-L-Phe were synthesized at the SUNY Buffalo Microsequencing Lab by the solid phase method of Tam, et al. (1983). An N-terminal carbobenzoxy moiety was added to these peptides by the method of Bodanszky, (1984). A solution of the peptide to be derivatized in 3 ml water and 2 ml 5 N NaOH was stirred on ice. Benzyl chloroformate (1.1 moles per mole of peptide) and 5.5 ml 2 N NaOH were added, alternately, in 10 portions while the reaction mixture was being stirred at 10° C. Additions were completed within 90 min. The reaction mixture was stirred an additional 30 min. at 25° C after which the pH was adjusted to 10.0 with 1 N NaOH and the solution was extracted 3 times with 2 volumes of diethyl ether. The aqueous layer was acidified to pH 3.0 with HCl and extracted 3 times with 2 volumes of diethyl ether. Solvent was removed by lyophilization and the crude product was purified by reverse phase HPLC using an Altex Ultrasphere™ C_{18} column. The products were eluted over a period of 1 hr. with a linear gradient of 15%-100% acetonitrile in 0.1% trifluoroacetic acid. Z-L-Phe-L-Phe eluted at approximately 61% acetonitrile and Z-Gly-L-Phe-L-Phe eluted at approximately 65% acetonitrile.

The "fusion peptide" of measles (FAGVVLAGAALGVAAAAQI) was chemically synthesized using an Applied Biosystems 430A solid phase peptide synthesizer. Peptide was desalted and purified by HPLC as previously described (Richardson et al. 1985). Rechromatography showed one peak on HPLC. The threonine residue at position 15 was converted to alanine to make the synthesis simpler. The amino acid analysis of the purified peptide agreed with the above sequence.

---

[1]The abbreviations used are: N-methyl DOPE, N-methyl dioleoylphosphatidyl-ethanolamine; LUV, large unilamellar vesicles; ANTS, 1-aminonaphthalene-3,6,8-trisulfonic acid; DPX, p-xylene bis(pyridinium bromide); R_{18}, octadecylrhodamine B chloride; Z-D-Phe-L-PheGly, carbobenzoxy-D-Phenyl-L-PhenylGlycine; Z-L-Phe-L-Tyr, carbobenzoxy-L-Phenyl-L-Tyrosine and Z-Gly-L-Phe, Z-L-Phe-L-Phe, carbo-benzoxy-L-Phenylalanyl-L-Phenylalanine; Z-L-Phe, carbobenzoxy-L-Phenyl-alanine; Z-D-Phe, carbobenzoxy-D-Phenylalanine; Z-L-Phe-L-Tyr, carbobenzoxy-L-Phenylalanyl-L-Tyrosine; Z-Gly-L-Phe-L-Phe, carbobenzoxy-Glycyl-L-Phenyl-alanyl-L-Phenylalanine;carbobenzoxy-Glycyl-L-Phenylalanine; ^{31}P NMR, ^{31}P nuclear magnetic resonance; EDTA, ethylenediamine tetraacetic acid; PC phosphatidylcholine.

Membrane preparation for NMR

Multilamellar liposomes containing the measles peptide in the phospholipid bilayer were prepared in the following manner. The phospholipid and the peptide were co-solubilized in trifluoroacetic acid at room temperature in the indicated mole ratios. The trifluoroacetic acid was removed by evaporation under a stream of nitrogen gas followed by evaporation under high vacuum. The material was then hydrated with 50 mM histidine, 1 mM EDTA, pH 7.4. The pH was measured to detect residual trifluoroacetic acid. In the case of a significant pH change, the pH was adjusted with NaOH. Following hydration, the membranes were subjected to three freeze-thaw cycles using liquid nitrogen. The membranes were then warmed to the starting temperature of the experiment. Control experiments were also performed using LUV of pure N-methyl DOPE. No difference in the phase behavior was observed, using ^{31}P NMR, between the LUV and the multilamellar preparations (Ellens et al. 1989). The multilamellar preparation was used in the ^{31}P NMR experiments on the measles peptide because of limited amounts of the peptide available and because the yield of material after the extrusions was too low for the ^{31}P NMR experiments.

The preparations for the ^{31}P NMR experiments in the presence of peptide inhibitors were prepared identically to the LUV used for the fusion experiments.

Vesicle Preparation

Large unilamellar vesicles were prepared according to methods described by Szoka and Papahadjopoulos (1978) with further details described by Ellens, et al. (1989). N-methyl DOPE was hydrated for 3 hours on ice, under N_2 in either 25 mM ANTS, 45 mM NaCl, 10 mM glycine, pH 9.5, or 90 mM DPX, 10 mM glycine, pH 9.5 or 12.5 mM ANTS, 45 mM DPX, 22.5 mM NaCl, 10 mM glycine, pH 9.5. The lipid suspension was next subjected to 5 freeze-thaw cycles followed by 10 extrusions through a polycarbonate membrane with 0.1 μm pores (Nuclepore Corp., Pleasanton, CA). Encapsulated material was separated from unencapsulated material on a Sephadex G-50 column (Pharmacia) with 100 mM NaCl, 10 mM glycine, 0.1 mM EDTA, pH 9.5 used as the elution buffer. Vesicles were stored on ice, under N_2 and were used within 2 to 3 days. Vesicles were characterized by negative stain transmission electron microscopy and by gel chromatography as a function of the number of extrusions. After 10 extrusions, no further improvement in homogeneity of vesicle size was seen. Also, no evidence of multilamellar vesicles was observed. According to the electron microscopy, the LUV ranged in size from 200 nm to 900 nm, with most LUV near 400 nm.

Sendai Virus Preparation and Labeling

Sendai virus was grown in the chorioallantoic membrane of embryonated chicken eggs. The allantoic fluid was harvested 72 hrs. post-infection and the virus was partially purified by centrifugation through a 30 - 60% sucrose density gradient. The amount of virus used in the fusion assays was quantified by the number of μg of viral protein determined by dye binding assay (Bradford, 1976). Virus was labeled with Octadecylrhodamine B chloride (R_{18}) as described by Hoekstra, et al. (1984). Briefly, 20 nmol of R_{18} in 10 μl ethanol was added for each mg of viral protein in a total volume of 1 ml. The mixture was vortexed and allowed to incubate at room temperature for 1 hour. Labelled virus was separated from unincorporated R_{18} by passing the incubation mixture over a Sephadex G-75 column and eluting with 100 mM NaCl, 10 mM glycine, 0.1 mM EDTA, pH 7.4.

Fusion and Leakage Assays

All fluorescence measurements were made on an SLM 8000D fluorimeter. The ANTS/DPX fusion and leakage assays were carried out as described by Ellens, et al., 1985. Vesicles contained either 25 mM ANTS and 45 mM NaCl, or 90 mM DPX, or 12.5 mM ANTS, 45 mM DPX and 22.5 mM NaCl. Fluorescence intensity was monitored with an excitation wavelength of 380 nm and an emission wavelength of 510 nm. All assays were carried out in a total volume of 1 ml. The final lipid concentration was 0.4 μmole/ml for both fusion and leakage assays. Fusion or leakage was initiated by lowering the pH from 9.5 to 4.5 with 25 μl of 2 M sodium acetate/acetic acid buffer. For fusion assays a 9:1 molar ratio of DPX containing LUV to ANTS containing LUV was used. Fluorescence quenching due to contents mixing (resulting in an ANTS-DPX complex with reduced quantum yield) reflected the rate of LUV fusion. For vesicle-vesicle fusion assays baseline fluorescence was taken to be the level obtained with the shutters of the fluorimeter closed and 100% fluorescence was taken to be the initial fluorescence intensity before lowering the pH. Leakage was measured by dequenching of fluorescence due to leakage and dilution (and dissociation) of the ANTS-DPX complex. For leakage assays baseline fluorescence was taken as the initial level of fluorescence prior to lowering the pH, 100% fluorescence was determined by adding 25 μl of 5% deoxycholate to the vesicles.

The R_{18} lipid mixing fusion assay was used to monitor fusion of Sendai virus with target membranes. Ideally, the same fusion assay would have been used to study both vesicle-vesicle fusion and vesicle-virus fusion. Two obstacles prevented such experiments from being performed. First, since neither ANTS nor DPX can be easily encapsulated within the Sendai virion, the ANTS/DPX contents mixing assay could not be used to study virus-vesicle fusion. Second, the incorporation of 9 mole% R_{18} into N-methyl DOPE LUV seriously disrupts the phase behavior of the LUV as assessed by ^{31}P NMR (data not shown). Therefore the R_{18} lipid mixing assay could not be used to measure vesicle-vesicle fusion.

The R_{18} fusion assay for virus-vesicle fusion was carried out as described by Hoekstra, (1984). N-methyl DOPE LUV were diluted to 1 μmole/ml in 100mM NaCl, 10 mM glycine pH 9.5. A total volume of 1 ml was used for each assay. The LUV were first allowed to equilibrate to the appropriate temperature for 5 minutes. 25 μl of 2 M sodium acetate/acetic acid buffer and 50 μl of R_{18} labelled virus were added simultaneously to the vesicles. Fluorescence was monitored with an excitation wavelength of 560 nm and an emission wavelength of 586 nm. The fluorescence intensity obtained without the addition of 2 M sodium acetate/acetic acid buffer was taken as baseline. 100% fluorescence was determined by adding 100 μl of 10% Triton X-100 to the vesicle/virus mixture.

In order to assess the effect of Sendai virus on vesicle-vesicle fusion the same protocol was used as in the R_{18} assays except that the virus was unlabeled and the vesicles contained ANTS or DPX as for the ANTS/DPX fusion assay described above.

Preparation of Human Erythrocyte Ghosts
Human erythrocyte ghosts were prepared by the method of Clague, et al., (1990) except that resealing was done at 20° C. Fresh whole human blood was washed 3 times in phosphate-buffered saline (PBS), which consists of 137 mM NaCl, 2.7 mM KCl, 8.1 mM Na_2HPO_4, 1.5 mM KH_2PO_4, pH 7.4. Erythrocytes were lysed at 4° C in 20 volumes of 10 mM Tris, 1 mM $MgCl_2$, 1 mM $CaCl_2$, 0.1 mM EGTA, 0.1% BSA, pH 7.4. After 2 min. isotonicity was restored by addition of one tenth volume of hypertonic buffer consisting of 150 mM Na_2HPO_4, 50 mM KH_2PO_4, 1.22 M NaCl, 30 mM KCl, 1 mM $CaCl_2$ and 1 mM $MgCl_2$. The cells were stirred at 20° C for 40 min. followed by 3 washes in PBS.

Sonication
Sonication was performed with a Branson W350 probe sonicator with an ice bath. First, multilamellar liposomes containing the peptide in the phospholipid bilayer were prepared in the following manner. Egg phosphatidylcholine and Z-D-Phe-L-PheGly were co-solubilized in chloroform-methanol (2:1) at room temperature in the indicated mole ratios. The solvent was removed by evaporation under a stream of nitrogen gas followed by evaporation under high vacuum overnight. The material was then hydrated in D_2O with 50 mM NaCl, sealed under nitrogen gas and vortexed vigorously. The membranes were then sonicated with five minute sonications followed by a one minute rest period in which the light scattering of the sample was determined as effective absorbance at 400 nm.

RESULTS

Fusion of Sendai with N-methyl DOPE LUV

It has been previously shown that there is a correlation between the kinetics of fusion of N-methyl DOPE vesicles and the appearance of isotropic ^{31}P NMR resonances superimposed on the ^{31}P NMR bilayer powder patterns that reflect the phase behavior of the lipid (Ellens et al. 1989). That correlation was reproduced in the course of this study. The rate of fusion of N-methyl DOPE vesicles was measured using the ANTS/DPX contents mixing fusion assay as described in Methods. A marked increase in the initial rate of fusion can be observed around 35° - 40°C in the pure N-methyl DOPE vesicles, quantitatively similar to data published previously (Ellens et al. 1989). The rate of fusion was proportional to the incidence of the isotropic resonance in ^{31}P NMR spectra of identically prepared N-methyl DOPE LUV in this temperature range (data not shown).

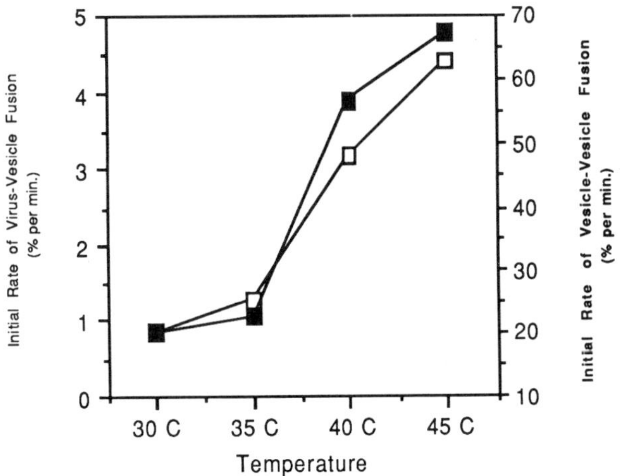

Fig. 1. The initial rates of virus–vesicle (—□—) and vesicle–vesicle (—■—) fusion are shown above. Virus–vesicle fusion was determined by the R_{18} lipid-mixing assay with 2.5 μg of labeled virus. Vesicle–vesicle fusion by the ANTS/DPX contents-mixing assay. The total lipid concentration in all assays was 0.4 μmol/mL. The data points for vesicle–vesicle fusion represent the average of six separate experiments. Data points for virus–vesicle fusion represent the average of three separate experiments.

The ability of Sendai virus to fuse with LUV of N-methylDOPE was examined with the R_{18} lipid mixing assay as described in Methods. Sendai virus - vesicle fusion occurred with a qualitatively similar increase in initial fusion rate that was seen in the case of vesicle-vesicle fusion. However, the initial rate of virus-vesicle fusion was lower than that of vesicle-vesicle fusion by about an order of magnitude at all temperatures. Representative data are shown in figure 1.

These data suggested that fusion of Sendai virus with the N-methyl DOPE LUV offered a simplified fusion pathway in that viral fusion occurs in this system in the absence of a receptor. Furthermore, viral fusion here is apparently facilitated in the presence of I_S just as is the fusion of these LUV with each other.

Table 1. Inhibition of fusion by hydrophobic peptides of N-methyl DOPE LUV, of Sendai with N-methyl DOPE LUV, and of Sendai with erythrocyte ghosts: concentration of added peptide which caused 50% inhibition.

Peptide	Vesicle-vesicle	Virus-vesicle	Virus-RBC Ghost
Z-D-Phe-L-PheGly	100 µM	200 µM	200 µM
Z-L-Phe-L-Phe	200 µM	200 µM	N.D.
Z-L-PheGly	200 µM	>200 µM	N.D.
Z-L-Phe-L-Tyr	200 µM	>200 µM	N.D.
Z-L-Phe	>200 µM	>200 µM	N.D.
Z-D-Phe	>200 µM	>200 µM	N.D.
Z-Gly-L-Phe	DNI	DNI	DNI
Z-Gly-L-Phe-L-Phe	DNI	N.D.	N.D.

">" at 200 µM inhibition was observed but 50% inhibition was not achieved;
DNI   no inhibition detected up to 200 µM;
N.D.   not determined.

Richardson, et al., (1980) demonstrated the effectiveness of small peptides at inhibiting the infectivity of enveloped viruses. In the present study the effectiveness of eight of the same peptides at inhibiting fusion of N-methyl DOPE LUV was tested. The goal was to achieve a better understanding of the inhibition process by using a well defined membrane fusion system. Experiments were carried out in which the initial rate of vesicle fusion was measured after the addition of the indicated peptide. A summary of these results is presented in Table 1.

The ability of these peptides to inhibit viral fusion with N-methyl DOPE LUV were also examined. Figure 2 shows a representative experiment. 100 µM Z-D-Phe-L-PheGly was effective at inhibiting virus-vesicle fusion whereas Z-Gly-L-Phe did not show any inhibition of fusion in a similar concentration range. The results from all eight peptides are summarized in Table 1. The relative activity towards inhibition of vesicle fusion of the peptides is the same as was observed by Richardson, et al., (Richardson, Scheid, and Choppin 1980) towards

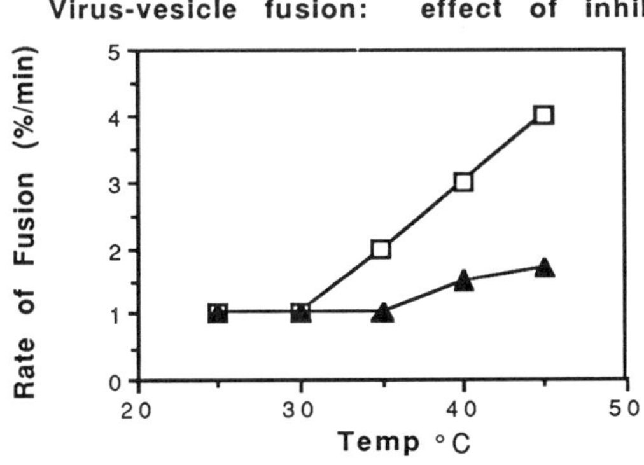

Fig. 2.   Effect of Z-D-Phe-L-PheGly on the fusion of Sendai virus with N-methyl DOPE LUV. (——□——), no inhibitor; (——▲——), 100 µM Z-D-Phe-L-PheGly.

inhibition of measles viral infectivity measured using plaque assays. Vesicle-vesicle fusion was inhibited in the presence of Sendai virus (data not shown). The initial rate of fusion of virus with N-methyl DOPE LUV increased with increasing amounts of virus (data not shown). No dequenching of the R_{18} was observed when R_{18}-labeled virus was incubated with unlabeled virus, confirming that Sendai virions did not fuse with themselves when labelled with R_{18} (data not shown). Sendai virus was not able to fuse with egg PC LUV on the same time scale as the fusion with N-methyl DOPE LUV. These data also helped to rule out artifacts due to spontaneous transfer of the R_{18} probe from one membrane to another without a fusion event. This latter result was in good agreement with previous results (Hoekstra et al. 1985).

Effect of peptides on vesicle leakage.

The same peptides were assayed for their ability to inhibit leakage of vesicle contents (data not shown). Experiments were carried out in which the initial rate of vesicle leakage (determined with the dequenching leakage assay described in Methods) was measured after the addition of the indicated peptide. The relative activity of the three peptides towards inhibition of vesicle leakage exhibited same trend was seen in the capacity of these peptides to inhibit leakage as was seen above for their ability to inhibit fusion.

Effect of peptides on fusion of Sendai virus with human erythrocyte ghosts

The question of whether or not peptides that inhibited vesicle-vesicle or virus-vesicle fusion exhibit similar activities with respect to viral fusion with a biological membrane was examined. Ghosts were preincubated at 37° C for 5 min. in the presence or absence of various concentrations of Z-D-Phe-L-PheGly. R_{18} labelled virus was then added to the assay mixture and fusion was measured as an increase in fluorescence as a function of time. The results show that significant inhibition first occurs at 100-200 µM Z-D-Phe-L-PheGly. This is consistent with results obtained by (Asano and Asano 1985) using an hemolysis assay. An identical experiment using 200 µM Z-Gly-L-Phe showed no inhibition of Sendai fusion with erythrocyte ghosts. These results are summarized in Table 1.

In the above experiments erythrocyte ghosts were exposed to the hydrophobic peptides prior to the addition of virus. In order to test the possibility that a different order of addition would result in a different level of inhibitory activity, virus was pre-bound to erythrocytes using a low temperature incubation. The ability of Z-D-Phe-L-PheGly to inhibit virus-erythrocyte fusion was abolished if R_{18} labelled virus was incubated with erythrocyte ghosts for 15 min. on ice and fusion was initiated by injecting the virus-ghost mixture into buffer containing 200 µM Z-D-Phe-L-PheGly that was pre-warmed to 37° C.

Studies on the mechanism of inhibition of fusion by these hydrophobic peptides

^{31}P NMR spectra of N-methyl DOPE LUV show the development of an isotropic resonance near 35 °C. The proportion of the phospholipid contributing to this isotropic resonance increased with an increase in temperature. Figure 3 shows typical ^{31}P NMR spectra.

The presence of structures in the phospholipid giving rise to this isotropic resonance has been correlated with the initial rate of fusion of these N-methyl DOPE LUV (Ellens et al. 1989). It was found that addition of Z-D-Phe-L-PheGly broadened this resonance in a dose dependent manner and the non-inhibitory peptide did not significantly broaden the resonance in the same concentration range (see figure 4).

The isotropic ^{31}P NMR resonance in the N-methyl DOPE LUV was likely generated by structures in the membrane with small radii of curvature. The broadening of the isotropic ^{31}P NMR resonance was consistent with an increase in that radius of curvature. Therefore, we examined the effect of the peptide Z-D-Phe-L-PheGly on the formation of structures of small radii of curvature. Z-D-Phe-L-PheGly was chosen for study as the most potent inhibitor of fusion among the

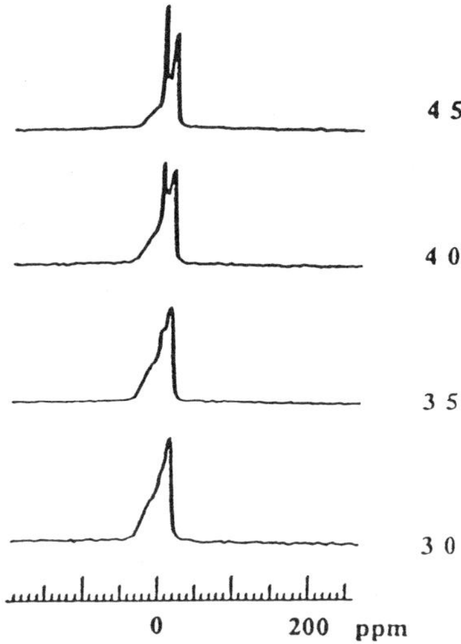

Fig. 3. ^{31}P NMR spectra of N-methyl DOPE LUV as a function of temperature.

peptides that were examined in this study. Sonicated phospholipid vesicles were chosen for examination because these structures had a radius of curvature small enough to produce an isotropic resonance of similar linewidth to that observed in the N-methyl DOPE LUV (Yeagle et al. 1975). The procedure described in Methods was employed. Egg phosphatidylcholine was chosen for sonication because PE and its derivatives did not sonicate at neutral pH due to poor headgroup hydration and

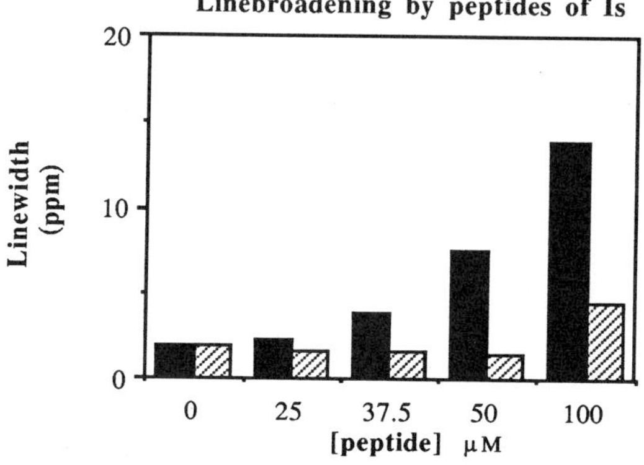

Fig. 4. Broadening of isotropic ^{31}P NMR isotropic resonance of N-methyl DOPE by Z-D-Phe-L-PheGly (solid bars) and Z-Gly-Phe (hatched bars) as a function of concentration of peptide.

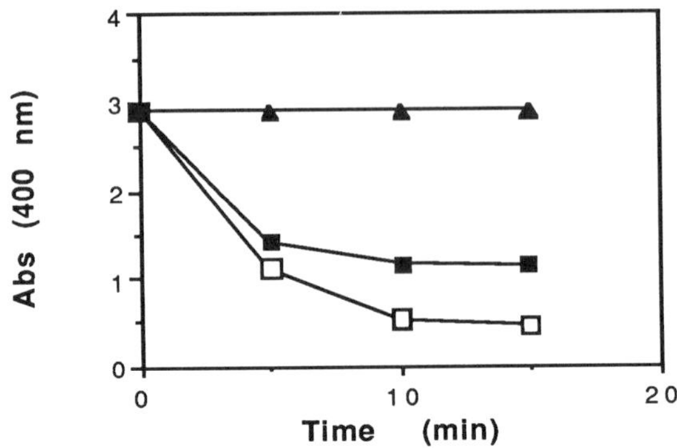

Fig. 5.   Effects of inhibitory and non-inhibitory peptides on
the ability of egg phosphatidylcholine to form small
unilamellar vesicles by sonication.   (——□——), pure
egg phosphatidylcholine; (——■——) in the presence
of Z-Gly-Phe; and (——▲——) in the presence of 100
μM Z-D-Phe-L-Phe-Gly.

because the physical properties of egg phosphatidylcholine sonicated vesicles were
very well characterized (Huang 1969).   The normal product from such sonication (of
pure egg phosphatidylcholine) was formation of a suspension that was nearly clear
from a suspension that was at first highly turbid (due to the large multilamellar
liposomes).   Figure 5 shows quantitatively the reduction in light scattering
observed when small radii of curvature were achieved through sonication.
As also shown in figure 5, in the presence of the Z-D-Phe-L-PheGly at a 4:1 mole
ratio with the phospholipid (i.e., at a ratio that caused substantial inhibition of
fusion and caused a very strong broadening of the isotropic ^{31}P NMR resonance),
the light scattering of the suspension did not significantly change upon sonication
for a period much longer than that needed to completely clarify a suspension of the
same lipid without the peptide.   Thus it was not possible to form small vesicles by
sonication when Z-D-Phe-L-PheGly was incorporated into the membrane.

DISCUSSION

These experiments identified a common feature among the following
membrane fusion systems: N-methyl DOPE LUV fusion, Sendai virus fusion with N-
methyl DOPE LUV, Sendai virus fusion with erythrocyte ghosts, and Sendai virus and
measles virus infection of cells.   The common feature was the ability of a specific set
of hydrophobic, amino-terminal modified small peptides to inhibit these events.
The same relative potency at inhibition of each of these events was exhibited by the
eight peptides studied.   Thus the same structural specificity for the inhibition was
operative for each of the fusion and infection events that were inhibited.

The commonality among these events suggested first that the inhibition by the
peptides was due to inhibition of membrane fusion.   In three of the systems,
membrane fusion is the event being measured.   In the studies of viral infection,
membrane fusion is an essential element of the infection process by enveloped
viruses.   Therefore it is reasonable to suggest that in viral infection, the step that is
inhibited by the hydrophobic peptides is the membrane fusion step of that
infection pathway.

This conclusion raises the possibility that inhibition of membrane fusion may
provide an effective target for anti-viral research.   A serious issue in such a line of

172

research is the specificity of the inhibition. Membrane fusion occurs in many essential intracellular processes. Some of these fusions are calcium stimulated, such as secretion. Others, such as fusions involving transport of membraneous material between endoplasmic reticulum and Golgi, are regulated by different factors. In preliminary studies on the fusion between rod outer segment disk membranes and rod outer segment plasma membranes (a calcium-dependent intracellular fusion event), these inhibitory peptides proved ineffective at inhibiting the fusion event (Boesze-Battaglia & Yeagle, unpublished observations). This result is encouraging for the possible use of this line of research for the development of anti-viral agents.

The question that is raised by these data and the conclusions described above is whether there is any step in common between the pathway of fusion of virus with cells and the pathway for fusion of LUV with LUV or Sendai with LUV. This question cannot be adequately answered by these data. However, there is a coincidence among the results that encourages the suggestion of a hypothesis for future testing.

As summarized in the introduction, a general pathway for fusion can be divided into four steps. Step three of this codification calls for a destabilization of the bilayer to facilitate the mixing of the membraneous structures allowing an intermediate to be formed that can evolve into a fused product. Such a destabilization might be expected to take the form of some non-lamellar structure formed from phospholipids. A suitable candidate has been described for the fusion of N-methyl DOPE LUV (Ellens et al. 1989). This was characterized by the appearance of an isotropic resonance in the ^{31}P NMR spectra of the LUV, whose intensity was directly proportional to the initial rate of fusion in this system. Under the same conditions, freeze fracture electron microscopy identified the presence of lipidic particles in these LUV (Ellens et al. 1989). Theoretical analysis has suggested possible structures for fusion intermediates in this system (Siegel 1987) and cryo-electron microscopy has provided evidence in support of that theoretical analysis (Siegel et al. 1989b).

In this context, the observation of fusion between the N-methyl DOPE LUV and Sendai virus is of interest. Normally membrane fusion between Sendai and membranes made only from zwitterionic and/or neutral lipids (i.e., without receptors for HN of Sendai) occurred slowly if at all. Data in this report showed fusion of Sendai with N-methyl DOPE vesicles, in the absence of any receptors for HN in the target membrane. Therefore, some property or properties of these N-methyl DOPE vesicles made them highly fusogenic towards Sendai.

The fusion activity of Sendai with the N-methyl DOPE vesicles increases significantly in approximately the same temperature range that pure vesicle-vesicle fusion has been observed to dramatically increase. This is the temperature range in which non-lamellar structures begin to appear in the bilayer, as described above. It would appear that the same non-lamellar structures that may facilitate the fusion of N-methyl DOPE LUV may also facilitate the fusion of Sendai virus with N-methyl DOPE LUV.

It was observed in this report that the peptide inhibitors target the non-lamellar structures in their interaction with the lipid bilayer of the N-methyl DOPE LUV. If that represents their mode of action (see below) it is then interesting to contemplate the ability of the peptide inhibitors to inhibit N-methyl DOPE LUV fusion, Sendai virus fusion with N-methyl DOPE LUV, Sendai virus fusion with erythrocyte ghosts, and Sendai virus and measles virus infection of cells. Do all these fusion events involve as a part of their pathway the development of some non-lamellar structure in the membrane as an intermediate in the pathway of fusion that is sensitive to the peptide inhibitors?

It has been suggested that the "fusogenic sequence" of the viral fusion proteins must penetrate the target membrane for the facilitation of the fusion event (Asano and Asano 1984, Daniels et al. 1985, Doms, Helenius, and White 1985,

Novick and Hoekstra 1988, Ruigrok et al. 1988, Skehel et al. 1982). We have recently discovered that a peptide with the sequence corresponding to the "fusion peptide" of measles fusion protein can destabilize membrane structures. In particular, the presence of this peptide in the membrane facilitated the formation of structures that gave rise to the isotropic ^{31}P NMR resonance in the spectra from these membranes and facilitated viral fusion with N-methyl DOPE LUV (Yeagle et al. 1991).

One question that can be addressed directly with the data in this report is the question of the target of the inhibitor peptides. The most potent of the inhibitory peptides, Z-D-Phe-L-PheGly, had a strong affect on the structure of the putative fusion intermediates in N-methyl DOPE LUV fusion, as indicated by the broadening of the isotropic ^{31}P NMR resonance. Therefore the target of the inhibitory peptide may be a putative fusion intermediate in the lipid of the involved membranes. In addition it was reported previously that Z-D-Phe-L-PheGly increased the transition temperature of phosphatidylethanolamine from the lamellar to the hexagonal II phase (Epand, 1986). We have focussed here on the isotropic structures since they appear in the temperature range where fusion is also observed, whereas in this N-methyl DOPE system, the transition temperature to the hexagonal II phase occurs at a temperature much higher than the temperatures at which membrane fusion can be measured.

The most straightforward interpretation of the increase in the ^{31}P NMR linewidth was an increase in the effective radius of curvature of the structure giving rise to that resonance. For example, ^{31}P linewidths of LUV increase with increasing size (Burnell, Cullis, and deKruijff 1980). Therefore, we propose as the most simple, but not the only, interpretation that the inhibitory peptides increase the radius of curvature of a putative intermediate in the membrane fusion process.

To test this hypothesis, we examined the effect of these peptide inhibitors on the formation of small vesicles by sonication. We used phosphatidylcholine in this experiment, since N-methyl DOPE would not sonicate into small vesicles at neutral pH (Yeagle, et al., data not shown). The vesicles that formed from the egg phosphatidylcholine had a small enough radius of curvature (Huang, 1969) to exhibit a ^{31}P NMR resonance that is nearly identical in linewidth to the isotropic ^{31}P NMR resonance in the N-methyl DOPE LUV (Hutton, Yeagle, and Martin 1977). It should be pointed out that the theory of the NMR motional averaging does not require that the structures in the LUV that gave rise to the isotropic ^{31}P NMR resonances need be of the size of the sonicated vesicles or vesicular at all; only that they formed structures that covered about one quadrant of a sphere and had a similar radius of curvature to the sonicated vesicles for that portion of a sphere. The data showed that the inhibitory peptides inhibit the formation of structures of phospholipids of small radii of curvature.

The last question that could be addressed in this discussion was what role small radii of curvature might play in membrane fusion. It was argued elsewhere that small radii of curvature were required to facilitate the fusion of two membranes (Siegel 1986, Siegel 1987, Siegel et al. 1989). Models of the fusion promoted by fusion proteins of enveloped viruses have suggested a role for a similar intermediate (Daniels et al. 1985, Doms, Helenius, and White 1985, Skehel et al. 1982). Therefore the ability of Z-D-Phe-L-PheGly to inhibit the formation of structures with small radii of curvature would be expected, on the basis of the present state of knowledge of membrane fusion, to inhibit the fusion event. How these fusion inhibitory peptides may inhibit the formation of structures in the membrane with small radii of curvature is under further investigation in this laboratory.

Acknowledgment: We thank E. Johnson for the preparation of this virus and acknowledge the use of the Anatomical Sciences Electron Microscope Facility. This work was supported by National Institutes of Health Grant AI26800.

REFERENCES

Asano, A., and K. Asano. 1984. Molecular mechanism of virus entry to target cells. Tumor Res. 19 : 1-20.

Asano, K., and A. Asano. 1985. Why is a specific amino acid sequence of F glycoproteins required for the membrane fusion reaction between envelope of HVJ (Sendai virus) and target cell membranes? Biochem. Int. 10 : 115-122.

Bentz, J., and H. Ellens. 1988. Membrane fusion: Kinetics and mechanisms. Colloids and Surfaces 30 : 65-112.

Brasseur, R., M. Vandenbranden, B. Cornet, A. Burny, and J.-M. Ruysscshaert. 1990. Orientation into the lipid bilayer of an asymmetric amphipathic helical peptide located at the N-terminue of viral fusion proteins. Biochim. Biophys. Acta 1029 : 267-273.

Cai, W., B. Gu, and S. Person. 1988. Role of glycoprotein B of herpes eimplex virus type 1 in viral entry and cell fusion. J. Virol. 62 : 2596-2604.

Daniels, R. S., J. C. Downie, A. J. Hay, M. Knossow, J. J. Skehel, M. L. Wang, and D. C. Wiley. 1985. Fusion mutants of the influenza virus hemagglutinin glycoprotein. Cell 40 : 431-439.

Doms, R. W., A. Helenius, and J. White. 1985. Membrane fusion activity of the influenza virus hemagglutinin. The low pH-induced conformational change. J. Biol. Chem. 260 : 2973-2981.

Elango, N., M. Satake, J. E. Coligan, E. Norrby, E. Camargo, and S. Venkatesan. 1985. Respiratory syncytial virus fusion glycoprotein: nucleotide sequence of mRNA, identification of cleavage activation site and amino acid sequence of N-terminus of F1 subunit. Nucleic Acids Res. 13 : 1559-1574.

Ellens, H., J. Bentz, and F. C. Szoka. 1986. Fusion of phosphatidylethanolamine-containing liposomes and mechanism of the L-alpha to H_{II} phase transition. Biochemistry 25 : 4141-4147.

Ellens, H., D. P. Siegel, D. Alford, P. L. Yeagle, L. Boni, L.J. Lis, P. J. Quinn, and J. Bentz. 1989. Membrane fusion and inverted phases. Biochemistry 28 : 3692-3703.

Gagne, J., L. Stamatatos, T. Diacovo, S. W. Hui, P. L. Yeagle, and J. Silvius. 1985. Physical properties and surface interactions of bilayer membranes containing N-methylated phosphatidylethanolamines. Biochemistry 24 : 4400-4408.

Haywood, A. M. 1988. 'Entry' of enveloped viruses into liposomes. In Molecular Mechanisms of Membrane Fusion. Edited by S. Ohki, D. Doyle, T. Flanagan, S. W. Hui and E. Meyhew. 427-440. NY: Plenum Publishing Corp.

Hoekstra, D. 1982a. Fluorescence method for measuring the kinetics of Ca^{2+}-induced phase separations in phosphatidylserine-containing lipid vesicles. Biochemistry 21 : 1055-1061.

Hoekstra, D. 1982b. Role of lipid phase separations and membrane hydration in phospholipid vesicle fusion. Biochemistry 21 : 2833-2840.

Hoekstra, D., K. Klappe, T. de Boer, and J. Wilschut. 1985. Characterization of the fusogenic properties of Sendai virus: Kinetics of fusion with erythrocyte membranes. Biochemistry 24 : 4739-4745.

Hsu, M.C., A. Scheid, and P.W. Choppin. 1979. Reconstitution of membranes with individual paramyxovirus glycoproteins and phospholipid in cholate solution. Virology 95 : 476-491.

Huang, C. 1969. Studies on phosphatidylcholine vesicles. Formation and physical characteristics. Biochemistry 8 : 344-352.

Leventis, R., J. Gagne, N. Fuller, R. P. Rand, and J. R. Silvius. 1986. Divalent cation induced fusion and lipid lateral segregation in phosphatidylcholine-phosphatidic acid vesicles. Biochemistry 25 : 6978-6987.

Manservigi, R., P. Spear, and A. Buchan. 1977. Cell fusion induced by herpes simplex virus is promoted and suppressed by different glycoproteins. Proc. Natl. Acad. Sci. USA 74 : 3913-3917.

Nakanishi, M., T. Uchida, and Y. Okada. 1982. Glycoproteins of Sendai virus (HVJ) have a critical ratio for fusion between virus envelopes and cell membranes. Exp. Cell Res. 142 : 95-101.

Nicolaides, E., H. de Wald, R. Westland, M. Lipnik, and J. Posler. 1968. Potential antiviral agents. Carbobenzoxy di- and tripeptides active against measles and herpes viruses. J. Med. Chem. 11 : 74-79.

Novick, S.L., and D. Hoekstra. 1988. Membrane penetration of Sendai virus glycoproteins during the early stages of fusion with liposomes as determined by hydrophobic photoaffinity labeling. Proc. Natl. Acad. Sci. USA 85 : 7433-7437.

Ohnishi, S., and M. Murata. 1988. Molecular mechanism of protein-mediated low pH-induced membrane fusions. In Molecular Mechanisms of Membrane Fusion. Edited by S. Ohki, D. Doyle, T. Flanagan, S. W. Hui and E. Meyhew. 357-366. NY: Plenum Publishing Corp.

Papahadjopoulos, D., W. J. Vail, C. Newton, S. Nir, K. Jacobson, G. Poste, and R. Lazo. 1977. Studies on membrane fusion. III. The role of calcium-induced phase changes. Biochim. Biophys. Acta 465 : 579-598.

Richardson, C.D., A. Berkovich, S. Rozenblatt, and W.J. Bellini. 1985. Use of antibodies directed against synthetic peptides for identifying cDNA clones, establishing reading frames, and deducing the gene order of measle virus. J. Virology 154 : 186-193.

Richardson, C.D., A. Scheid, and P.W. Choppin. 1980. Specific inhibition of paramyxovirus and myxovirus replication by oligopeptides with amino acid sequences similar to those at the N-termini of the F-1 or HA-2 viral polypeptides. Virology 105 : 205-222.

Roizman, B., and A. E. Sears. 1991. Herpes Simplex Viruses and Their Replication. In Fundamental Virology. Edited by B. N. Fields and D. M. Knipe. 849-896. New York: Raven Press.

Ruigrok, R.W.H., A. Aitken, L.J. Calder, S.R. Martin, J.J. Skehel, S.A. Wharton, W. Weis, and D.C. Wiley. 1988. Studies on the structure of the influenza virus Haemagglutinin at the pH of membrane fusion. J. Gen. Vir. 69 : 2785-2795.

Sarmiento, M., M. Haffey, and P. Spear. 1979. Membrane proteins specified by herpes simplex viruses. II. Role of glycoprotein VP7 (B2) in virion infectivity. J. Virol. 29 : 1149-1158.

Scheid, A., and P.W. Choppin. 1974. Identification of biological activities of paramyxovirus glycoproteins. Activation of cell fusion, hemolysis, and infectivity by proteolytic cleavage of an inactive precurser protein of Sendai virus. Virology 57 : 475-490.

Siegel, D.P. 1987. Inverted micellar intermediates and the transitions between lamellar, cubic, and inverted hexagonal phases. 3. Isotropic and inverted cubic state formation via intermediates in transitions between L_α and H_{II} phases. Chem. Phys. Lipids 42 : 279-301.

Siegel, D. P., J. Banschbach, D. Alford, H. Ellens, L. Lis, P. J. Quinn, P. L. Yeagle, and J. Bentz. 1989a. Physiological levels of diacylglycerols in phospholipid membranes induce membrane fusion and stabilize inverted phases. Biochemistry 28 : 3703-3709.

Siegel, D. P., J. L. Burns, M. H. Chestnut, and Y. Talmon. 1989b. Intermediates in membrane fusion and bilayer/non-bilayer phase transitions imaged by time-resolved cryo-transmission electron microscopy. Biophys. J. 56 : 161-169.

Silvius, J. R., and J. Gagne. 1984. Lipid phase behavior and calcium-induced fusion of phosphatidylethanolamine-phosphatidylserine vesicles. Calorimetric and fusion studies. Biochemistry 23 : 3232-3240.

Skehel, J.J., P.M. Bayley, E. B. Brown, S. R. Martin, M. D. Waterfield, J. White, I. A. Wilson, and D. C. Wiley. 1982. Changes in the conformation of influenza virus hemagglutinin at the pH optimum of virus-mediated membrane fusion. Proc. Natl. Acad. Sci. USA 79 : 968-972.

Stegmann, T., D. Hoekstra, G. Scherphof, and J. Wilschut. 1985. Kinetics of pH-dependent fusion between influenza virus and liposomes. Biochemistry 24 : 3107-3133.

White, J., L. Kartenbeck, and A. Helenius. 1982. Membrane fusion activity of influenza virus. <u>EMBO J.</u> 1 : 217-222.

Yeagle, P.L., 1987, "The Membranes of Cells", Academic Press, San Diego.

Yeagle, P. L., R. M. Epand, C. D. Richardson, and T. D. Flanagan. 1991. Effects of the "fusion peptide" from Measles Virus on the Structure of N-methyl Dioleoylphosphatidylethanolalmine Membranes and their Fusion with Sendai Virus. <u>Biochim. Biophys. Acta</u> : in press.

Yeagle, P. L., W. C. Hutton, C. Huang, and R. B. Martin, 1975, Headgroup conformation and lipid-cholesterol association in phosphatidylcholine vesicles: a ^{31}P $\{^{1}H\}$ nuclear Overhauser effect study, <u>Proc. Nat. Acad. Sci USA</u> ,72 : 3477-3481.

FUSION OF INFLUENZA, SENDAI AND SIMIAN IMMUNODEFICIENCY VIRUSES WITH CELL MEMBRANES AND LIPOSOMES

Nejat Düzgüneş[1,2,3*], Maria C. Pedroso de Lima[1,5],
Charles E. Larsen[2‡], Leonidas Stamatatos[1,2], Diana
Flasher[1,2], Dennis R. Alford[2‡], Daniel S. Friend[4]
and Shlomo Nir[6]

[1]Department of Microbiology, University of the
Pacific, School of Dentistry, San Francisco,
California 94115; [2]Cancer Research Institute and
Departments of [3]Pharmaceutical Chemistry and
[4]Pathology, University of California, San
Francisco, California 94143; [5]Center for Cell
Biology, and the Department of Chemistry,
University of Coimbra, 3049 Coimbra, Portugal;
[6]Department of Soil and Water Sciences, The Hebrew
University of Jerusalem, Rehovot 76100, Israel

INTRODUCTION

The cell entry routes of various lipid enveloped viruses
have been studied extensively. Influenza, Semliki Forest and
vesicular stomatitis viruses infect their host cells by fusing
with the endosome membrane after endocytosis of the virion and
acidification of the endosome lumen (Matlin et al., 1981;
Yoshimura et al., 1982; White et al., 1983; Ohnishi, 1988;
Marsh and Helenius, 1989). Sendai virus fuses with the plasma
membrane at neutral pH (Asano and Asano, 1984; White et al.,
1983; Okada, 1988; Hoekstra & Kok, 1989), and human immunodeficiency virus is also believed to fuse in a similar manner
(Marsh and Helenius, 1989; Sattentau, 1990; Stein and Engleman,
1991).

Fluorescence Assays for Fusion

The kinetics and extent of fusion of these viruses with
liposomes, erythrocyte ghosts and cultured cells can be
monitored with fluorescence assays (Düzgüneş and Bentz, 1988;

---

* Address for correspondence: Department of Microbiology,
University of the Pacific, School of Dentistry, 2155 Webster
Street, San Francisco, CA 94115.
‡ Present address: Department of Bioscience and Biotechnology,
Drexel University, Philadelphia, PA 19104

Cell and Model Membrane Interactions
Edited by S. Ohki, Plenum Press, New York, 1991

Hoekstra, 1991; Loyter et al., 1988). Studies utilizing these
assays have provided insights into the mechanisms of membrane
fusion and the role of target membrane components in adhesion
and fusion. The most convenient and widely used fluorescence
assay monitoring virus-target membrane fusion utilizes the
amphiphilic fluorophore octadecyl rhodamine (R-18), which
partitions into the virus membrane when added from an ethanolic
solution (Hoekstra et al., 1984, 1985; Hoekstra and Klappe,
1991). The overlap in the excitation and emission spectra of
R-18 results in self-quenching when the probe is inserted into
a membrane at a concentration above a few mole percent of the
lipid. When the probe is incorporated into liposomes composed
of dioleoylphosphatidylcholine at different mole ratios, the
percent quenching is a linear function of the mol % of the
fluorophore (Hoekstra et al., 1984). To monitor virus fusion
with a target membrane, R-18 is initially incorporated in the
viral membrane at 3-5 mol % of the total membrane lipid. When
the viral membrane fuses with the target membrane, the surface
concentration of the probe is diluted due to rapid lateral
diffusion, and the fluorescence intensity increases due to the
relief of self-quenching.

Viral Fusion Proteins

Viral envelope proteins mediate both the binding and
fusion steps of the cell entry process (Düzgüneş, 1985; White,
1990; Ohnishi, 1988; Hoekstra and Kok, 1989; Stegmann et al.,
1989a). The molecular basis for the fusion step is far less
understood than for binding. The interaction of a hydrophobic
segment of the fusion protein with the target membrane bilayer
is thought to destabilize the target membrane and increase the
hydrophobicity of the membrane surface, thereby changing the
water structure of the region of adhesion between the viral and
target membranes and altering the energy of interaction between
the two membranes (Ohki, 1988; Stegmann et al., 1989a; Düzgüneş
and Shavnin, 1991). We have proposed that the region of the
viral envelope bilayer in contact with the modified target
membrane cannot maintain its integrity when confronted with a
highly hydrophobic surface with an altered water structure; its
bilayer configuration is forced into non-bilayer fluctuations
that expose the hydrophobic membrane interior. If the two
hydrophobic regions on apposed membranes coincide they coalesce
to form a single membrane (Düzgüneş and Shavnin, 1991).

Influenza Virus

Using erythrocyte ghosts as a biological target membrane
for influenza virus labeled with R-18, Stegmann et al. (1986)
have shown that the pH-dependence of fusion is similar to that
of the cell-cell fusion activity of the virus (White et al.,
1981). These studies have also found correlations with the
"inactivation" of the hemolytic activity of influenza virus
following incubation at low pH in the absence of target
membranes (Sato et al., 1983; Junankar and Cherry, 1986;
Stegmann et al., 1986). When the target membranes used for
determining the fusion activity of the virus were cardiolipin
liposomes, however, significant inactivation was not observed
(Stegmann et al., 1986).

Sendai Virus

The rate of fusion of Sendai virus with erythrocyte
ghosts and mouse spleen cells is optimal at neutral pH
(Hoekstra et al., 1985; Hoekstra and Kok, 1989). Cell-cell
fusion induced by Sendai virus also occurs at lower or higher
pH values, which has led to the interpretation that the fusion
of Sendai virus with cells is pH independent (White et al.,
1983). The rates and final extents of fusion of Sendai virus
with liposomes of several compositions, such as phosphatidylserine, phosphatidylglycerol, cardiolipin and cardiolipin/
dioleoylphosphatidylcholine (DOPC) are enhanced at mildly
acidic pH, i.e., 4-5 (Amselem et al., 1986; Klappe et al.,
1986; Nir et al., 1986b). With liposomes composed of DOPC/
dioleoylphosphatidylethanolamine/cholesterol/disialoganglioside
(G_{D1a}), the rate of fusion has been shown to be considerably
slower than in the case of the compositions mentioned above, to
have a local optimum at pH 7, and to increase greatly below pH
5 (Klappe et al.,1986). Pre-incubation of Sendai virus at low
pH has been shown to lead to some inactivation of the fusion
capacity of the virus towards liposomes and erythrocyte ghosts
(Hsu et al., 1982; Hoekstra et al., 1985; Amselem et al.,
1986). In contrast, exposure of the virus to basic pH enhances
its hemolytic and cell-cell fusion activity (Hsu et al., 1982).

Simian Immunodeficiency Virus

Simian immunodeficiency virus (SIV) causes simian acquired
immunodeficiency syndrome (AIDS; Letvin et al., 1985;
Desrosiers, 1988). Infection of rhesus macaques by SIV
represents the best animal model for human AIDS (Desrosiers and
Letvin, 1987). The molecular and cellular mechanisms of entry
of SIV, as well as of the human immunodeficiency virus type 1
(HIV-1), into their host cells are not well understood.
Although some controversy remains (Maddon et al., 1986; Pauza
and Price, 1988; Pauza, 1991), it is generally believed that
HIV fuses with the plasma membrane and does not require
endocytosis or acidification of endosomes (Maddon et al., 1988;
Stein et al., 1987; McClure et al., 1988; Marsh and Dalgleish,
1987). The CD4 molecule, found on circulating T helper cells,
monocytes and macrophages, is recognized as the primary
cellular receptor for HIV and SIV_{mac} (SIV originally isolated
from macaques) (Dalgleish et al., 1984; Klatzmann et al., 1984;
McDougal et al., 1985; Kannagi et al., 1985), but several cell
types that do not express this protein can be infected by these
viruses (Cheng-Mayer et al., 1987; Clapham et al., 1989; Tateno
et al., 1989; Harouse et al., 1989; Weber et al., 1989; Konopka
et al., 1991a). The expression of CD4 on the cell surface does
not appear to be sufficient for HIV or SIV_{mac} infection (Maddon
et al., 1986; Koenig et al., 1989). The soluble ectodomain of
the CD4 molecule (rCD4), at concentrations in the range of 1
µg/ml, can inhibit the infectivity of laboratory strains of HIV
type 1 (HIV-1) (Fisher et al., 1987; Hussey et al., 1988; Byrn
et al., 1989), but much higher concentrations are necessary to
block primary isolates of the virus (Daar et al., 1990).

The binding and fusion of immunodeficiency viruses are
mediated by two envelope-associated glycoproteins. The
ectoplasmic gp120 is non-covalently attached to the transmembrane gp41 in HIV-1 (Lifson et al., 1986; Sodroski et al.,

1986; McDougal et al., 1986) or gp32/gp40 in HIV-2 and SIV
(Chakrabarti et al., 1987). Gp120 is thought to mediate
binding to CD4, and gp41 (gp32/40) to facilitate membrane
fusion (Kowalski et al., 1987). The N-terminal amino acid
sequence of gp41 (gp32/40) bears some resemblance to that of
other viral fusion proteins (Marsh & Dalgleish, 1987; Gallaher,
1987; Gonzalez-Scarano et al., 1987; Bosch et al., 1989). This
segment of the protein is thought to interact with the target
membrane bilayer and possibly bridge the viral and cellular
lipid bilayers. It is possible, however, that certain regions
on gp120 are also involved in fusion (Freed et al., 1991).

FUSION OF INFLUENZA VIRUS WITH CULTURED CELLS

Although the kinetics of fusion of influenza virus with
erythrocyte ghosts and liposomes has been examined extensively,
few studies have investigated the kinetics of fusion with
cellular membranes. We have used several cell lines grown in
suspension culture as targets for influenza virus fusion, and
investigated the kinetics of low pH-induced fusion of the virus
with the plasma membrane (Düzgüneş et al., 1991a). Suspension
cells can be grown in large numbers to provide ample target
membranes without limitations of binding sites for the virus,
particularly since microgram quantities of virus must be used
to provide a sufficient fluorescence signal for reliable
analysis. Since they can be transferred readily, and without
harsh manipulations, into fluorometer cuvettes, suspension
cells are amenable to continuous fluorometric analysis. We
have used the human promyelocytic leukemia cell line HL-60, the
human T lymphoblastic leukemia cell line CEM, and the murine
lymphoma cell line S49. These cells do not internalize the
virus to a significant extent during the relatively short times
of pre-incubation before the reduction of the pH of the medium.

Figure 1 shows the time course of fusion of the A/PR/8/34
strain of influenza virus with HL-60 cells when the pH was

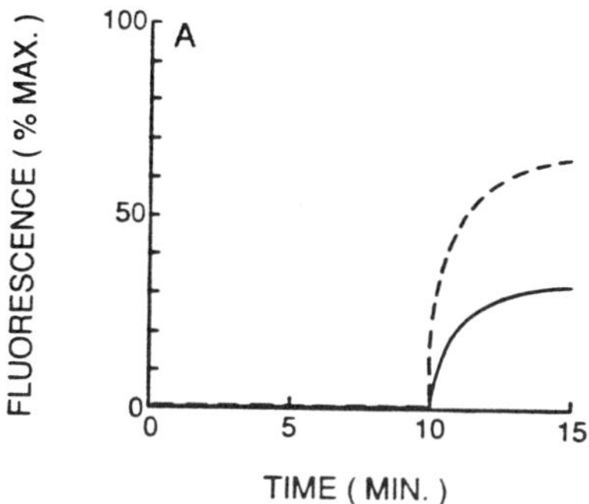

Figure 1.   The time-course of fusion of influenza virus with
            2×10^7/ml HL-60 cells (solid line: 5 µg/ml virus;
            dashed line: 2 µg/ml virus). The pH was lowered
            from 7.6 to 5.0 at t = 10 min. The fluorescence
            scale was set to 100% by lysing the virus and cell
            suspension with detergent. The initial fluorescence
            of the suspension was taken as 0%. Data reproduced
            from Düzgüneş et al., 1991a, with permission.

lowered to 5.0, following a 10 min incubation at neutral pH.
No fusion took place at neutral pH, or from within endosomes.
Electron microscopy of the cell and virus suspension at this
time point indicated numerous viral particles adhering to the
plasma membrane, and no significant endocytosis of virus
(Düzgüneş et al., 1991a,b). The figure indicates that the
virus/cell ratio affects the initial rate and extent of fusion,
with a lower ratio resulting in a higher initial rate and
extent. This suggests that a higher percentage of the virus
has associated and fused with the cells, and that the number of
favorable fusion sites with the cell may be limited. Although
a higher extent of fusion was obtained with a different
preparation of the virus, when the concentration of cells was
reduced while keeping the virus concentration constant, the
initial rate and and the extent of fusion were again reduced
(Düzgüneş et al., 1991a). Electron microscopy of influenza
virus and HL-60 cells fixed 0.5 min after lowering the pH,
showed fusion of virions with the plasma membrane, as well as
the amorphous morphology of the viral spike glycoproteins
(Figure 2).

When R-18-labeled virus was incubated with cells in the
presence of a 20- or 50-fold excess of unlabeled virus, the
extent of fusion at the end of 4 min was reduced to 25% or 18%
of the control, respectively (Figure 3). In the case of the
50-fold excess unlabeled virus, binding of the labeled virus
was reduced to 43% of the control (where only labeled virus was
present) (Düzgüneş et al., 1991a,b). This observation suggests
that not only are the number of binding sites for the virus
limited under these conditions, but also the number of fusion
sites, since fusion was reduced even more (to 18% of the
control).

Fusion and Aggregation Rate Constants: The Effect of pH

The separation of the overall fusion process into the
adhesion and the actual fusion stages, expressed as the rate
constants of adhesion and fusion, can produce a quantitative
description of the mechanism of membrane fusion (Nir et al.,

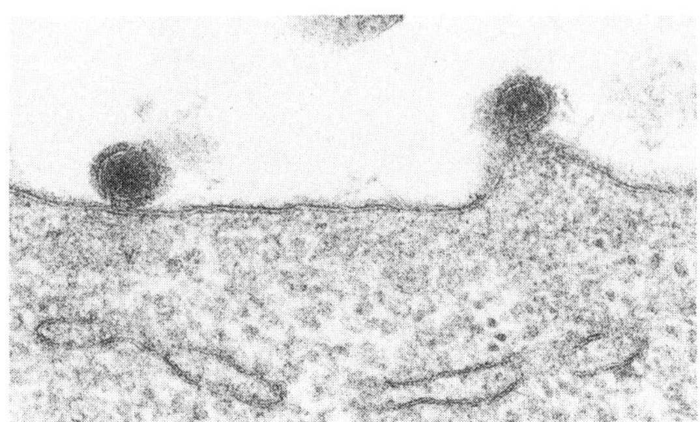

Figure 2.   Electron micrograph showing the fusion of influenza
            virions with an HL-60 cell following the reduction
            of the pH from 7.6 to 5.0. Magnification: 110,000x.
            Data reproduced from Düzgüneş et al., 1991a, with
            permission.

1983, 1986a-c, 1990). When the final extents of fusion as well
as the rate constants of adhesion and fusion in virus-cell
systems are compared with those of virus-liposome systems, the
role of various membrane components in the fusion process can
be elucidated. The adhesion rate constant for the influenza
virus-HL-60 cell system was found to be greater than 3×10^{10}
$M^{-1}s^{-1}$, which is higher than the upper limit on diffusion-
controlled processes (Düzgüneş et al., 1991a, Smoluchowski,
1917). This observation can be explained by the unequal sizes
of the adhering particles (viruses and cells), whose adhesion
rate constants can be 25-fold higher than diffusion-controlled
processes (Berg and von Hippel, 1985).

Optimal fusion was observed around pH 5, with a rather
sharp dependence on pH. The fusion rate constant increased
from 7×10^{-4} s^{-1} at pH 5.8 to 18×10^{-3} s^{-1} at pH 5.0, and
decreased to 2.5×10^{-3} s^{-1} at pH 4.5 (Düzgüneş et al., 1991a,b).
Fusion was also highly dependent on the temperature (Figure 4).
The rotational mobility of the hemagglutinin within the
membrane is reduced at lower temperature (Junankar & Cherry,
1986), and may be one of reasons for the temperature dependence
of fusion. In contrast, the pH-dependent binding of the
ectodomain of the hemagglutinin to liposomes is temperature
independent (Doms et al., 1985). Therefore, the low pH
conformation of hemagglutinin, by itself, does not appear to be
sufficient to mediate rapid membrane fusion. We have proposed
that the process of conformational change of hemagglutinin
while the virus membrane is in close proximity to the target
membrane, and not merely the low-pH conformation, may be
essential for membrane fusion (Düzgüneş and Gambale, 1988;
Düzgüneş and Shavnin, 1991; Düzgüneş et al., 1991a).

Analysis of the final extents of fluorescence increase
demonstrated that, at pH 5, all the virus particles were
capable of fusing with the plasma membranes of suspension
cells, as well as erythrocyte ghost membranes (Düzgüneş et al.,
1991a). In contrast, only partial fusion activity (20-45%) was
observed towards liposomes of several compositions, such as
phosphatidylcholine/phosphatidylethanolamine with or without
G_{D1a} (Stegmann et al., 1989b).

Pre-treatment of Influenza Virus at Low pH

Previous reports have indicated that the hemolytic
activity of the PR/8 strain of influenza virus and of its
hemagglutinin is abolished when the the virus alone, or the
hemagglutinin, is pre-exposed to pH below 5.4 (Sato et al.,
1983). The fusion of the virus with HL-60 or CEM cells,
however, was inhibited by only 21-41% (measured as the extent
of fusion within 5 min), although fusion with erythrocyte
ghosts was inhibited by about 95% (Düzgüneş et al., 1991a).
Radiation inactivation studies have suggested that while the
fusion activity of the PR/8 strain of influenza virus is
mediated by a single monomer of hemagglutinin, the hemolysis
and leakage activities require a larger and more complex
functional unit (Gibson et al., 1986). Although this may
explain the differential effect of low pH pre-treatment on
hemolysis or fusion with cultured cells, why fusion with
erythrocyte ghosts is inhibited under similar conditions is not

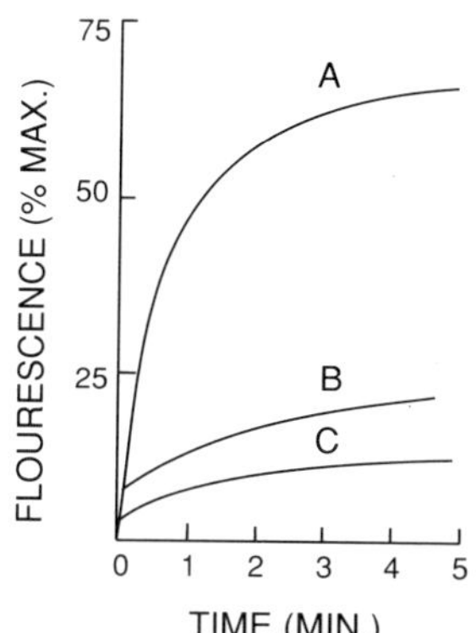

Figure 3.   The fusion of R-18-labeled influenza virus with HL-
60 cells in the presence of excess unlabeled
virions.   The virus (2 μg protein/ml) was incubated
with $2x10^7$/ml HL-60 cells for 5 min at pH 7.4, and
the pH was lowered to 5 at time = 0 min.   A: Labeled
virus only.   B: Labeled virus in the presence of a
20-fold excess of unlabeled virus.   C: Labeled virus
in the presence of a 50-fold excess of unlabeled
virus.   The fluorescence scale was set as in Figure
1.   Data from Düzgüneş et al., 1991a.

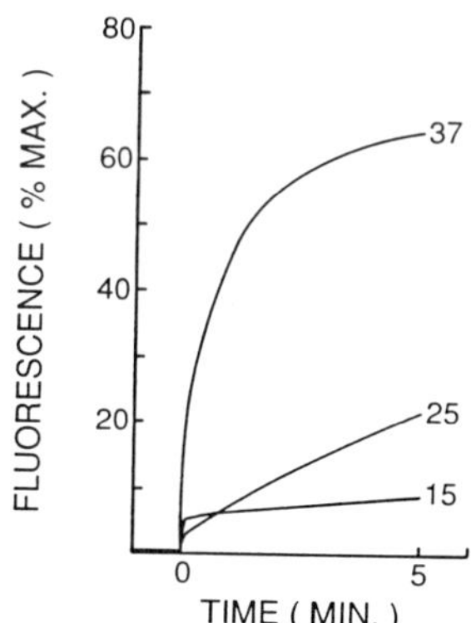

Figure 4.   The effect of temperature on the fusion of
influenza virus with HL-60 cells.   After a pre-
incubation of 2 μg/ml of the virus (A/PR/8/34
strain) with $2x10^7$/ml cells at the indicated
temperatures for 10 min, the pH was lowered to 5.
The fluorescence scale was set as in Figure 1.   Data
from Düzgünes et al., 1991a.

clear. It is possible that adhesion of the virus to the
surface of a cultured cell brings about the proper, i.e.
fusogenic, orientation of the hemagglutinin that has previously
undergone the low pH conformational change.

FUSION OF SENDAI VIRUS WITH CULTURED CELLS

When Sendai virus was added to a suspension of HL-60 or
CEM cells at neutral pH, fusion started immediately at a
relatively slow rate, which began to increases after about a
minute (Figure 5; Pedroso de Lima et al., 1990, 1991; Düzgüneş
et al., 1991b). If the virus was added to the cells at pH 5,
the initial slow rate was maintained (Curve B). However, if
the pH was lowered to 5 after an initial incubation at 7.5,
rapid fusion was initiated, but the rate gradually leveled off.

When Sendai virus was pre-incubated at pH 5 or 9 in the
absence of target membranes, and then added to HL-60 cells at
neutral pH, the rate and extent of fusion did not change
significantly (Pedroso de Lima et al., 1991). These
observations are in contrast to those of Hsu et al. (1982), who
reported that previous exposure of Sendai virus to high pH
enhanced the hemolytic and cell-cell fusion activity of the
virus. They also differ from the observations reported by
Hoekstra and Kok (1989), that pre-incubation of Sendai virus at
pH 5-6 or pH 9 reduced the initial rate of fusion with mouse
spleen cells at pH 7.4. These differences may be due to the
different strains of virus used for these studies, and the
different methods used to detect fusion activity, i.e.
hemolysis, cell-cell fusion, virus-cell fusion. The molecular
requirements for Sendai virus-cell fusion and virus-induced
cell-cell fusion may be different (Aroeti and Henis, 1988).
Different isolates of the same strain of virus may also show
differences in fusion behavior. For example, the Z/G isolate

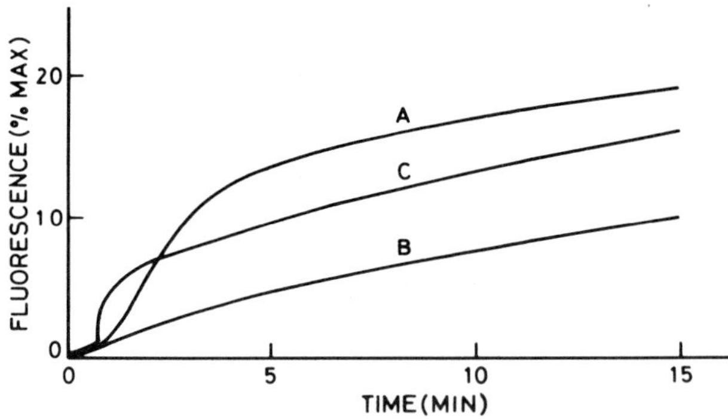

Figure 5. Fusion of Sendai virus with CEM cells. Five µg/ml
of the Z/SF strain of Sendai virus was incubated
with 2x10⁷/ml CEM cells at 37° C in RPMI 1640
without phenol red and containing 25 mM HEPES
buffer. Curve A: pH 7.5. Curve B: pH 5, adjusted
with citrate buffer before the virus was added to
the cells. Curve C: The experiment was started at
pH 7.5, but the pH was lowered to 5 after 45 s. The
fluorescence scale was set as in Figure 1. Data
from Pedroso de Lima et al., 1991.

of Sendai virus displayed enhanced fusion when added to cells
at mildly acidic pH (Pedroso de Lima et al., 1990, 1991;
Düzgünes et al., 1991b).

FUSION OF SIV$_{mac}$ WITH LIPOSOMES AND ERYTHROCYTE GHOSTS

Liposomes as Target Membranes

Liposomes have been used as target membranes for lipid
enveloped viruses in order to investigate the types of
molecules with which the viral membrane can interact and
undergo fusion. The demonstration that the fusion of Semliki
Forest virus with target membranes was dependent on the
presence of cholesterol in the latter was based on the use of
liposomes (White and Helenius, 1980; Kielian and Helenius,
1984; Düzgüneş, 1988). That acidic pH was required for the
induction of fusion between Semliki Forest virus and target
membranes was shown by the use of liposomes (White and
Helenius, 1980; Helenius et al., 1980). A similar observation
was made with influenza virus (Maeda et al., 1981; White et
al., 1982). Influenza virus hemagglutinin was shown to have
amphiphilic properties by studying its interaction with
liposomes (Doms et al., 1985). The demonstration of the
penetration of segments of viral envelope proteins into the
target membrane during fusion, by means of photo-activatable
probes, also required the use of target membranes composed only
of phospholipids (Novick and Hoekstra, 1988; Harter et al.,
1989).

*Fusion of SIV with Liposomes: The Effects of Lipid Composition
and Soluble CD4*

Although immunodeficiency viruses are generally believed
to require the cell surface CD4 molecule as a receptor, we have
found that both SIV$_{mac}$ and HIV-1 undergo fusion with liposomes
not containing CD4 (Larsen et al., 1990, 1991; Konopka et al.,
1990; Düzgüneş et al., 1991c). Figure 6 shows the time-course
of fusion of SIV$_{mac}$ with cardiolipin liposomes at pH 7.5.
Fusion was inhibited by rCD4 at concentrations of 10-100 µg/ml,
but inhibition was not complete. Watanabe et al. (1989) have
reported that 12.5 µg/ml rCD4 inhibited SIV$_{mac}$ infection of
rhesus monkey peripheral blood lymphocytes by only about 50%,
while 125 µg/ml rCD4 inhibited infectivity by 10-fold. Varying
levels of rCD4 were necessary to inhibit the infectivity of
different HIV-1 isolates, with primary isolates from patients
requiring much higher concentrations than laboratory strains
(Clapham et al., 1989; Daar et al., 1990). Although the fusion
of SIV$_{mac}$ and cardiolipin liposomes proceeds without the
specific attachment of gp120 to a CD4 receptor, binding of rCD4
to gp120 probably limits the close approach of the envelope
glycoprotein to the liposome surface. This may, in turn,
prevent the efficient penetration of the fusogenic domains of
the protein (possibly the N-terminal of the gp32/40) into the
liposome membrane. We should point out, however, that bovine
serum albumin, at high concentrations (i.e. 100 µg/ml), also
inhibited SIV$_{mac}$-liposome fusion to an extent similar to that
obtained with rCD4 (Düzgüneş et al., 1991c). This observation
suggests that the inhibition by rCD4 may be non-specific.

A puzzling effect of rCD4 is that it enhances the infectivity of SIV_{agm} (isolated from African green monkeys) (Allan et al., 1990; Werner et al., 1990). Furthermore, addition of rCD4 to HIV-1, or cells infected with HIV-1, causes the release of gp120 from the gp41, and thus from the membrane (Moore et al., 1990; Hart et al., 1991). This release induces the exposure of a region of gp41 that may be involved in membrane fusion (Hart et al., 1991). Applying this scenario to SIV, if binding of CD4 to gp120 is necessary for the exposure of the fusogenic domain of gp32/40, the fusion of SIV_{mac} with liposomes without the CD4 molecule would be difficult to explain. It is possible that some of the gp32/40 is already exposed on the virus surface due to shedding of gp120 during virus isolation and purification. In the studies on gp120 shedding cited above, cells or the virus are incubated with rCD4 for extended periods of time, while in our experiments the rCD4 is added shortly before the measurement of fusion. It is therefore likely that gp120 is shed only minimally from the SIV membrane in our experiments with rCD4.

The initial rate and extent of fusion of SIV_{mac} was highly dependent on the phospholipid composition of the liposomes (Larsen et al., 1990; Düzgüneş et al., 1991c). Fusion was fastest with pure cardiolipin vesicles, among the lipid compositions tested. When 70% of the cardiolipin was replaced with DOPC, the rate and extent of fusion were decreased precipitously. Fusion was slower with pure phosphatidylserine liposomes, than with cardiolipin-containing vesicles, and even slower with pure DOPC vesicles. Similar observations were made with HIV-1 (Larsen et al., 1991). Cardiolipin liposomes were also found to specifically inhibit the infectivity of HIV-1 (Konopka et al., 1990, 1991b).

Fusion of SIV with Liposomes: The Effects of pH and Calcium

Lowering the pH enhanced the initial rate and extent of fusion of SIV_{mac}, particularly with pure cardiolipin liposomes (Figure 7), similar to observations made with Sendai and influenza viruses (Stegmann et al., 1986; Klappe et al., 1986; Düzgüneş et al., 1991c). When 15 mol% disialoganglioside (G_{D1a}) was incorporated in DOPC liposomes, the rate and the extent of fusion of SIV_{mac} were similar to that obtained with phosphatidylserine liposomes. Low pH did not significantly alter the kinetics of fusion with this target membrane (Larsen et al., 1990). In contrast, gangliosides can act as receptors for influenza and Sendai viruses (Stegmann et al., 1989b; Haywood and Boyer, 1984, 1985).

The rate and extent of fusion of SIV_{mac} with cardiolipin liposomes increased significantly in the presence of Ca^{2+} in the medium; for example, the initial rate increased 3-fold in the presence of 1 mM Ca^{2+} (Larsen et al., 1990). This observation contrasts with the Ca^{2+}-independence of the fusion of Sendai and influenza viruses (Stegmann et al., 1985; D. Hoekstra, personal communication). Fusion in the presence of 0.5 mM Ca^{2+} was also inhibited by rCD4, the relative inhibition at 100 µg/ml being considerably greater than in the absence of Ca^{2+} (Düzgünes et al., 1991c). Calcium up to 5 mM had no

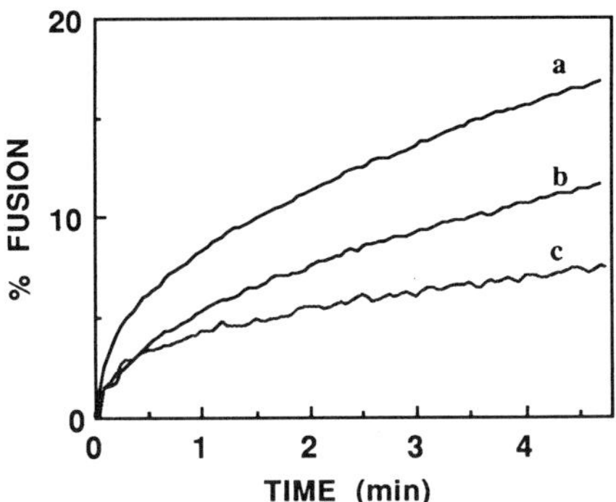

Figure 6. Kinetics of fusion of SIV_{mac} with cardiolipin liposomes, measured by the dequenching of R-18 incorporated in the viral membrane, and the effect of recombinant soluble CD4. Curve a: SIV_{mac} (1 μg protein/ml) incubated with large unilamellar cardiolipin liposomes (0.05 μmol lipid/ml) at pH 7.5. Curve b: As in a, but in the presence of 10 μg/ml rCD4. Curve c: With 100 μg/ml rCD4.

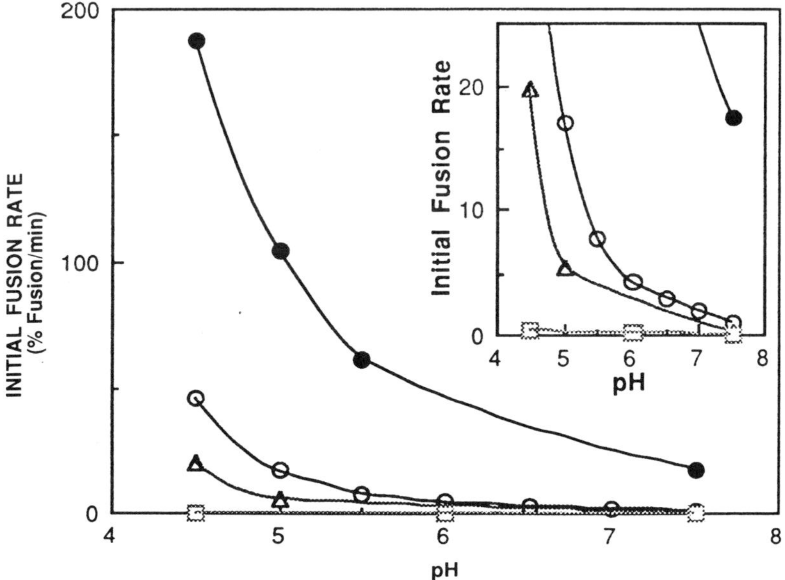

Figure 7. The pH-dependence of the initial rate of fusion of SIV_{mac} with liposomes of various phospholipid composition. Cardiolipin: Closed circles. Cardiolipin/DOPC: Open circles. Phosphatidylserine: Triangles. DOPC: Squares. Reproduced with permission from Larsen et al., 1990a.

effect on the fusion of the virus with DOPC liposomes. Thus,
Ca^{2+} did not activate SIV_{mac} to fuse rapidly with a membrane
with which the virus did not fuse appreciably in the absence of
the cation.

Fusion of SIV with Cationic Liposomes

SIV_{mac} could also fuse with positively charged liposomes
composed of N[1-(2,3-dioleyloxypropyl]-N,N,N-trimethylammonium
(DOTMA) and cholesterol, but fusion was not enhanced by low pH
or Ca^{2+} (C. Larsen, D. Alford, L. Young, T. McGraw, N.
Düzgünes, unpublished results). Fusion was also not inhibited
by rCD4. Since DOTMA-containing liposomes fuse with negatively
charged liposomes (Düzgüneş et al., 1989) it is possible that
DOTMA interacts directly with an acidic lipid component of the
SIV_{mac} membrane or with acidic amino acids on the viral
proteins. The infectivity of HIV-1 was enhanced drastically in
the presence of DOTMA liposomes, which also fused with HIV-1
(Konopka et al., 1990, 1991c; Düzgüneş et al., 1991c).

Fusion of SIV with Erythrocyte Ghosts

Various lipid-enveloped viruses, including influenza
(Stegmann et al., 1986) and Sendai (Hoekstra et al., 1985;
Hoekstra and Klappe, 1986) fuse with erythrocyte ghost
membranes. Fusion of SIV_{mac} with erythrocyte ghosts was very
slow at neutral pH, but was enhanced at acidic pH. The rate of
fusion increased further when calcium was also present in the
low pH medium, but not at neutral pH (Düzgüneş et al., 1991c).
Thus, the enhancement of SIV_{mac} fusion activity by calcium or
low pH is not restricted to its fusion with liposomes as target
membranes. These experiments also indicate that SIV_{mac} can
fuse with biological membranes without the CD4 receptor
molecule. Although low pH is not required for the fusion
activity of SIV_{mac}, endocytosed virions (Pauza, 1991) that
encounter acidic pH in the endosomes of lymphocytes or
monocyte/macrophages may fuse more efficiently with the
endosome membrane than with the plasma membrane. In contrast,
the fusion of HIV-1 with $CD4^{+}$ CEM cells, detected with the R-18
assay, was reported to be optimal at pH 7, but also occured at
pH 5 (Sinangil et al.,1988).

ACKNOWLEDGMENTS

This work was supported by Grant AI-25534 (N.D.) and
Fellowship AI-08117 (C.E.L.) from the National Institute of
Allergy and Infectious Diseases, Fellowship PF-3394 from the
American Cancer Society (C.E.L.), Grant 86SF017 from the State
of California Universitywide AIDS Research Program (N.D), U.S.-
Israel Binational Science Foundation Grant 86-00010 (S.N. and
N.D.), and NATO Collaborative Research Grant CRG 900333
(M.C.P.de L. and N.D.).

REFERENCES

Allan, J. S., Strauss, J. and Buck, D. W., 1990, Enhancement of
    SIV infection with soluble receptor molecules, <u>Science</u>
    247:1084-1088.

Amselem, S., Barenholz, Y., Loyter, A., Nir, S. and
     Lichtenberg, D., 1986, Fusion of Sendai virus with
     negatively charged liposomes as studied by pyrene-
     labelled phospholipid liposomes, _Biochim. Biophys. Acta_
     860:301-313.
Aroeti, B. and Henis, Y. I., 1988, Effects of fusion
     temperature on the lateral mobility of Sendai virus
     glycoproteins in erythrocyte membranes and on cell fusion
     indicate that glycoprotein mobilization is required for
     cell fusion, _Biochemistry_ 27:5654-5661.
Asano, A., and Asano, K., 1984, Molecular mechanism of virus
     entry to target cells, _Tumor Res._ 19:1-20.
Berg, O. G. and von Hippel, P. H., 1985, Diffusion-controlled
     macromolecular interactions, _Ann. Rev. Biophys. Chem._
     14:131-160.
Bosch, M. L., Earl, P. L., Fargnoli, K., Picciafuoco, S.,
     Giombini, F., Wong-Staal, F. and Franchini, G., 1989,
     Identification of the fusion peptide of primate im-
     munodeficiency viruses, _Science_ 244:694-697.
Byrn, R. A., Sekigawa, I., Chamow, S. M., Johnson, J. S.,
     Gregory, T. J., Capon, D. J. and Groopman, J. E., 1989,
     Characterization of in vitro inhibition of human
     immunodeficiency virus by purified recombinant CD4, _J.
     Virol_ 63:4370-4375.
Chakrabarti, L., Guyader, M., Alizon, M., Daniel, M. D.,
     Desrosiers, R. C., Tiollais, P. and Sonigo, P., 1987,
     Sequence of simian immunodeficiency virus from macaque
     and its relationship to other human and simian
     retroviruses, _Nature_ 328:543-547.
Cheng-Mayer, C., Rutka, J. T., Rosenblum, M. L., McHugh, T.,
     Stites, D. P. and Levy, J. A., 1987, Human
     immunodeficiency virus can productively infect cultured
     human glial cells, _Proc. Natl. Acad. Sci. USA_ 84:3526-
     3530.
Clapham, P. R., Weber, J. N., Whitby, D., McIntosh, K.,
     Dalgleish, A. G., Maddon, P. J., Deen, K. C., Sweet, R.
     W. and Weiss, R. A., 1989, Soluble CD4 blocks the
     infectivity of diverse strains of HIV and SIV for T cells
     and monocytes but not for brain and muscle cells, _Nature_,
     337:368-370.
Daar, E. S., Li, X. L., Moudgil, T., Ho, D. D., 1990, High
     concentrations of recombinant soluble CD4 are required to
     neutralize primary human immunodeficiency virus type 1
     isolates, _Proc. Natl. Acad. Sci. USA_ 87:6574-6578.
Dalgleish, A. G., Beverly, P. C. L., Clapham, P. R., Crawford,
     D. H., Greaves, M. F., and Weiss, R. A., 1984, The CD4
     (T-4) antigen is an essential component of the receptor
     for the AIDS retrovirus, _Nature_ 312:763-766.
Desrosiers, R. C., 1988, Simian immunodeficiency viruses.
     Annu. Rev. Microbiol. 42:607-625.
Desrosiers, R. C. and Letvin, N. L., 1987, Animal models for
     acquired immunodeficiency syndrome, _Rev. Infect. Dis._
     9:438-446.
Doms, R. W., Helenius, A. and White, J., 1985, Membrane fusion
     activity of the influenza virus hemagglutinin: The low
     pH-induced conformational change, _J. Biol. Chem._
     260:2973-2981.

Düzgüneş, N., 1985,   Membrane fusion,   <u>Subcell. Biochem</u>. 11:195-286.

Düzgüneş, N., 1988, Cholesterol and membrane fusion, <u>in</u>: "Biology of Cholesterol," pp. 197-212, P. L. Yeagle, ed., CRC Press, Boca Raton, Florida.

Düzgüneş, N. and Bentz, J., 1988, Fluorescence assays for membrane fusion, in: "Spectroscopic Membrane Probes, Vol. I," pp. 117-159, L. M. Loew, ed., CRC Press, Boca Raton, Florida.

Düzgüneş, N. and Gambale, F., 1988, Membrane action of synthetic peptides from influenza virus hemagglutinin and its mutants, <u>FEBS Lett</u>. 227:110-114.

Düzgüneş, N. and Shavnin, S. A., 1991, N-terminal peptides of viral fusion proteins: Interaction with phospholipid vesicles, <u>Biochim. Biophys. Acta</u> (submitted).

Düzgüneş, N., Goldstein, J. A., Friend, D. S. and Felgner, P. L., 1989, Fusion of liposomes containing a novel cationic lipid, N[1-(2,3-dioleyloxypropyl]-N,N,N-trimethylammonium: Induction by multivalent anions and asymmetric fusion with acidic phospholipid vesicles, <u>Biochemistry</u> 28:9179-9184.

Düzgüneş, N., Pedroso de Lima, M. C., Stamatatos, L., Flasher, D., Alford, D., Friend, D. S. and Nir, S., 1991a, Fusion activity and inactivation of influenza virus: Kinetics of low-pH induced fusion with cultured cells, <u>J. Gen. Virol</u>. (submitted).

Düzgüneş, N., Pedroso de Lima, M. C., Stamatatos, L., Flasher, D., Friend, D. S., Alford, D., Klappe, K., Hoekstra, D. and Nir, S., 1991b, Fusion of influenza and Sendai viruses with cultured cells, <u>Biophys. J</u>. 59:206a.

Düzgüneş, N., Larsen, C. E., Konopka, K., Alford, D. R., Young, L. J. T., McGraw, T. P., Davis, B. R., Nir, S. and Jennings, M., 1991c, Fusion of HIV-1 and SIV$_{mac}$ with liposomes and modulation of HIV-1 infectivity, <u>in</u>: "Mechanisms and Specificity of HIV Entry into Host Cells," N. Düzgüneş, ed., Plenum Press, New York (in press).

Fisher, R. A., Bertonis, J. M., Meier, W., Johnson, V. A., Costopoulos, D. S., Liu, T., Tizard, R., Walker, B. D., Hirsch, M. S., Schooley, R. T. and Flavell, R. A., 1988, HIV infection is blocked in vitro by recombinant soluble CD4, <u>Nature</u> 331:76-78.

Freed, E. O., Myers, D. J. and Risser, R., 1991, Identification of the principal neutralizing determinant of human immunodeficiency virus type 1 as a fusion domain, <u>J. Virol</u>. 65:190-194.

Gallaher, W. R., 1987, Detection of a fusion peptide sequence in the transmembrane protein of human immunodeficiency virus, <u>Cell</u> 50:327-328.

Gibson, S., Jung, C. Y., Takahashi, M. and Lenard, J., 1986, Radiation inactivation analysis of influenza virus reveals different target sizes for fusion, leakage and neuraminidase activities, <u>Biochemistry</u> 25:6264-6268.

Gonzalez-Scarano, F., Waxham, M. N., Ross, A. M. and Hoxie, J. A., 1987, Sequence similarities between human immunodeficiency virus gp41 and paramyxovirus fusion proteins, <u>AIDS Res. Human Retrovir</u>. 3:245-252.

Harouse, J. M., Kunsch, C., Hartle, H. T., Laughlin, M. A., Hoxie, J. A., Wigdahl, B., and Gonzalez-Scarano, F., 1989, CD4-independent infection of human neural cells by human immunodeficiency virus type 1, J. Virol. 63:2527-2533.

Hart, T. K., Kirsh, R., Ellens, H., Sweet, R. W., Lambert, D. M., Petteway, S. R. Jr., Leary, J. and Bugelski, P. J., 1991, Binding of soluble CD4 proteins to human immunodeficiency virus type 1 and infected cells induces release of envelope glycoprotein gp120, Proc. Natl. Acad. Sci. USA 88:2189-2193.

Harter, C., James, P., Bächi, T., Semenza, G. and Brunner, J., 1989, Hydrophobic binding of the ectodomain of influenza hemagglutinin to membranes occurs through the "fusion peptide." J. Biol. Chem. 264:6459-6464.

Haywood, A. M. and Boyer, B. P., 1984, Effect of lipid composition upon fusion of liposomes with Sendai virus membranes, Biochemistry 23:4161-4166.

Haywood, A. M. and Boyer, B. P., 1985, Fusion of influenza virus membranes with liposomes at pH 7.5, Proc. Natl. Acad. Sci. USA 82:4611-4615.

Helenius, A., Kartenbeck, J., Simons, K. and Fries, E., 1980, On the entry of Semliki Forest virus into BHK-21 cells, J. Cell Biol. 84:404-420.

Hoekstra, D., 1991, Fusion of enveloped viruses: From microscopic observation to kinetic simulation, in: "Membrane Fusion," pp. 289-311, J. Wilschut and D. Hoekstra, eds., Marcel Dekker, New York.

Hoekstra, D. and Klappe, K., 1986, Sendai virus-erythrocyte membrane interaction: Quantitative and kinetic analysis of viral binding, dissociation and fusion, J. Virol. 58:87-95.

Hoekstra, D. and Kok, J. W., 1989, Entry mechanisms of enveloped viruses. Implications for fusion of intracellular membranes, Biosci. Rep. 9:273-305.

Hoekstra, D. and Klappe, K., 1991, Fluorescence assays to monitor fusion of enveloped viruses, in: "Membrane Fusion Techniques," N. Düzgüneş, ed., Academic Press, San Diego (in press).

Hoekstra, D., de Boer, T., Klappe, K., and Wilschut, J., 1984, Fluorescence method for measuring the kinetics of fusion between biological membranes, Biochemistry 23:5675-5681.

Hoekstra, D., Klappe, K., de Boer, T., and Wilschut, J., 1985, Characterization of the fusogenic properties of Sendai virus: Kinetics of fusion with erythrocyte membranes, Biochemistry 24:4739-4745.

Hsu, M.-C., Scheid, A. and Choppin, P. W., 1982, Enhancement of membrane-fusing activity of Sendai virus by exposure of the virus to basic pH is correlated with a conformational change in the fusion protein, Proc. Natl. Acad. Sci. USA 79:5862-5866.

Hussey, R. E., Richardson, N. E., Kowalski, M., Brown, N. R., Chang, H. C., Siliciano, R. F., Dorfman, T., Walker, B., Sodroski, J. and Reinherz, E. L., 1988, A soluble CD4 protein selectively inhibits HIV replication and syncytium formation, Nature (London) 331:788-81.

Junankar, P. R. and Cherry, R. J., 1986, Temperature and pH
    dependence of the haemolytic activity of influenza virus
    and of the rotational mobility of the spike glyco-
    proteins, _Biochim. Biophys. Acta_ 854:198-206.
Kannagi, M., Yetz, J. M., and Letvin, N. L, 1985, _In vitro_
    growth characteristics of simian T-lymphotropic virus
    type III, _Proc. Natl. Acad. Sci. USA_ 82:7053-7057.
Kielian, M., and Helenius, A., 1984, Role of cholesterol in
    fusion of Semliki Forest virus with membranes, _J. Virol_.
    52:281-283.
Klappe, K., Wilschut, J., Nir, S. and Hoekstra, D., 1986,
    Parameters affecting fusion between Sendai virus and
    liposomes. Role of viral proteins, liposome composition,
    and pH, _Biochemistry_ 25:8252-8260.
Klatzmann, D., Champagne, E., Chamaret, S., Gruest, J.,
    Guétard, D., Hercend, T., Gluckman, J.-C., and
    Montagnier, L., 1984, T-lymphocyte T4 molecule behaves as
    the receptor for human retrovirus LAV, _Nature_ 312:767-
    768.
Koenig, S., Hirsch, V. M., Olmsted, R. A., Powell, D., Maury,
    W., Rabson, A., Fauci, A. S., Purcell, R. H. and Johnson,
    P. R., 1989, Selective infection of human CD4$^+$ cells by
    simian immunodeficiency virus: Productive infection
    associated with envelope glycoprotein-induced fusion,
    Proc. Natl. Acad. Sci. USA 86:2443-2447.
Konopka, K., Davis, B. R., Larsen, C. E., Alford, D. R., Debs,
    R. J. and Düzgüneş, N., 1990, Liposomes modulate human
    immunodeficiency virus infectivity, _J. Gen. Virol_.
    71:2899-2907.
Konopka, K., Davis, B. R. and Düzgüneş, D., 1991a, HIV-1
    infection of a non-CD4-expressing variant of HUT-78
    cells: Lack of inhibition by Leu3a antibodies and
    enhancement by cationic DOTMA liposomes, _in_: "Mechanisms
    and Specificity of HIV Entry into Host Cells," N.
    Düzgüneş, ed., Plenum Press, New York (in press).
Konopka, K., Davis, B. R., Larsen, E. and Düzgüneş, N., 1991b,
    Cardiolipin liposomes specifically inhibit HIV-1
    infectivity, _Antiviral Res_. Suppl. I:68.
Konopka, K., Stamatatos, L., Larsen, C. E., Davis, B. R. and
    Düzgüneş, N., 1991c, Enhancement of HIV-1 infection by
    cationic liposomes: The role of CD4, serum and liposome-
    cell interactions. _J. Gen. Virol_. (submitted).
Kowalski, M., Potz, J., Basiripour, L., Dorfman, T., Goh, W.
    C., Terwilliger, E., Dayton, A., Rosen, C., Haseltine, W.
    and Sodroski, J., 1987, Functional regions of the
    envelope glycoprotein of human immunodeficiency virus
    type 1, _Science_ 237:1351-1355.
Larsen, C. E., Alford, D. R., Young, L. J. T., McGraw, T. P.
    and Düzgüneş, N., 1990, Fusion of simian immunodeficiency
    virus (SIV$_{mac}$) with liposomes and erythrocyte ghosts:
    Effect of liposome composition, low pH and calcium, _J.
    Gen. Virol_. 71:1947-1955.
Larsen, C. E., Alford, D. R., Nir, S., Jennings, M., Lee, K.-D.
    and Düzgüneş, N., 1991, Fusion of HIV-1 with liposomes
    and erythrocyte ghosts, _Biophys. J_. 59:131a.
Letvin, N. L., Daniel, M. D., Sehgal, P. K., Desrosiers, R. C.,
    Hunt, R. D., Waldron, L. M., MacKey, J. J., Schmidt, D.
    K., Chalifoux, L. V., and King, N. W., 1985, Induction of
    AIDS-like disease in macaque monkeys with T-cell tropic
    retrovirus STLV-III, _Science_ 230:71-73.

Lifson, J. D., Feinberg, M. B., Reyes, G. R., Rabin, L., Banapour, B., Chakrabarti, S., Moss, B., Wong-Staal, F., Steimer, K. S. and Engleman, E. G., 1986, Induction of CD4-dependent cell fusion by the HTLV-III/LAV envelope glycoprotein, _Nature_ 323:725-728.

Loyter, A., Citovsky, V., and Blumenthal, R., 1988, The use of fluorescence dequenching measurements to follow viral membrane fusion events, _Methods Biochem. Analysis_ 33:129-164.

Maddon, P. J., Dalgleish, A. G., McDougal, J. S., Clapham, P. R., Weiss, R. A., and Axel, R., 1986, The T4 gene encodes the AIDS virus receptor and is expressed in the immune system and the brain, _Cell_ 47:333-348.

Maddon, P. J., McDougal, J. S., Clapham, P. R., Dalgleish, A. G., Jamal, S., Weiss, R. A., and Axel, R., 1988, HIV infection does not require endocytosis of its receptor, CD4, _Cell_ 54:865-874.

Maeda, T., Kawasaki, K., and Ohnishi, S., 1981, Interaction of influenza virus hemagglutinin with target membrane lipids is a key step in virus-induced hemolysis and fusion at pH 5.2, _Proc. Natl. Acad. Sci. USA_ 78:4133-4137.

Marsh, M., and Dalgleish, A., 1987, How do human immuno-deficiency viruses enter cells? _Immunol. Today_ 8:369-371.

Marsh, M. and Helenius, A., 1989, Virus entry into animal cells. _Adv. Virus Res_. 36:107-151.

Matlin, K., Reggio, H., Helenius, A. and Simons, K., 1981, The infective entry of influenza virus into MDCK cells, _J. Cell Biol_. 91:601-613.

McClure, M. O., Marsh, M., and Weiss, R. A., 1988, Human immunodeficiency virus infection of CD4-bearing cells occurs by a pH-independent mechanism, _EMBO J_. 7:513-518.

McDougal, J. S., Mawle, A., Cort, S. P., Nicholson, J. K. A., Cross, G. D., Scheppler-Campbell, J. A., Hicks, D., and Sligh, J., 1985, Cellular tropism of the human retrovirus HTLV-III/LAV, _J. Immunol_. 135:3151-3162.

McDougal, J. S., Kennedy, M. S., Sligh, J. M., Cort, S. P., Mawle, A. and Nicholson, J. K. A., 1986, Binding of HTLV-III/LAV to T4+ T cells by a complex of the 110K viral protein and the T4 molecule, _Science_ 231:382-385.

Moore, J. P., McKeating, J. A., Weiss, R. A. and Sattentau, Q. J, 1990, Dissociation of gp120 from HIV-1 virions induced by soluble CD4, _Science_ 250:1139-1142.

Nir, S., Bentz, J., Wilschut, J. and Düzgüneş, N., 1983, Aggregation and fusion of phospholipid vesicles, _Prog. Surface Sci_. 13:1-124.

Nir, S., Stegmann, T. and Wilschut, J., 1986a, Fusion of influenza virus with cardiolipin liposomes at low pH: Mass action analysis of kinetics and fusion, _Biochemistry_ 25:257-266.

Nir, S., Klappe, K. and Hoekstra, D., 1986b, Kinetics and extent of fusion between Sendai virus and erythrocyte ghosts: Application of a mass action kinetic model, _Biochemistry_ 25:2155-2161.

Nir, S., Klappe, K. and Hoekstra, D., 1986c, Mass action analysis of kinetics and extent of fusion between Sendai virus and liposomes, _Biochemistry_ 25:8261-8266.

Nir, S., Düzgüneş, N., Pedroso de Lima, M. C. and Hoekstra, D., 1990, Fusion of enveloped viruses with cells and liposomes: Activation and inactivation, _Cell Biophys_. 17:181-201.

Novick, S. L. and Hoekstra, D., 1988, Membrane penetration of
     Sendai virus glycoproteins during the early stages of
     fusion with liposomes as determined by hydrophobic
     photoaffinity labeling, Proc. Natl. Acad. Sci. USA
     85:7433-7437.
Ohki, S., 1988, Surface tension, hydration energy and membrane
     fusion, in: "Molecular Mechanisms of Membrane Fusion,"
     pp. 123-138, S. Ohki, D. Doyle, T. D. Flanagan, S. K.
     Hui and E. Mayhew, eds., Plenum Press, New York.
Ohnishi, S.-I., 1988, Fusion of viral envelopes with cellular
     membranes, in: "Membrane Fusion in Fertilization,
     Cellular Transport and Viral Infection," pp. 257-296, N.
     Düzgüneş and F. Bronner, eds., Academic Press, San Diego.
Okada, Y., 1988, Sendai virus-mediated cell fusion, in:
     "Membrane Fusion in Fertilization, Cellular Transport and
     Viral Infection," pp. 297-336, N. Düzgüneş and F.
     Bronner, eds., Academic Press, San Diego.
Pauza, C. D., and Price, T. M., 1988, Human immunodeficiency
     virus infection of T cells and monocytes proceeds via
     receptor-mediated endocytosis, J. Cell Biol. 107:959-968.
Pauza, C. D, 1991, The endocytic pathway for human
     immunodeficiency virus infection, in: "Mechanisms and
     Specificity of HIV Entry into Host Cells," N. Düzgüneş,
     ed., Plenum Press, New York (in press).
Pedroso de Lima, M. C., Ramalho-Santos, J., Martins, M. F. and
     Düzgüneş, N., 1990, Fusion of Sendai and influenza
     viruses with cultured cells: Insights toward cell
     infection, Abst. 20th Mtg. Fed. Eur. Biochem. Soc.,
     Budapest, Hungary.
Pedroso de Lima, M. C., Nir, S., Flasher, D., Klappe, K.,
     Hoekstra, D. and Düzgüneş, N., 1991, Kinetics of fusion
     of Sendai virus with human HL-60 and CEM cells, Biochim.
     Biophys. Acta (submitted).
Sato, S. B., Kawasaki, K. and Ohnishi, S.-I., 1983, Hemolytic
     activity of influenza virus hemagglutinin glycoproteins
     activated in mildly acidic environments, Proc. Natl.
     Acad. Sci. USA 80:3153-3157.
Sattentau, Q., 1990, Molecular interactions between CD4 and the
     HIV envelope glycoproteins, in: "AIDS and the New
     Viruses," pp. 41-53, A. G. Dalgleish and R. A. Weiss,
     eds., Academic Press, London.
Sinangil, F., Loyter, A., and Volsky, D. J., 1988, Quantitative
     measurement of fusion between human immunodeficiency
     virus and cultured cells using membrane fluorescence
     dequenching, FEBS Lett. 239:88-92.
Smoluchowski, M., 1917, Investigation into the mathematical
     theory of the kinetics of coagulation of colloidal
     solutions, Z. physik. Chem., Abt. A 92:129-168.
Sodroski, J., Goh, W. C., Rosen, C., Campbell, K. and
     Haseltine, W., 1986, Role of the HTLV-III/LAV envelope in
     syncytium formation and cytopathicity, Nature 322:470-
     474.
Stegmann, T., Hoekstra, D., Scherphof, G. and Wilschut, J.,
     1985, Kinetics of pH-dependent fusion between influenza
     virus and liposomes, Biochemistry 24:3107-3113.
Stegmann, T., Hoekstra, D., Scherphof, G. and Wilschut, J.,
     1986, Fusion activity of influenza virus. A comparison
     between biological and artificial target membrane
     vesicles, J. Biol. Chem. 261:10966-10969.

Stegmann, T., Doms, R.W. and Helenius, A., 1989a, Protein-mediated membrane fusion, <u>Annu. Rev. Biophys. Chem</u>. 18:187-211.

Stegmann, T., Nir, S., and Wilschut, J., 1989b, Membrane fusion activity of influenza virus. Effects of gangliosides and negatively charged phospholipids in target liposomes, <u>Biochemistry</u> 28:1698-1704.

Stein, B. S. and Engleman, E. G., 1991, Mechanism of HIV-1 entry into CD4+ T cells, <u>in</u>: "Mechanisms and Specificity of HIV Entry into Host Cells," N. Düzgüneş, ed., Plenum Press, New York (in press).

Stein, B. S., Gowda, S. D., Lifson, J. D., Penhallow, R. C., Bensch, K. G., and Engleman, E. G., 1987, pH-Independent HIV entry into CD4-positive T cells via virus envelope fusion to the plasma membrane, <u>Cell</u> 49:659-668.

Tateno, M., Gonzalez-Scarano, F., and Levy, J. A., 1989, Human immunodeficiency virus can infect CD4-negative human fibroblastoid cells, <u>Proc. Natl. Acad. Sci. USA</u> 86:4287-4290.

Watanabe, M., Reimann, K. A., DeLong, P. A., Liu, T., Fisher, R. A. and Letvin, N. L., 1989, Effect of recombinant soluble CD4 in rhesus monkeys infected with simian immunodeficiency virus of macaques, <u>Nature</u> 337:267-270.

Weber, J., Clapham, P., McKeating, J. Stratton, M., Robey, E. and Weiss, R., 1989, Infection of brain cells by diverse human immunodeficiency virus isolates: Role of CD4 as receptor, <u>J. Gen. Virol</u>. 70:2653-2660.

Werner, A, Winskowsky, G and Kurth, R., 1990, Soluble CD4 enhances simian immunodeficiency virus SIV_{agm} infection, <u>J. Virol</u>. 64:6252-6256.

White, J., 1990, Viral and cellular membrane fusion proteins, <u>Annu. Rev. Physiol</u>. 52:675-697.

White, J. and Helenius, A., 1980, pH-Dependent fusion between the Semliki Forest virus membrane and liposomes, <u>Proc. Natl. Acad. Sci. USA</u> 77:3273-3277.

White, J., Matlin, K. and Helenius, A., 1981, Cell fusion by Semliki Forest, influenza and vesicular stomatitis viruses, <u>J. Cell Biol</u>. 89:674-679.

White, J., Kartenbeck, J. and Helenius, A., 1982, Membrane fusion activity of influenza virus, <u>EMBO J</u>. 1:217-222.

White, J., Kielian, M. and Helenius, A., 1983, Membrane fusion proteins of enveloped animal viruses, <u>Q. Rev. Biophys</u>. 16:151-195.

Yoshimura, A., Kuroda, K., Kawasaki, K., Yamashina, S., Maeda, T. and Ohnishi, S., 1982, Infectious entry mechanism of influenza virus, <u>J. Virol</u>. 43:284-293.

RED BLOOD CELL INTERACTION WITH A GLASS SURFACE

J. K. Angarska[♣], K. D. Tachev[♣], I. B. Ivanov[*], P. A. Kralchevsky[*] and E. F. Leonard[**]

[♣]Department of Chemistry, University of Shumen, Shumen, 9700 Bulgaria
[*]Laboratory of Thermodynamics and Physico-Chemical Hydrodynamics, University of Sofia, Faculty of Chemistry, Sofia, 1126 Bulgaria
[**]Artificial Organs Research Laboratory, Department of Chemical Engineering, Columbia University, New York, N.Y. 10027, U.S.A.

ABSTRACT

Adhesion films formed between red blood cells (RBC) and a glass plate have been studied. The work of adhesion per unit of film area was determined from its relation to the contact angle of the adhesion film. This angle was measured using an interferometric method in conjunction with a specially designed experimental cell. The RBC investigated were swollen to a nearly or fully spherical shape in hypotonic buffer solutions. The adhesion films were classified into three groups according to the area of the adhesion film, which was probably determined by differences within the cell population. Each group was analyzed separately. It was confirmed that adhesion film thickness decreases with increasing ionic strength. Variation of the total solution osmolarity at constant ionic strength did not affect the film thickness but did change the contact angle. The latter effect can be attributed to the action of a positive line tension at the periphery of the adhesion film. The experimental method is applicable to other solid surfaces, non-spherical RBC, and other biological cells.

1. INTRODUCTION

Over the last decade, interest in cell adhesion has grown because of its biological and medical importance. Cell adhesion has been found to play a considerable role in the embryonic development of organisms (Gershey and d'Alisa, 1980). It also plays an important role in some stages of cell recognition and in the immune response (Nicol and Garrod, 1979), as well as in some physiologic functions of blood cells (Leonard et al., 1987). Knowledge and rudimentary control of cell adhesion has allowed the competent and successful application of a number of artificial materials in artificial kidneys, hearts, cardiac valves, blood vessels and limbs. Cell adhesion has also turned out to be important for some biotechnological processes such as tissue culture and microbiological culture.

Both the experimental study and the theoretical interpretation of cell adhesion data are complex problems due to the multiple physicochemical processes involved as well as the intricate structural features and variable properties of cell membranes (Bongrand et al., 1988). The following parameters are currently used to quantify cell adhesion:

F - the minimum force required for detachment of the adhering surfaces;

A - the area of contact between the two adhering surfaces;

H - the distance between the two adhering surfaces.

Irrespective of its different possible definitions, the distance H is an important characteristic of the adhesion process (Leonard et al., 1987). The knowledge of H enables one to calculate both the long-range (Parsegian and Gingell, 1973); Gingell and Vince, 1980) and the short-range (Bell, 1988) forces of interaction. Experimentally H can be determined interferometrically (Gingell and Todd, 1979), ellipsometrically (Izzard and Lochner, 1976) or by using TIRFM (Leonard et al., 1987).

The following physical stages of cell-cell (cell-surface) adhesion can be distinguished depending on the magnitude of H (see Fig. 1):

(I) H > 1000 Å: hydrodynamic interactions, resulting from macroscopic fluid mechanical conditions as fixed by the flow rate and dimensions of the apparatus.

(II) 200 < H < 1000 Å: long-range intermolecular forces (van der Waals and electrostatic) and hydrodynamic interactions;

(III) H < 200 Å: short-range intermolecular forces (steric, hydration, others) as well as formation of chemical or hydrogen bounds with the substratum.

The DLVO-theory of coagulation in colloids (Derjaguin and Landau, 1941; Verwey and Overbeek, 1948) takes into account the van der Waals and electrostatic forces. Further development and application of this theory to cell adhesion has been given by Parsegian and Gingell (1973), Dolowy (1980), Lerche (1983) and Donath and Voigt (1983, 1988). Antigen - antibody bonds and ligand-donor interactions have been taken into account by Bongrand et al. (1982) and Bell (1988). The forces governing the process of adhesion during its different stages have been investigated experimentally by Gingell and Vince (1980), Wolf and Gingell (1983) and Todd and Gingell (1980).

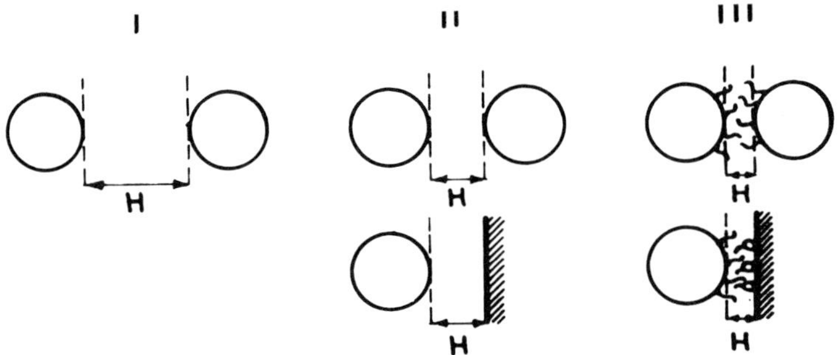

Fig. 1.   Stages of the cell adhesion: (I) hydrodynamic interactions only; (II) hydrodynamic and long range interactions; (III) short range and chemical interactions.

The present work is concerned with the second stage of cell adhesion (200 < H < 1000 Å). The principal aim is to determine the specific work of adhesion from its connection with the contact angle. Such an approach enables one to adapt and utilize the known methodology of the theory of thin liquid films (see e.g. Ivanov, 1988) to describe cell adhesion. In particular, the work of adhesion and other physical properties can be calculated from the measured contact angle, radius and other purely geometrical parameters.

To simplify considerations, we studied the adhesion of spherical (swollen) red blood cells (RBC) to a plane glass surface. In Section 2 the physicochemical aspect of the problem is discussed. The experimental interferometric method for contact angle measurements is presented in Section 3. The preparation of the RBC and buffer solutions used is described in Section 4. The experimental results for the film thickness, contact angle and radius are given and discussed in Section 5.

This study could be extended to adhering surfaces that have been modified, for example, glass on which poly-L-lysine had been pre-adsorbed, possibly causing enhanced adhesion, or glass on which serum albumin had been pre-adsorbed, possibly suppressing adhesion (Wolf and Gingell, 1983). The RBC surface can be modified by enzyme treatment to remove the hydrophilic parts of its glycoproteins (Wolf and Gingell, 1983) or by artificial enrichment with glycophorin (Arvinte et al., 1989). Such modifications studied as outlined here are a "symbiosis" of biological membrane science and thin liquid film theory that could lead to a deeper understanding of cell adhesion.

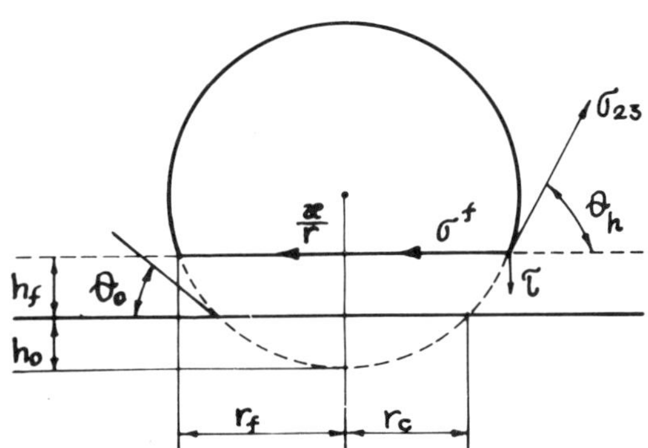

Fig. 2. Sketch of the thin film formed between a biological cell and a solid surface.

2. CONTACT ANGLE AND WORK OF ADHESION

As pointed out by Evans and Skalak (1979) the mechanical and thermodynamical properties of biomembranes can be successfully described by means of a two- dimensional continuum, analogous to a Gibbs dividing surface. The latter concept is currently used in the physical chemistry of interfaces (Rowlinson and Widom, 1982) and thin liquid films (Ivanov and Kralchevsky, 1988). In this sense, biomembranes can be described with the concepts and formalism of the theory of capillarity. As an example we consider below the work of adhesion of either liquid drops or biological cells to a solid surface.

The work of adhesion between a liquid (phase 2) and a solid (phase 1) which have been initially separated by another fluid (phase 3) is

$$W = \sigma_{13} + \sigma_{23} - \sigma_{12} \tag{1}$$

(e.g. Alexander and Johnson (1950)). Here the subscripts 1, 2 and 3 refer to the respective phases and the σ's denote the surface free energies per unit area of the respective three surfaces. If liquid 1 has a finite contact angle, α, against the solid then by using Young's equation:

$$\sigma_{13} = \sigma_{12} + \sigma_{23} \cos \alpha \tag{2}$$

one can represent the work of adhesion in the form

$$W = \sigma_{23}(1 + \cos\alpha) = \sigma_{23}(1 - \cos\theta) \tag{3}$$

Here θ (equal to $180° - \alpha$) is the contact angle measured through the medium phase, 3. Thus the work of adhesion can be determined knowing σ_{23} and θ. This equation is applied below to measurements on red blood cells.

When a RBC is attached to a solid surface the analysis must be more complicated because a thin film intervenes between the cell and the solid (See Fig. 2.). Two different but equivalent approaches are used in the macroscopic theory of thin films; they are called the "membrane" and the "detailed" approach. See, for example, Kralchevsky and Ivanov (1985). In the membrane approach the cell surface is extrapolated until it intersects the solid surface. The contact angle θ_0 thus defined allows calculation of the work of adhesion by using a counterpart of Eq. (3).

In the detailed approach the film is treated as a layer of finite thickness h_f. As shown in Fig. 2, the contact angle θ_h is then slightly different from θ_0. In addition, the film surface tension σ^f (Fig. 2) is different from σ_{23}. This difference is due to specific interactions in the thin film. The counterpart of Young's equation in this case reads

$$\sigma^f = \sigma_{23}\cos\theta_h \tag{4}$$

(See Toshev and Ivanov (1975).) The respective specific work of adhesion can be defined as

$$\Delta\sigma = \sigma_{23} - \sigma^f = \sigma_{23}(1 - \cos\theta_h) \tag{5}$$

The force balance equations (2) and (4) are more complicated when the line tension effect turns out to be important (Pethica, 1977, 1980; Kralchevsky and Ivanov, 1985). If such is the case, Eq. (4) reads

$$\sigma^f + \frac{\kappa}{r_f} = \sigma_{23}\cos\theta_h \tag{6}$$

where κ is the line tension and r_f is the radius of curvature of the contact line shown in Fig. 2. Since $\Delta\sigma = \sigma_{23} - \sigma^f$ is again the specific work of adhesion, Eq. (8) can be transformed to read

$$\cos\theta_h = \frac{\kappa}{\sigma_{23}}\frac{1}{r_f} + (1 - \frac{\Delta\sigma}{\sigma_{23}}) \tag{7}$$

If $\Delta\sigma$, σ_{23} and κ do not depend on r_f, then the specific work of adhesion, $\Delta\sigma$, and the line tension, κ, can be in principle determined from the intercept and slope of the plot of $\cos\theta_h$ vs. $1/r_f$. (Both θ_h and r_f are geometrical parameters that are experimentally accessible, v.i.).

As known (Ivanov and Toshev, 1975) both θ_h and σ_{23} depend on the choice of the mathematical dividing surfaces, representing the surfaces of the thin film. Once this choice has been made, all parameters have well defined fixed values for a given physical state of the real system. In the case of a red blood cell attached to a solid we have chosen to treat the adhesion film as non-symmetrical. The lipid bilayer of the cell membrane is encircled from the outside by a charged glycocalyx of thickness 50 - 100 Å. The thickness of the adhesion film can be determined from the interference of light waves reflected from the attached cell and from the flat solid surface (see below). Thus it is expedient to define the film surfaces as the two surfaces on which the gradient of the refractive index is maximum. These are the surface of the solid and the lipid bilayer of the cell membrane.

In summary, Eq. (5) or (7) allows calculation of the adhesion energy $\Delta\sigma$ at known σ_{23} by measuring optically some geometrical parameters (the radii of interference fringes) and determining θ_h and r_f. (See Section 3.)

3. EXPERIMENTAL METHOD

In our experiments we used RBC swollen to spherical shape in hypotonic solutions of different osmolarity and ionic strength. The cells were allowed to attach to a horizontal plate of optical glass (the experimental cell and procedure are described in the next section). At the lower osmolarities the attached RBC were almost perfect spheres - Fig. 3[a] and they remained free and non-adhered in contact with the glass. At the higher osmolarities flat, thin adhesion films were formed between the cells and the glass - Fig. 3[b].

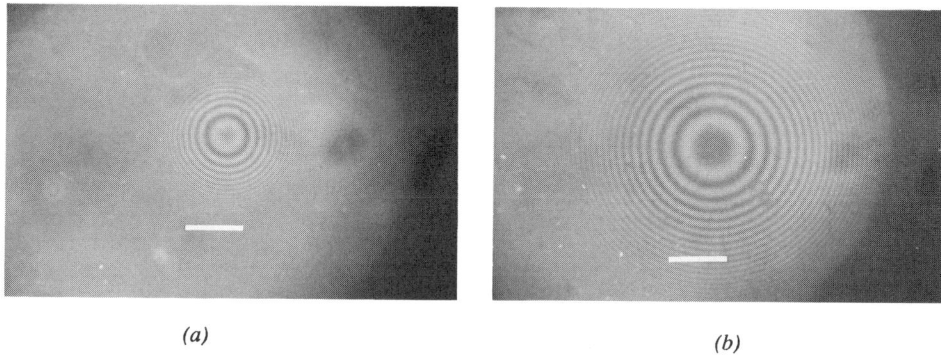

(a)   *(b)*

Fig. 3. Sketch of a swollen RBC attached to a horizontal glass surface without (a) and with (b) formation of an adhesion thin-film.

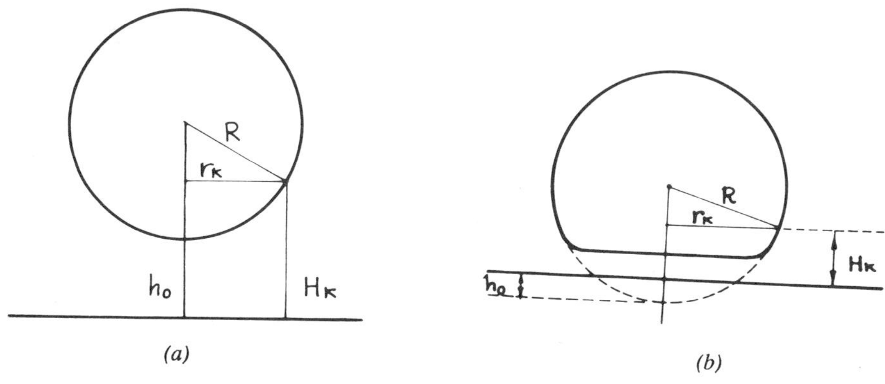

(a)   *(b)*

Fig. 4. Interference pattern created by two glass beads of radii 148±5 μm (a) and 384±6 μm (b), placed on the flat glass bottom of a container filled with water. The scale bar represents 20 μm.

Cells were observed from below, through the glass plate, by means of a metalographic microscope ("Epitip 2", Carl Zeiss, Jena). An immersion objective of magnification 100 and numerical aperture 1.30 was used. The cells were illuminated through the microscope objective. The light source was a high-pressure mercury lamp (HBO-101), from which only the green line of the mercury spectrum (546 nm) was extracted using an interference filter.

Under these conditions one observes Newton interference rings around the zone of contact between the RBC and the glass. A similar interference pattern is observed with small glass beads (Duke Scientific Co.) placed on the glass plate. (See Fig. 4.) The fringes are due to the interference of light beams reflected from the bead-water and water-glass interfaces. If H is the thickness of the liquid meniscus along the vertical (see Fig. 3^a), the fringes of maximum (and minimum) intensity correspond to

$$H_k = k \frac{\lambda}{4n}, \qquad k = 0,1,2,\dots \; , \tag{8}$$

where n is the refractive index of the liquid and k is the order of interference. (For more details - see Dimitrov et al., 1990.) In particular, the odd k correspond to bright rings and the even k to dark rings. The radii, r_k, of maximum intensity (brightness or darkness) were measured from photomicrographs of the interference pattern using a moving stage micrometer with a minimum resolution of 0.2 μm. The position of the rings was measured along three different diameters and the resulting r_k values were averaged.

The experimental points (r_k, H_k), k = 1, 2, 3,..., contain information about the profile of the liquid meniscus between the sphere and the plate. Indeed, r_k and H_k satisfy the equation

$$(H_k - R - h_0)^2 + r_k^2 = R^2 \tag{9}$$

where R is the radius of the sphere and h_0 is the distance from the lower point of the sphere to the liquid-glass interface. (See Fig. 3.) Eq. (9) can be transformed:

$$H_k^2 + r_k^2 = 2(R + h_0)H_k - (2R + h_0)h_0 \tag{10}$$

suggesting that the slope and intercept of a plot of $(H_k^2 + r_k^2)$ vs. H_k can be used to determine R and h_0.

Fig. 5 shows the interference pattern obtained with RBC replacing glass beads. A comparison of Figs. 4 and 5 shows less contrast in the interference fringes created by the RBC. This difference is expected because of the small difference between the refractive indices of the RBC membrane and the surrounding solution. The small number of RBC fringes is due to the size of the cells. In fact, the fourth interference ring (second dark ring) corresponds to a distance of about 0.45 μm. It is hard to observe fringes arising from higher orders of interference because of the greater slope of the cell (the fringes are too narrow and manifest even smaller contrast). To avoid optical difficulties connected with the great aperture of the objective (Gingell and Todd, 1979) we used an illuminating aperture of appropriate diameter.

The dark spots in the centers of the RBC interference patterns (Fig. 5) indicate the presence of thin adhesion films. Since the remaining part of the RBC surface is spherical (Fig. 3^b) one can process the data (r_k, H_k) for the interference rings by using Eq. (10). However, in this case h_0 is found to be negative. As indicated in Fig. 3$_b$, a negative h_0 means that the lower point of the (extrapolated) sphere is located below the interface solution-glass plate.

The contact radius r_c (Fig. 2) can be determined by setting H_k equal to 0 in Eq. (10):

$$r_c^2 = -(2R + h_0)h_0 \tag{11}$$

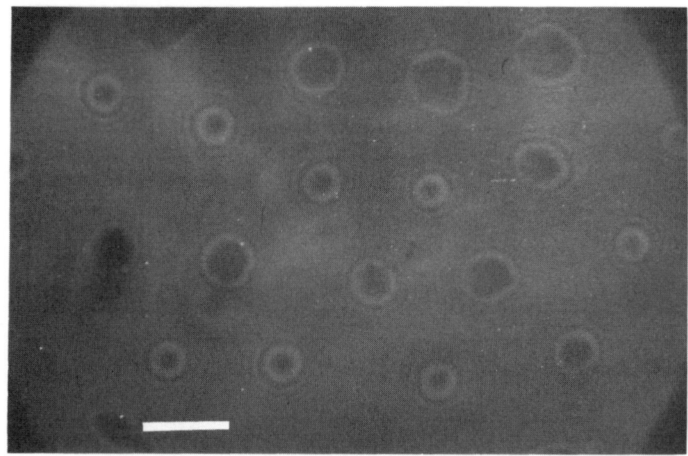

Fig. 5. Interference pattern due to RBC adhering to the flat glass bottom of the experimental cell when filled with a hypotonic solution of 151 mosM, 0.045 mol/l NaCl. Immersion objective: 100X, NA 1.30. The scale bar represents 5 μm.

Simple geometrical considerations (see Fig. 3^b) yield an equation for calculating the contact angle θ_0:

$$\tan\theta_0 = \frac{r_c}{R+h_0} \tag{12}$$

Usually $h_0 \ll R$ and Eqs. (11-12) may be simplified:

$$r_c = \sqrt{2R|h_0|}, \qquad \tan\theta_0 = \sqrt{\frac{2|h_0|}{R}} \tag{13}$$

The parameters r_c and θ_0 refer to the "membrane" model of the film. To apply these measurements to the "detailed" model, one can determine a film thickness, h_f, by measuring the radius, r_f, of the dark spot (the adhesion film) in the center of the interference pattern. Indeed, by substituting $r_k = r_f$ and $H_k = h_f$ in Eq. (9) one finds

$$h_f = R + h_0 - \sqrt{R^2 - r_f^2} \tag{14}$$

A modification of Eq. (12) allows calculation of the contact angle θ_h in the detailed approach:

$$\tan\theta_h = \frac{r_f}{R + h_o - h_f} \tag{15}$$

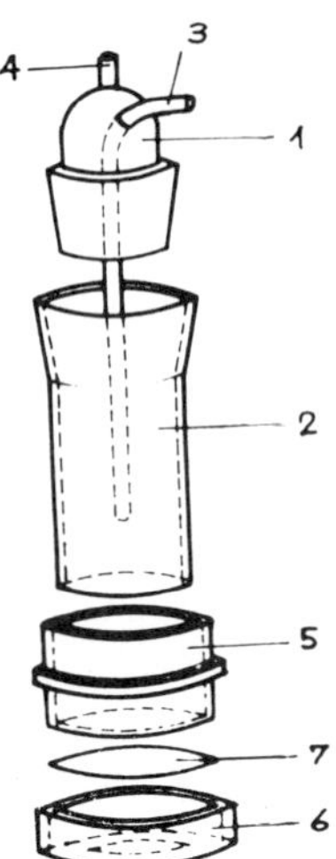

Fig. 6. Glass experimental cell for studying the RBC adhesion: 1 - upper part; 2 - lower (cell) part; 3 - inflow tube; 4 - outflow tube; 5 and 6 - teflon rings; 7 - cover glass (plate).

The area of the dark spots (the films) in Fig. 5 is different for the different cells. This is probably due to the different age and membrane characteristics of the RBC. The procedure by which these cells were separated from blood plasma, the composition of the hypotonic solution used, and the construction of the experimental cell are described in the next section.

4. EXPERIMENTAL PROCEDURE AND MATERIALS

Venous blood was drawn from a normal young female and stabilized with heparin. The red blood cells were separated from the plasma by centrifugation for 20 minutes at 1000g. The serum was removed by an aspirator. The RBC were washed and centrifuged three times, each time with a five volume isotonic solution, containing 0.153 mol/l NaCl, 0.0147 mol/l Trizma HCl and 0.023 mol/l Trizma base, pH = 7.4. After the washing the RBC were suspended in the same isotonic solution at a hematocrit of 50% and were kept in ice for use during the same day. The RBC suspensions were adjusted to a hematocrit of 0.04% by dilution of the base suspension with hypotonic solutions. The osmolarity and the ionic strength were varied independently by adding different amounts of NaCl and saccharose to 0.017 mol/l Tris buffer. When preparing the solutions, analytically pure NaCl (Fluka) and saccharose (Merck) and doubly distilled water of specific dc resistance greater than 1 MΩ cm. Trizma HCl and Trizma base were used as obtained from Sigma. All experiments described below were carried out at 20°C.

The cell in which RBC adhesion was studied is shown schematically in Fig. 6. It consists of two glass parts (1 and 2) connected by a lapped joint. To allow fluid to enter and exit the cell two glass tubes (3 and 4) are used. A circular cover glass (Fisher Sci. Co) with a diameter of 18 mm and thickness of 0.17 mm (7) is fixed at the lower part of the experimental cell by means of two teflon rings (5 and 6). The RBC adhere on the glass plate, thus allowing their photomicrography.

RBC suspensions were supplied to the experimental cell by an infusion pump at a rate of 99 ml/min. The glass cell was kept on the stage for two hours. After that it was turned so that the lower side faced upward and was kept in this position for another two hours to allow non-adhered RBC to be detached from the cover glass. Then the suspension was gradually replaced by infusion of buffer solution at a rate of 20 ml/min (a rate small enough to prevent detachment or irreversible deformation of the adhering RBC). Finally the glass cell was restored in its initial position and placed carefully on the microscope stage. Photomicrographs of the interference patterns were taken and processed as described in the previous section.

5. RESULTS AND DISCUSSION

a) Preliminary experiments with glass beads

As mentioned earlier, small glass beads of known radii were used to check the applicability of the interferometric method. The glass cell depicted in Fig. 6 was filled with water and glass spheres were allowed to settle onto its bottom (the plate (7) in Fig. 6). Fig. 4 illustrates the observed interference pattern.

Fig. 7[a] represents the experimental data for H_k vs. r_k for a glass bead with radius of 148 μm. The values of H_k were determined from the interference order by using Eq. (8), while the r_k were directly

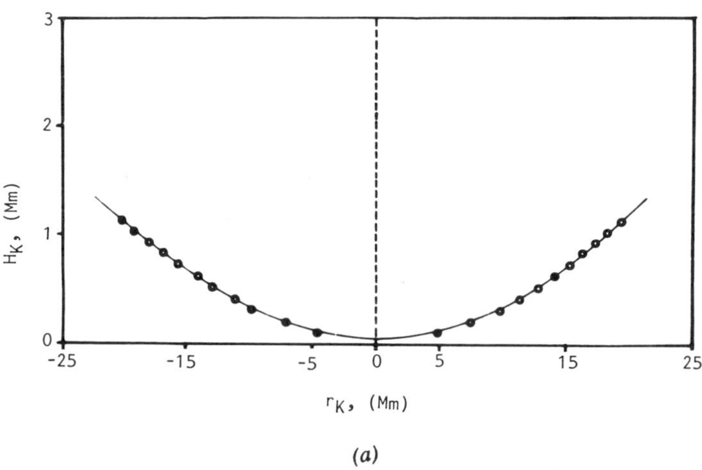

(a)

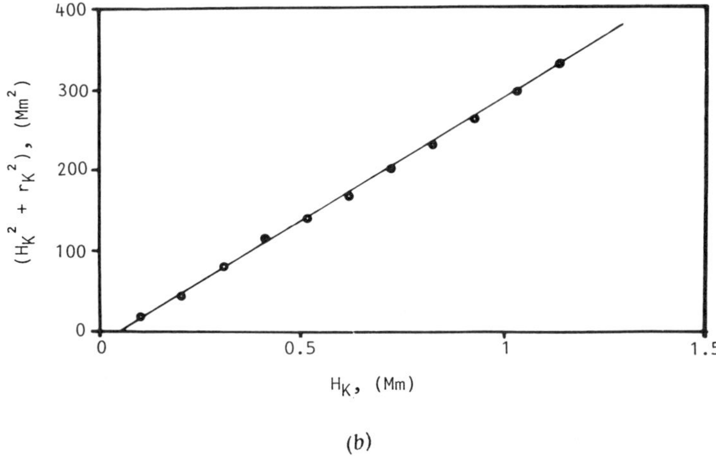

(b)

Fig. 7. Plots of H_k vs r_k (a), and $(H_k^2 + r_k^2)$ vs H_k (b) for a glass bead of radius 148 μm.

Table 1. Data for two glass beads of different radii measured directly and interferometrically.

R [μm] directly	R [μm] interferometry	h_0 [nm] directly	h_0 [nm] interferometry
148 ± 5	151 ± 15	—	63 ± 13
384 ± 6	383 ± 13	—	11 ± 60

measured from the photomicrographs as explained in Section 3. Fig. 7^b represents the same data (averaged from the r_k measurements along different radii) plotted in accordance with Eq. (10). The straight line (correlation coefficient 0.999) shows that the shape of the bead agrees well with that of a sphere.

Table 1 contains values of R measured directly and interferometrically (in accordance with Eq.(10)) from the photomicrographs in Fig. 4. It is not possible to measure h_0 directly. The R values measured directly are obtained by microscopic observations of the equator of the bead in transmitted light. These measurements revealed that the beads are not in fact perfect spheres. This fact may explain, at least in part, the greater standard deviation of the interferometrically determined values of R and h_0. Indeed, the interference pattern covers 0.2 - 0.4% of the bead area and a local deviation of the curvature in this region can affect the calculated values of R and h.

Notwithstanding these details, we conclude that the experimental verification of the interferometric technique with glass beads shows that it is precise enough for measurements with spherical RBC.

b) Effect of Ionic Strength on RBC adhesion

Due to the different age and membrane properties of the RBC they form adhesion films of different area at the same conditions as Fig. 5 shows. One can approximately divide the RBC into three types depending on their adhesion area: Type A - small adhesion area;   Type B - intermediate adhesion area; Type C - large adhesion area.

Within each type, the interferometric data agree well with spherical profile. Fig. 8 shows the data for RBC classified as type B. For all types the correlation coefficient of the respective straight line was greater than 0.95.

To check the effect of ionic strength on RBC adhesion the concentration of NaCl was varied at constant osmolarity. The latter was achieved by corresponding changes in the concentration of saccharose in the solution. Thus the osmolarity was fixed to be 151 mosM at NaCl concentrations varying between 1×10^{-2} and 6.1×10^{-2} mol/l.

At 1×10^{-2} mol/l NaCl the RBC are not firmly adherent: one observes their Brownian motion. At concentrations of $3 - 6.1 \times 10^{-2}$ mol/l NaCl, the RBC form stable adhesion films whose area increases with electrolyte concentration.

Table 2 shows data obtained for RBC adhesion films at a fixed osmolarity of 151 mosM, but at different NaCl concentrations. To compare the interferometric data we chose RBC with almost equal radii. The three RBC that were compared were of type B. The values of R and h_0 were calculated from the slope and the intercept of the plot of $(H_k + r_k)$ vs H_k (cf. Eq. 10 and Fig. 8^b). The radius of the dark spot (the film radius) was measured directly from the photographs. Then θ_0, h_f and θ_h were calculated from Eqs. (12), (14) and (15).

For 3×10^{-2} mol/l NaCl the (extrapolated) spherical surface does not intersect the plane of the glass surface and the contact angle θ_0 is not defined. The film thickness h_f decreases and the contact angle θ_h increases with the increase of the NaCl concentration. (See Table 2.) The differences are more pronounced when extrapolated parameters h_0, θ_0 and r_c are compared. These findings agree with the DLVO theory predicting suppression of the electrostatic repulsion and thinner equilibrium films for higher electrolyte concentrations. The results for h_f agree with the results of Wolf and Gingell, (1983).

Table 2. Effect of electrolyte concentration on RBC adhesion at fixed osmolarity (153 mosM)

$c \times 10^2$ [mol/l]	R [μm]	h_0 [nm]	θ_0 [deg]	r_c [μm]	h_f [nm]	θ_h [deg]
3.0	4.372	7.6	—	—	89.6	9.7
4.5	4.036	-6.3	3.2	0.225	73.8	11.4
6.1	4.294	-15.7	4.9	0.367	73.6	11.7

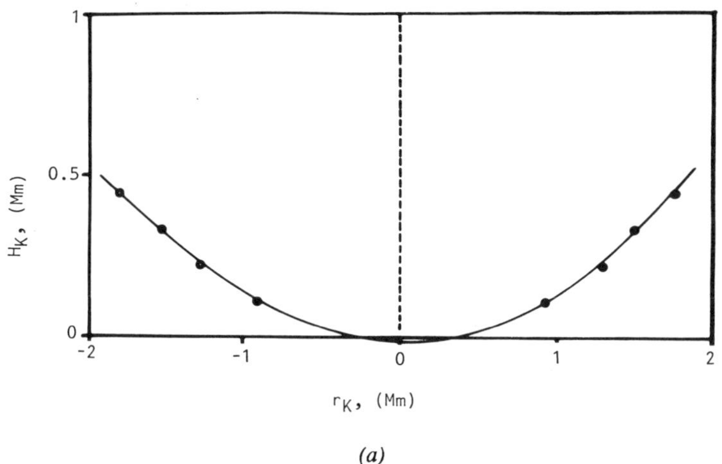

(a)

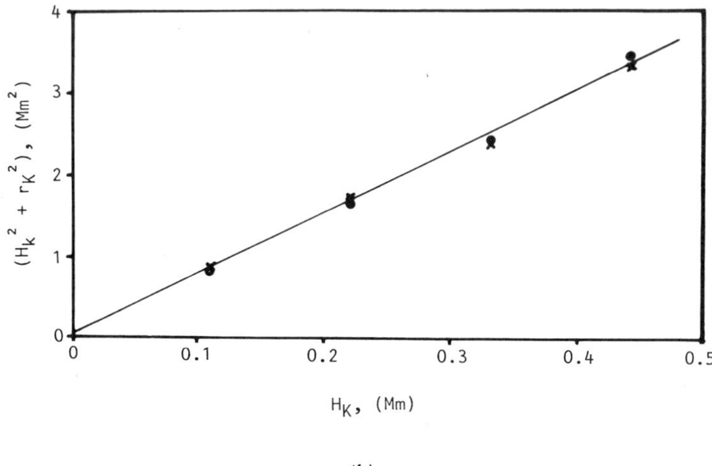

(b)

Fig. 8. Plots of H_k vs r_k (a) and $(H_k^2 + r_k^2)$ vs H_k (b) for a RBC type B (R = 4.036 μm, h_0 = -6.3 nm). (×) - left side of cell profile; (○) - right side of cell profile.

Table 3. Effect of solution osmolarity on the RBC adhesion at 4.5×10^{-2} mol/l NaCl

osmol. [mosM]	cell type	slope	inter-sect	R [μm]	h_0 [nm]	θ_0 [deg]	r_c [μm]	h_f [nm]	θ_h [deg]	r_f [μm]
	A	7.731	-0.164	3.845	21.2	—	—	75.0	9.7	0.641
151	B	7.158	-0.120	3.562	16.8	—	—	68.3	9.8	0.604
	C	7.558	-0.164	7.515	21.8	—	—	61.5	8.4	0.545
	A	7.498	0.012	3.750	-1.6	1.7	0.110	75.1	11.6	0.755
153	B	8.059	0.050	4.036	-6.3	3.2	0.225	73.8	11.4	0.797
	C	7.731	0.213	3.893	-27.5	6.8	0.462	22.0	9.1	0.619
	A	9.157	0.042	4.584	-4.6	2.6	0.204	71.0	10.4	0.829
155	B	10.692	0.410	5.384	-38.2	6.8	0.641	72.5	11.6	1.086
	C	8.413	0.762	4.296	-87.3	11.7	0.873	22.4	12.8	0.974

c) Effect of Osmolarity on the RBC Adhesion

To study separately the effect of osmolarity we fixed the NaCl concentration at 4.5×10^{-2} mol/l and varied the solution osmotic pressure by changing the concentration of saccharose. The osmolarity variation was limited to the range 151 to 156 mosM where spherical (swollen) RBC were observed. At osmolarities higher than 156 mosM (e.g. 161 mosM) cells began to shift from a spherical to a discoid shape; at osmolarities lower than 151 mosM (e.g. 145 mosM) lysis was observed at the experimental temperature. (In comparison at 25°C lysis occurred at an osmolarity of 140 mosM).

The results for the three types of RBC (A, B and C) and for three different osmolarities, namely 151, 153 and 156 mosM are presented in Table 3. The parameters R, h_0 and r_f were determined as explained above and then r_c, θ_0, h_f and θ_h were calculated from Eqs. (13-15).

At 151 mosM θ_0 and r_c are not defined since the (extrapolated) spherical surface of adhering cell did not intersect the plane of the glass plate surface. At 153 and 156 mosM, both r_c and θ_0 increased with the size of the RBC. The increase of the contact angle θ_0 could be due (i) to a reduction of film thickness

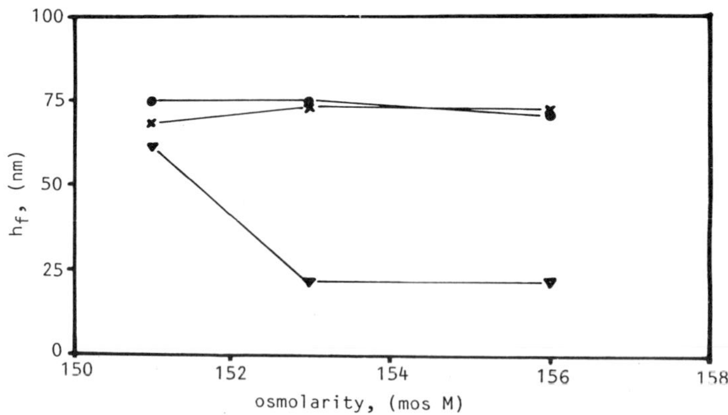

Fig. 9. Plot of the adhesion film thickness h_f vs the osmolarity of the buffer solution for the three types of RBC: A (○), B (×) and C (▽).

and an increase of the adhesion energy and (ii) to the action of line tension at the periphery of the adhesion film. In the framework of the "membrane" model (when h_f is neglected) one cannot decide which of these two effects prevails.

The "detailed" film model provides additional possibilities. The data for the adhesion film thickness h_f is visualized in Fig. 9. One sees that (except the last two points for the RBC type C) h_f practically does not depend on osmolarity. This fact is in agreement with the DLVO theory because the changes in the saccha rose concentration do not affect neither the electrostatic, nor the van der Waals interactions in the adhesion film. The anomalous data for some RBC type C deserve additional study.

The constancy of the film thickness h_f at different osmolarities implies that the observed variation of the contact angle can be explained by the action of a positive line tension, κ, rather than by the changes in the energy of adhesion. To strengthen this interpretation, we plotted $\cos\theta_h$ vs $1/r_f$ using the data in Table 3. (See Fig. 10.) The experimental points are fitted by a straight line with a correlation coefficient 0.89. If the RBC tension, σ_{23}, and the line tension, κ, are assumed to be constant, from Eq. (7) and Fig. 10, one can determine $\Delta\sigma/\sigma$ to be 0.035 and κ/σ to be 1.3×10^{-6} cm.

It is well known (see e.g. Kralchevsky and Ivanov, 1985) that the line tension is an excess quantity accounting for the existence of a narrow transition region between the thin film and the liquid meniscus. In the present case, the shape of the RBC changes from planar to spherical across this transition region (cf. Fig. 3b)). In addition, the RBC tension changes from σ_f to σ_{23}. The ratio $\kappa/\sigma = 13$ nm can be considered as an estimate for the width of the transition zone.

The tension of the RBC, σ_{23}, is a quantity which is difficult to measure directly. One way to determine σ_{23} is to use the Laplace equation

$$\frac{2\sigma_{23}}{R} = \Delta P \tag{16}$$

where ΔP is the pressure drop across the RBC membrane. (The contribution of the shearing elasticity in Eq. (16) is zero because the two radii of curvature of the membrane surface are equal - see, for example, Kralchevsky, 1990). In principle ΔP can be calculated from the condition for Donnan equilibrium between the RBC interior and exterior. However the effective concentration of the macro-ions (which can not penetrate the membrane) inside the cell is not known. Thus, to estimate the specific work of adhesion, $\Delta\sigma$, and the line tension, κ, we used the value $\sigma_{23} = 6$ mN/m determined by Evans et al. (1976). From the slope and the intersect of Fig. 10 we determined $\kappa = 7.8 \times 10^{-11}$ N and $\Delta\sigma = 0.21$ mN/m.

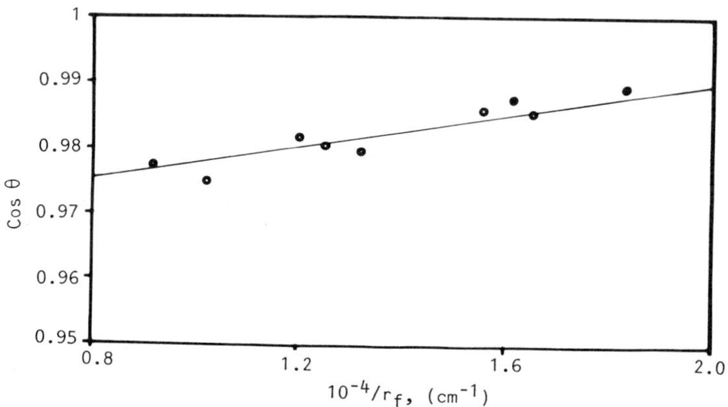

Fig.10. Plot of $\cos\theta_h$ vs $1/r_f$ at 0.045 mol/l NaCl and temperature 20°C.

It should be noted that the observed dependence of the contact angle θ_h on r_f could well be due to changes of the RBC tension, σ_{23}, connected with changes in the osmotic pressure. However, the verification of such a hypothesis needs a method for direct measurement of σ_{23} which is still not available.

6. CONCLUDING REMARKS

The main purpose of the present study is to determine the specific work of adhesion of RBC to solid surfaces by using its relation to the contact angle. In our experiments we used spherically-shaped, swollen RBC. The solid substratum was a horizontal plane-parallel glass plate. A specially designed experimental cell (Fig. 6) was used, providing both for the supply of buffer solution and the possibility of making interferometric measurements.

It was established experimentally that three types of adhesion films between the RBC and the glass surface (called type A, B and C) are formed - see Fig. 5. They differ in the area of their adhesion films, presumably caused by the different age and membrane properties of the individual RBC in cell suspension.

By processing photomicrographs of the interference pattern induced by the cell and glass surfaces, the radius of the spherical RBC surface, R, the adhesion film thickness, h_f, the contact angle, θ_h, and other geometrical parameters were determined. It turns out that h_f decreases when increasing the electrolyte concentration at constant total osmolarity of the buffer solution. On the other hand, h_f is almost constant when the total osmolarity is varied at constant electrolyte concentration. This is an indication of the importance of the electrostatic interactions in the adhesion film. The observed dependence of the contact angle θ_h on the film radius at constant film thickness can be attributed to the action of a positive line tension, κ, at the film periphery. Using models discussed in the text, we determined $\kappa = 7.8 \times 10^{-11}$ N and specific work of RBC adhesion $\Delta\sigma = 0.21$ mN/m.

There are several possibilities for further experiments and development of the above method: (i) The glass substratum can be changed with other modified solid surfaces. (ii) The adhesion of non-spherical RBC can be studied by varying the solution content (osmolarity). (iii) The percentage of cells falling into categories A, B and C may be investigated for diagnostic purposes. (iv) The adhesion of other biological cells can be studied. We hope the results of the present work will be useful for all branches of biological and medical science and practice where cell adhesion to solid surfaces is important.

7. REFERENCES

Arvinte, T., Cuold, A., Schulz, B., and Nicolau, C., 1989, Low-pH association of proteins with the membranes of intact red blood cells. II. Studies of the mechanism, Biochim. Biophys. Acta, 981: 51.

Bell, G. I., 1988, Models of cell adhesion involving specific binding, In: "Physical Basis of Cell-Cell Adhesion", P. Bongrand, ed., CRC Press, Inc. Boca Raton, Florida.

Bongrand, P., Capo, C., Mege, J. L., and Benoliel, A. M., 1988, Surface physics and cell adhesion, In: "Physical Basis of Cell-Cell Adhesion", P. Bongrand, ed., CRC Press, Inc. Boca Raton, Florida.

Bongrand, P., Capo, C., and Depieds, R., 1982, Physics of cell adhesion, Progr. Surface Sci., 12: 217.

Derjaguin, B. V., and Landau, L. D., 1941, Theory of the stability of strongly charged lyophobic sols and of the adhesion of strongly charged particles in solutions of electrolytes, Acta Physicochemica USSR, 14: 633.

Dimitrov, A. S., Kralchevsky, P. A., Nikolov, A. D., and Wasan, D. T., 1990, Contact Angles of Thin Liquid Films — Interferometric Determination. Colloid Surfaces, 47: 299.

Dolowy K., 1980, A physical theory of cell-cell and cell-substratum interactions, In: "Cell Adhesion and Motility", A. S. G. Curtis and J. D. Pitts, eds., Cambridge Univ. Press, Cambridge.

Donath, E., and Voigt, A., 1983, Charge distribution within cell surface coats of single cell and interacting surfaces - a minimum free electrostatic approach. Conclusions for electrophoretic mobility measurements, J. Theor. Biol., 101: 569.

Evans, E. A., and Skalak, R., 1979, CRC Critical Rev. Bioeng., 3: 181.

Evans, E. A., Waugh, R. and Melnik, L., 1976, Elastic area compressibility modulus of red cell membrane, Biophys. J., 16: 585.

Gershey, E. L., and d'Alisa, R. M., 1980, Characterization of a CV: I. Cell cycle; II. The role of cell-substrate attachment, J. Cell Sci., 42:357.

Gingell, D., and Todd, I., 1979, Interference reflection microscopy: A quantitative theory for image interpretation and its application to cell-substratum separation measurement, Biophys. J., 26: 507.

Gingell, D., and Vince, S., 1980, Long-range forces and adhesion: An analysis of cell-substratum studies, In: "Cell Adhesion and Motitlity", A. S. G. Curtis and J. D. Pitts, eds., Cambridge Univ. Press, Cambridge.

Ivanov, I. B., 1988, "Thin Liquid Films", ed., M. Dekker, New York.

Ivanov, I. B., and Kralchevsky, P. A., 1988, Mechanics and Thermodynamics of Curved Thin Liquid Films. In: "Thin Liquid Films", I. B. Ivanov, ed., M. Dekker, New York.

Ivanov, I. B., and Toshev, B. V., 1975, Thermodynamics of thin liquid films. II. Film thickness and its relation to the surface tension and the contact angle, Colloid Polymer Sci., 253: 558.

Izzard, C. S., and Lochner, L., 1976, Cell-to-substrate contacts in living fibroblasts: an interference reflection study with an evaluation of the technique, J. Cell Sci., 21: 129.

Kralchevsky, P. A., 1990, Micromechanical Description of Curved Interfaces, Thin Films and Membranes. J. Colloid Interface Sci., 137: 217.

Lerche, D., 1983, A biophysical model for interaction of cells with a surface coat (glycocalyx). I. Electrostatic interaction profile, J. Theor. Biol., 104: 231.

Leonard, E. F., Rahmin, I., Angarska, J. K., Vassilieff, C. S., 1987, The close approach of cells to surfaces, Ann. N.Y. Acad. Sci., 516:502.

Nicol, A., and Garrod, D. R., 1979, The sorting out of embryonic cells in monolayer, the differential adhesion hypothesis and the non-specifity of cell adhesion, J. Cell Sci., 38: 249.

Parsegian, V. A., and Gingell, D., 1973, A physical force model of biological membrane interaction, Proc. Amer. Chem. Soc., In: "Recent Advances in Adhesion", L. Lee, ed., Gordon and Breach, London.

Pethica, B. A., 1977, The Contact Angle Equilibrium. J. Colloid Interface Sci., 62: 567.

Pethica, B. A., 1980, In: "Microbial Adhesion to Surfaces", R. C. W. Berkeley, et al., eds., Ellis Horwood, Chichester.

Rowlinson, J. S., and Widom, P., 1982, "Molecular Theory of Capillarity", Univ. Press, Oxford.

Todd, I., and Gingell, D., 1980, Red blood cell adhesion. I. Determination of the ionic conditions for adhesion to an oil-water interface, J. Cell Sci., 41: 125.

Verwey, E. J. W., and Overbeek, J. Th. G., 1948, "Theory of the Stability of Lyophobic Colloids", Elsevier, Amsterdam.

Wolf, H., and Gingell, D., 1983, Conformational response of the glycocalyx to ionic strength and interaction with modified glass surfaces: Study of live red cells by interferometry, J. Cell Sci., 63: 101.

ON THE MECHANISM OF MEMBRANE FUSION:

USE OF SYNTHETIC SURFACTANT VESICLES AS A NOVEL MODEL SYSTEM

Tino A.A. Fonteijn, Jan B.F.N. Engberts and Dick Hoekstra[*]

Laboratories of Organic Chemistry and Physiological
Chemistry, University of Groningen, Groningen, The
Netherlands

INTRODUCTION

The structural and physical properties of membranes affect many
functional and dynamic processes that take place in and between membranes.
Therefore, selective modulation of these properties may reveal fundamental
insight at the molecular level as to how a certain function, involving a
membrane, is accomplished mechanistically. To gain such an insight, the
model of choice may be particularly relevant, as one membrane model may be
more amenable and susceptible to modulation than another. This appears
particularly true in case of studies aimed at understanding and revealing
the mechanism of membrane fusion.

Early work involved studies using natural membranes such as those
derived from erythrocytes. The induction of fusion could be accomplished
by using, for example, chemical agents or viruses (Blumenthal, 1988;
Hoekstra and Wilschut, 1989; Hoekstra and Kok, 1989 and references
therein). These studies provided limited insight as far as the molecular
mechanism of fusion was concerned. In that respect, i.e., from the
fundamental point of view, the application of artificial membranes has
been more beneficial, particularly since such membranes allow a systematic
study of the effect of a great variety of lipid compositions, including
the effects of incorporation of non-bilayer forming lipids. Such systems
also provide possibilities for selective reconstitution, involving studies
of effects of membrane proteins in and on fusion.

More recently we have extended our studies of the molecular
mechanism(s) of fusion by employing a model system involving membrane
vesicles composed of synthetic amphiphiles (figure 1). In many respects
these amphiphiles appear to mimic the properties of bilayers consisting of
phospholipids, including polymorphic properties and the ability to undergo
membrane fusion. A particular advantage of synthetic amphiphiles is the
possibility to modify their structures in a well-defined manner, to an
almost unlimited degree. Accordingly, systematic and detailed studies can
be carried out to correlate structural and physical properties of these
amphiphilic molecules to functional aspects of membrane systems that they
form upon their suspension in an aqueous environment. This opportunity, in
turn, may contribute to a better fundamental understanding of the
properties and functioning of natural membranes. In the following we will

---

[*]Corresponding author: Laboratory of Physiological Chemistry, University
of Groningen, Bloemsingel 10, 9712 KZ Groningen, The Netherlands.

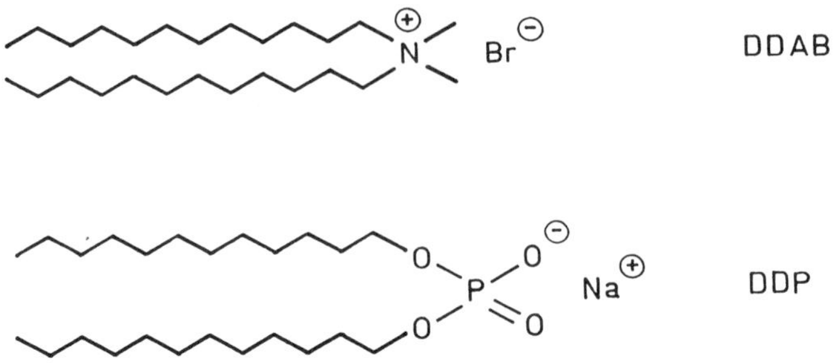

Fig. 1. Molecular structures of a cationic (DDAB) and anionic (DDP)
vesicle-forming synthetic amphiphile. DDAB, di-n-dodecyldimethyl-
ammonium bromide; DDP, di-n-dodecyl phosphate.

summarize some aspects of our studies dealing with membrane fusion
properties of bilayers composed of synthetic amphiphiles. Before
discussing experimental details, we will briefly describe some general
features of fusogens and membrane fusion.

SOME GENERAL FEATURES OF MEMBRANE FUSION AND MEMBRANE HYDRATION

Membrane fusion can be defined as the merging of two, initially
separated membrane spheres into one, which is accompanied by the mixing of
membranes and the contents of the vesicles. Whether artificial or
biological, in any membrane system the mechanism triggering the ultimate
merging of initially separated membranes may differ, but the overall
characteristics will be the same. This overall process entails that
adjacent membranes come into close approach and adhere or aggregate to
establish a zone of molecular contact. In a second stage, a perturbation
in this region of molecular contact has to take place in which the opposed
membranes destabilize. Obviously, for triggering the fusion step, the
bilayers will have to depart, albeit locally, from the bilayer structure.
Whether the structure thus formed, and actually representing a potential
fusion-intermediate structure, displays features of a defined, though
transient nonbilayer structure, is still a matter of debate. At any rate,
the result of this perturbation will be that membrane components and
aqueous contents, initially present in separate structures, intermix.
What is the nature of the parameters that determines whether fusion
occurs? Membranes do not fuse spontaneously since their mutual
interactions are governed by repulsive electrostatic and hydration forces
(Rand and Parsegian, 1989). In addition, membrane fusion, requiring
membrane perturbation, will likely depend on disruption of the combination
of hydration and hydrophobic stabilizing interactions, governing membrane
self-assembly and stability (Israelachivilli et al., 1980; Tanford, 1980).
The short-range repulsive hydration forces have been shown to be dominant
at interbilayer distances of less than 30Å (Marra and Israelachvili, 1985;
Rand and Parsegian, 1989). These forces, also known as 'structural' or
'solvation' forces arise from the tendency of the strongly polar part of
an amphiphilic molecule, packed in a bilayer configuration, to hold onto
the solvent in order to lower its energy. This tenacious property, which
in fact is a balance between the high energy of the hydrocarbon-water
interface and the energy-lowering interaction with solvent, is the origin
of hydration forces between amphiphilic surfaces.
One of the first theories of 'structural' forces was developed by

Marcelja and Radic (1976). They assumed that the hydrophilic bilayer surface affects the nearest water layers. The order parameter that describes this perturbed water structure, decays exponentially with increasing distance from the bilayer. Subsequently this order parameter has been described in terms of polarizability and free energy density (Cevc et al., 1982; Gruen and Marcelja, 1983). Finally, Jönsson and Wennerström (1983) have proposed a theory based on the interaction of discrete lipid molecule dipoles with their electrostatic images at the opposite bilayer. More recently, the solvent-structural origin of these hydration forces has been questioned. Instigated by the theory of Helfrich (1975), the possibility has been reconsidered that thermal mechanical fluctuations contribute to the repulsive forces between bilayers (Sornette and Ostrowsky, 1986; Israelachvilli and Wennerström, 1990). Israelachvilli and Sornette (1985) have demonstrated that there exists in fact a quantitative relationship between effective headgroup size and the magnitude of the hydration forces. In this context it is finally worth mentioning that Rand et al. (1988) have described a model which takes into account <u>attractive</u> hydration forces as well, resulting from intermembrane networks of hydrogen-bonded water. Although awaiting further experimental verification this model is consistent with some observations of intermembrane interactions (cf. Novick and Hoekstra, 1990).

Although much remains to be resolved, artificial membranes have been especially useful in the recognition and analysis of membrane hydration. In fact, it has become clear that for close apposition of bilayers, hydration forces represent a major barrier to be overcome before merging can take place. Any mechanism, attempting to clarify fusion, must therefore take into account the necessity of disruption of the hydration shells surrounding the polar headgroup of the amphiphile, as the ultimate prerequisite for fusion to occur.

Since fusion does not represent a 'spontaneous' event, particular fusogens are necessary to induce the process. In model systems a whole range of such agents have been used such as amphipathic fusogens, including lysophospholipids, long-chain alcohols and fatty acids; poly- and divalent cations; polypeptides and proteins; chemical agents, displaying dehydrating properties. The main features of these agents can be summarized as follows. Usually, they are capable of modulating the surface properties of the membranes, consisting of either phospholipids or synthetic amphiphiles. These changes include electrostatic neutralization, in case of charged membranes, and an ability to increase the (local) hydrophobicity of the surface. Enhanced hydrophobicity may lead to bilayer destabilization and may, when ocurring in closely apposed bilayers, eventually lead to fusion. Local destabilization may also facilitate the penetration and concomitant perturbation of bilayers occurring when polypeptides and/or proteins penetrate. These transiently perturbed structures may thus become highly susceptible to fusion although at the molecular level many details are still largely obscure. In fact the latter events may be applicable to current models of virus-induced membrane fusion (Hoekstra, 1990). The role of a viral envelope glycoprotein in this process can be seen as bringing viral and target membrane into close proximity to overcome the repulsive hydration forces and to induce formation of a nonbilayer fusion intermediate. Crucial to these events appears to be the penetration of a hydrophobic segment of the viral protein into the target membranes. This fusion process is strongly facilitated when such experiments are carried out in the presence of a dehydrating agent, such as poly(ethylene)glycol (Hoekstra et al., 1989; see also below).

SYNTHETIC AMPHIPHILES AS MODEL FOR MEMBRANE FUSION

In 1977, Kunitake and Okahata reported that synthetic amphiphiles, when suspended in an aqueous environment, could form a bilayer structure,

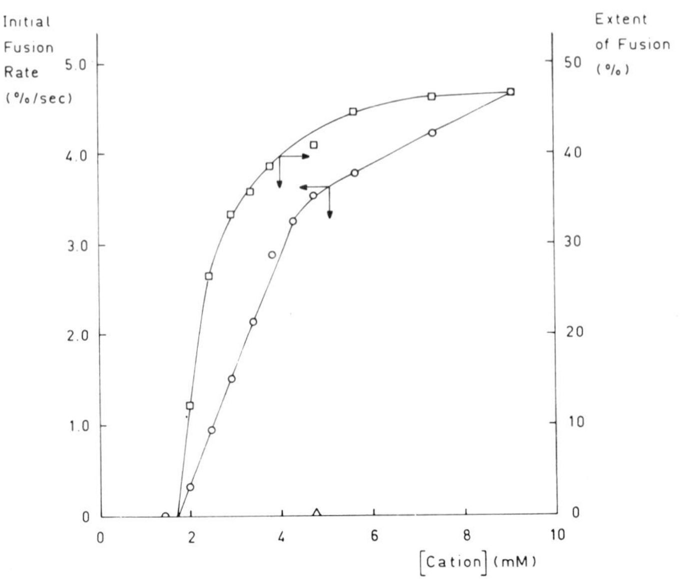

Fig. 2. Rate (o) and extent (□) of fusion of DDP vesicles, induced by Ca^{2+}. Note that Mg^{2+} fails to induce membrane merging (△). Fusion was monitored at 40°C, pH 7.4, using a lipid mixing assay. For details, see Rupert et al., 1987 (reproduced with permission).

very analogous to the organization of phospholipids in a similar environment. Ever since, the interest in synthetic amphiphiles as a novel membrane mimetic model, expanded rapidly (Fendler, 1982). In 1985, we (Rupert et al., 1985) observed that vesicles, composed of the cationic amphiphile didodecyldimethylammonium bromide (DDAB) fuse to giant vesicles with diameters of several μ's when vesicles with a size of approximately 3000Å (prepared by ethanol injection) were incubated in the presence of divalent anions, such as dipicolinic acid or sulphate, or with bi-functional organic anions consisting of a charged group and a hydrophobic moiety.

More recently, we have carried out detailed studies of factors that govern fusion of membrane vesicles, composed of n-dialkyl phosphates. Some of these amphiphiles exhibit polymorphic phase behavior similar to that of phospholipids. In addition, these systems appear to be particularly instructive in analyzing the role of membrane hydration in fusion.

Vesicles, formed from the anionic synthetic amphiphile didodecyl phosphate (DDP) readily aggregate and fuse in the presence of Ca^{2+}, while Mg^{2+} only causes aggregation (Fig. 2, Rupert et al., 1987. In this respect, the amphiphile displays similar properties as observed for large unilamellar vesicles, prepared from the phospholipid phosphatidyl-serine (PS). Interestingly, the threshold Ca^{2+} concentrations for aggregation of DDP vesicles is 1.0 mM, while fusion is triggered at 1.75 mM. This clear distinction in threshold concentrations, which is usually not seen for phospholipid systems, has been exploited to analyse ion binding behavior and physical properties of the amphiphile bilayer at conditions where aggregation has been established, but prior to fusion (Rupert et al., 1986, 1987, 1988a). The results of such studies demonstrated that the driving force for Ca^{2+}-induced fusion of DDP vesicles entails a dehydration of the membrane surface, without the necessity of propagation of such an effect into the bilayer interior. The same conclusion was reached in earlier studies of dipicolinic acid-induced fusion of DDAB vesicles (Rupert et al., 1986). Based on extensive studies, involving fluorescence polarization with ANS-(to report surface changes) and DPH-labeled vesicles (to report changes within the bilayer interior), we concluded that DPA^{2-}-induced fusion of DDAB vesicles does not require the induction of an isothermal phase transition (accompanying total removal of water). Rather, the data obtained for ANS-labeled vesicles indicated that the physical and structural alterations occurring upon DPA^{2-} binding to DDAB, are entirely restricted to changes at the membrane surface, i.e., local dehydration of the surface sufficed for driving the fusion reaction. These events occur primarily on separate vesicles, i.e., prior to aggregation. For the DDP system, the evidence supported the conclusion that after Ca^{2+} binding in a socalled 'cis'-mode, which caused aggregation of the vesicles, a Ca^{2+} complex between apposed vesicles ('trans'-mode) presumably triggered the actual fusion event, analogous to the hypothesis proposed for Ca^{2+}-induced fusion of phospholipid vesicles. The increase in Ca^{2+} binding, responsible for trans-Ca^{2+}/DDP complex formation amounted some 6-7%, relative to the amount bound at conditions of aggregation only (i.e., 'cis'-mode).

As noted above, Mg^{2+} did not trigger fusion although it caused DDP vesicles to aggregate. Furthermore, Mg^{2+} displays a higher binding constant for DDP than Ca^{2+}, which contrasts with phospholipid systems. Given the higher affinity constant of Mg^{2+} and the fact that Mg^{2+} competes for the same binding sites as Ca^{2+}, it is not surprising that in the presence of an equimolar mixture of Ca^{2+} and Mg^{2+}, fusion is almost completely suppressed (figure 3). Thus, in contrast to the synergism observed for the Ca^{2+}/Mg^{2+} system in case of PS vesicles (Hoekstra, 1982), no synergistic effect is seen for the DDP vesicles in the presence of both cations. Evidently, this difference is to be explained in terms of an inverse relationship in binding constants for Ca^{2+} and Mg^{2+} in both systems, possibly related to a difference in geometry of the cation binding sites in PS vs DDP (Kurland et al., 1979).

The significance of 'hydration forces' as a major barrier to overcome during membrane fusion, is also illustrated when comparing the hydration

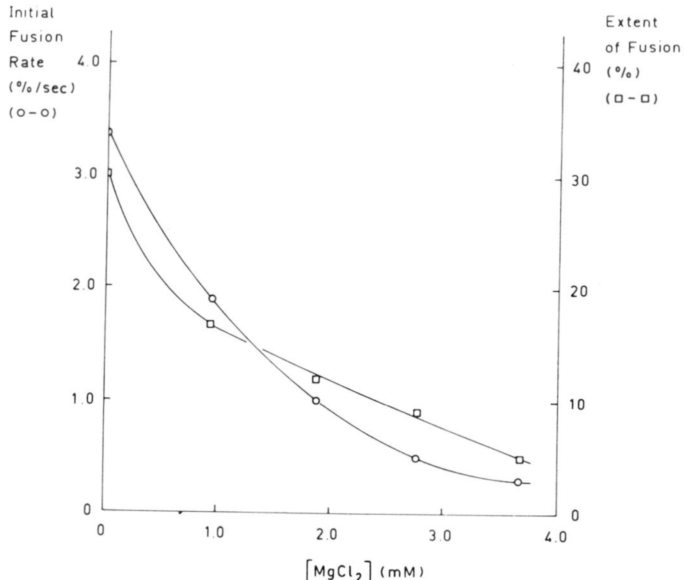

Fig. 3. Mg^{2+} inhibits Ca^{2+}-induced fusion of DDP vesicles. Fusion, induced by 4.7 mM Ca^{2+}, was measured in the presence of various concentrations of Mg^{2+}, as indicated. From Rupert et al., 1987, with permission.

properties of DDP versus those of DDAB. It can be shown that the phosphate
headgroup (of DDP) is more strongly hydrated than the cationic headgroup
of DDAB (Pullman et al., 1975; Rupert et al., 1985). This implies that
more free energy is required to dehydrate the phosphate headgroup and,
hence, to induce fusion. By the same token, since DDAB bilayers are
relatively poorly hydrated (a conclusion that can also be derived from
their high salt stability, relative to that of DDP (Carmona Ribeiro et
al., 1985)) it may be anticipated that the amount of thermodynamic work
needed for dehydration of this system and to induce fusion is relatively
limited. Indeed, divalent <u>anions</u> such as sulphate, phosphate and
dipicolinic acid readily perturb DDAB bilayer systems, resulting in fusion
(Rupert et al., 1986). Similar phenomena have been observed for other
cationic amphiphiles (Düzgünes et al., 1989). With these divalent anions
dehydration of the membrane surface of DDAB bilayers does occur at
concentrations up to 70 times lower than divalent cation concentrations,
needed to trigger fusion of DDP bilayers. In fact, with Mg^{2+} no fusion of
DDP bilayers is observed, even up to concentrations of at least 20 mM. The
much stronger hydration of DDP and PS bilayers would thus set forward also
particular criteria to the physical properties (in terms of dehydration)
of the divalent cations (Ca^{2+} versus Mg^{2+}).

Recently, we noted that synthetic amphiphiles exhibit polymorphic
phase behavior, similar to that of phospholipids (Rupert et al., 1987a).
Among the nonbilayer structures, proposed to act as intermediates in
membrane fusion, are those defined as 'inverted micellar intermediates'
(IMI, Siegel, 1986, 1986a). Such intermediates were proposed of being
capable to act as an interlamellar attachment that leads to fusion or,
depending on local density, could revert into hexagonal H_{II} phase
structures. In the presence of Ca^{2+}, DDP vesicles fuse to large vesicles
which subsequently collapse. Upon collapse, interbilayer contacts in
apposed bilayers occur which appear to lead to the formation of tubes that
possess the hexagonal H_{II} phase structure. These observations would be
consistent with the IMI concept (see Rupert et al., 1987a). It should also
be noted that the concept need not necessarily have a general
significance. Although DDAB vesicles can fuse into giant membrane
vesicles, no hexagonal phase formation is observed in this system. We are
currently examining molecular parameters that govern H_{II} phase formation
of the di-n-alkyl phosphates by employing phosphates that contain alkyl
chains of different lengths (Wagenaar et al., 1989).

EFFECT OF PEG ON FUSION OF AMPHIPHILES

High molecular weight polymers of poly(ethylene)glycol (PEG) have
been widely used to induced cell-cell fusion in cell hybridization
studies. PEG binds 2 to 3 water molecules per monomer unit (Arnold et al.,
1985) via hydrogen-bond interactions. At PEG concentrations above 3 wt%,
all water molecules present in bulk solution experience the presence of
PEG, while at 40 wt% essentially all water molecules are bound. At
concentrations above 30 wt% PEG is even capable of inducing fusion between
phosphatidylcholine (PC) vesicles (Boni et al., 1981, 1984; Parente and
Lentz, 1986) in spite of the fact that this lipid is particularly strongly
hydrated and usually interferes with Ca^{2+}-induced fusion of acidic
phospholipid vesicles that are supplemented with this zwitterionic lipid.
Evidently, the strong dehydrating properties of PEG facilitate in
overcoming the repulsive hydration barrier of this vesicle system.
We have studied in some detail the effect of PEG on Ca^{2+}-induced
fusion of DDP vesicles (Rupert et al., 1988). In this system we also noted
strong, molecular weight dependent effects of the polymer on fusion
(figure 4). The experiments were carried out at relatively low PEG
concentrations, such that medium viscosity did not interfere with mixing
of vesicle populations. In the presence of 10 wt% PEG, MW 8 kDa, no fusion
of DDP vesicles takes place, but as has been reported also for fusion of

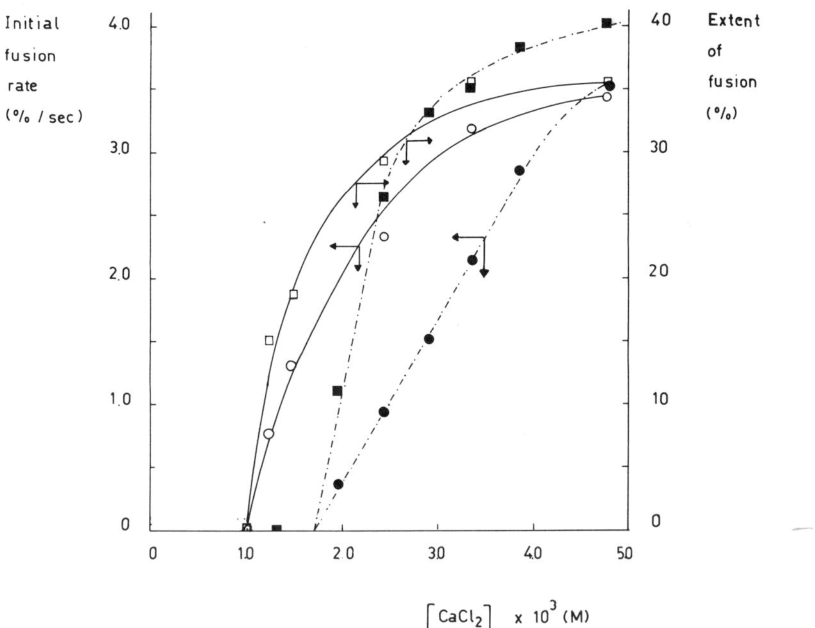

Fig. 4. Effect of PEG on fusion of DDP vesicles, as a function of the Ca^{2+} concentration.
Ca^{2+}-induced fusion was triggered in the presence (open symbols) or absence (filled symbols) of PEG (10 wt%, 8 kDa). Note that PEG facilitates Ca^{2+}-induced fusion at these conditions. For details, see Rupert et al., 1988.

acidic phospholipid vesicles (Hoekstra, 1982; MacDonald, 1986), Ca^{2+}-induced fusion is strongly facilitated in the presence of PEG. A substantial reduction in Ca^{2+}-threshold concentration can thus be obtained while in the presence of the polymer the initial fusion rates at lower Ca^{2+} concentrations increase more rapidly than in the absence of the polymer. A quite distinct effect on Ca^{2+}-induced fusion was exerted by PEG with a molecular weight of 20 kDa. At a 100-fold lower polymer concentration, but at the same Ca^{2+} concentration, the initial rate of DDP vesicle fusion was about 40-fold lower whereas the final extents at both conditions were essentially similar. Thus PEG 20 kDa affected the kinetics of fusion rather than fusion susceptibility. In fact, further experiments indicated that in contrast to PEG 8 kDa, PEG 20 kDa strongly interfered with interbilayer interactions between vesicles, thereby acting as a steric barrier for close fusion-susceptible membrane-membrane interactions. It was shown (Rupert et al., 1988) that in the combined presence of PEG 20 kDa and Ca^{2+}, Ca^{2+}-induced aggregation of DDP was inhibited when the PEG concentration was raised beyond a concentration of 1.5 x 10^{-3} wt% (figure 5). Thus PEG 20 kDa is capable of stabilizing the vesicles, presumaly eliminating formation of the required trans Ca^{2+}/DDP complex (Rupert et al., 1987). A likely and attractive explanation for this peculiar molecular weight dependent effect of PEG might be that differences occur in the clouding temperature of PEG molecules located near the bilayer-water interface. In water, PEG displays a lower critical solution temperature. Above this temperature a phase separation occurs with formation of a concentrated polymer phase (e.g., 20-40% PEG) and a dilute polymer phase (1% PEG). This temperature, known as the <u>clouding temperature</u>, depends on the molecular weight of PEG and on the presence of salts (Saeki et al., 1976; Florin et al., 1984; Ataman,

221

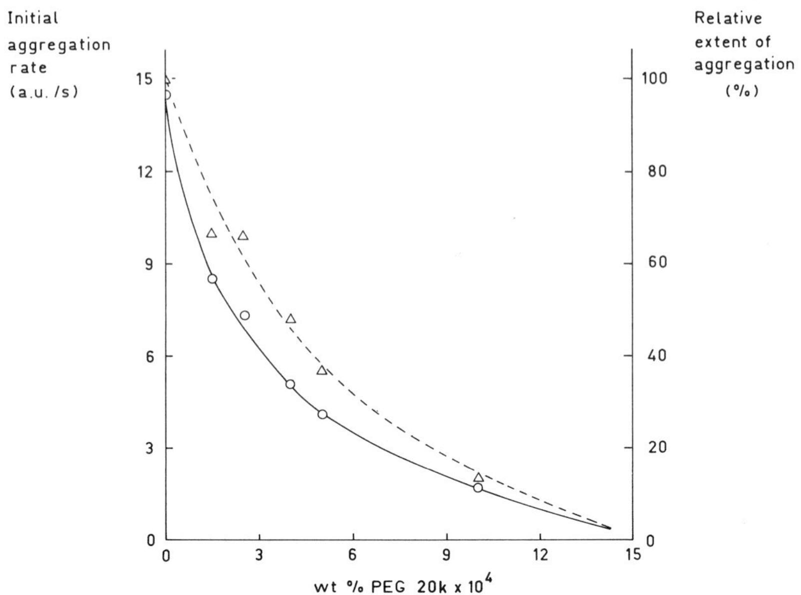

Fig. 5. Inhibition of Ca^{2+}-induced aggregation of DDP vesicles by PEG 20 kDa. DDP vesicles were suspended in PEG-containing media, followed by addition of Ca^{2+} (3.8 mm). The initial rate and relative extent of aggregation were determined by turbidity measurements. For details, see Rupert et al., 1988.

1987). Sodium and phosphate ions can lower the clouding temperature of PEG 20 kDa from 103°C (in water) to 43°C in a 1 M sodiumphosphate solution (Boucher and Hines, 1978; Ataman, 1987). These conditions are comparable to those near the DDP surface, which can be seen as a concentrated $(RO)_2PO_2Na$ solution. Since our experiments were carried out at 40°C it is quite conceivable that the local concentration of PEG 20 kDa near the bilayer surface was enhanced as a result of the phase separation from bulk solution.

NMR experiments revealed that this phenomenon is accompanied by a concomitant removal of hydration water from the bilayer headgroup, as evidenced by an upfield shift of the ^{31}P-signal. Furthermore, it could also be shown (Rupert et al., 1988) that the 'cis'-complexation constant for (lateral) Ca^{2+} binding to the DDP bilayer was not significantly affected by the presence of PEG, i.e., large parts of the vesicle surface are still freely accessible to Ca^{2+} and, hence, PEG 20 kDa does not appear to compete for Ca^{2+} binding sites. Yet, the polymer effectively prevents close approach and fusion of the DDP vesicles. Therefore, PEG 20 kDa must effectively shield the whole vesicle surface, which can be accounted for by an extension (with respect to the vesicle surface) of outer PEG loops or coils into the bulk water phase, which is a good solvent for PEG. Thus in this model (figure 6; for details see Rupert et al., 1988) the extending coils overlap unfavourably when the vesicles approach each other, thus resulting in steric stabilization. It should be noted in this respect that PEG 20 kDa coils possess dimensions comparable to the range of the Van der Waals attractions (11.0 nm), i.e., of the right dimensions to impart colloid stability.

Finally, the clouding temperature of PEG 8 kDa is (in water) about 13°C higher than that of the 20 kDa compound. Consistent with this notion is the observation that PEG 8kDa-inhibited fusion of DDP vesicles is observed around 52°C, thus providing further support for a role of

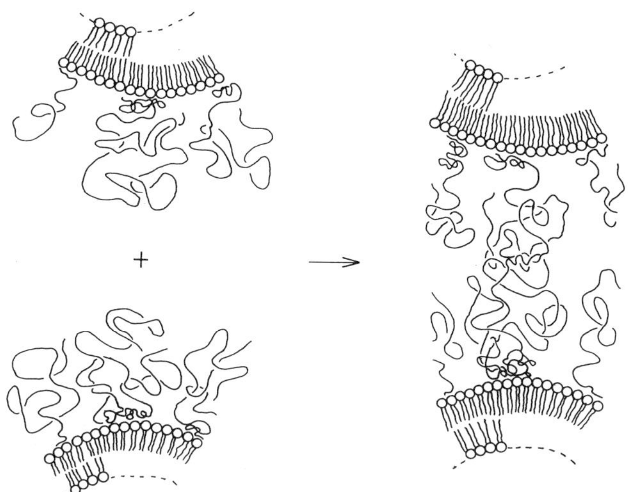

Fig. 6. Schematic representation of the interaction of PEG 20 kDa with DDP vesicles. The mode of interaction allows for 'cis' Ca^{2+}-DDP interactions. Due to the unfavorable overlap of the polymer coils, occurring when the vesicles approach each other, 'trans' Ca^{2+}-DDP interaction is prevented.

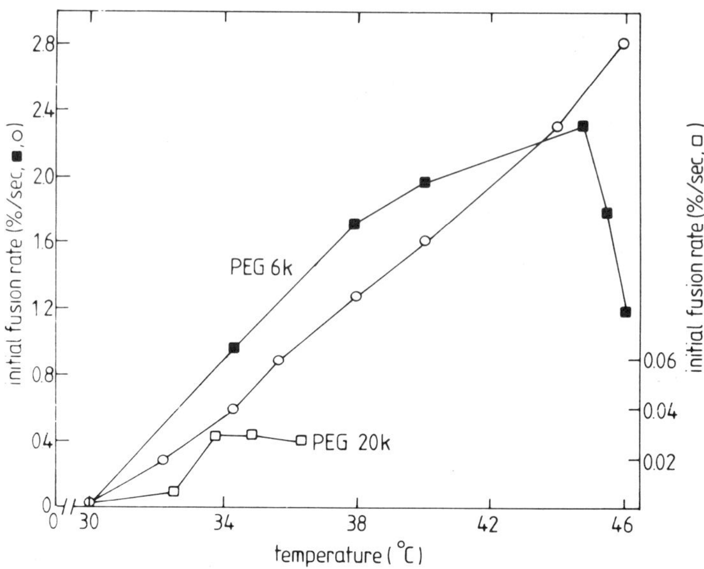

Fig. 7. Effect of temperature on Ca^{2+}/PEG-induced fusion of DDP vesicles. DDP vesicles were suspended in PEG-containing media (□, 0.3 wt% 20 kDa; ■, 10 wt% PEG 8 kDa). At the PEG concentrations used, spontaneous fusion did not occur. Membrane fusion was triggered upon addition of Ca^{2+} (O, in the absence of PEG) and initial rates were plotted as a function of temperature. Note that in the presence of PEG 20 kDa fusion becomes inhibited around 34°C while for PEG 8 kDa inhibition occurs around 45°C.

clouding in PEG-mediated fusion (figure 7). In a broader context, the present results may, at least in part, provide an explanation for the superior properties of PEG 8kDa over those of the 20 kDa polymer to induce cell-cell fusion, often used as a technique in the generation of hybrid cells. Possibly, the clouding phenomenon described here in the context of steric interference during intermembrane interaction, may also bear relevance to the interpretation of the stealth properties of PEG-derived lipids seen in studies that employ liposomes as drug carriers. Liposomes, containing such PEG-lipid analogs, have been shown to remain in the blood circulation of animals for extended periods of time after intravenous injection (Klibanov et al., 1990), implying their refractoriness to cellular uptake which could be due to steric stabilization.

SYMMETRIC AND ASYMMETRIC FUSION

Fusion between alike membranes can be defined as 'symmetric' fusion. Virtually all studies carried out thusfar on fusion of liposomes, deal with 'symmetric' fusion. Recently, we found (Fonteijn et al., 1990) that under appropriate experimental conditions vesicles prepared from synthetic amphiphiles can actually engage in 'asymmetric' fusion (figure 8). These conditions involve that vesicles prepared from didodecylphosphate are incubated with, for example, phospholipid vesicles, below the gel-liquid crystalline temperature (T_c, phase transition temperature) of the DDP bilayer. Below T_c (30°C), the DDP vesicles do not fuse upon addition of

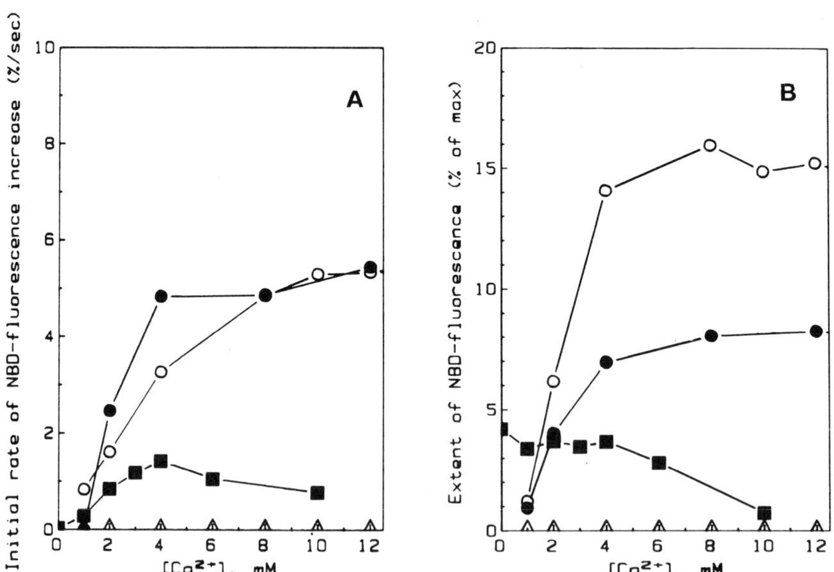

Fig. 8. Symmetric and asymmetric fusion of DDP vesicles. The initial rates (A) and extent of fusion (B) are indicated as a function of the Ca^{2+} concentration for the following systems: △, DDP-DDP; 0, PS-PS; ●, PS-DDP; □, DOPC-DDP. Fusion was monitored at 25°C, pH 7.4 with a lipid mixing assay. Note that symmetric fusion of DDP vesicles (△) does not occur below the gel/liquid crystalline phase transition temperature. Reproduced from Fonteijn et al., 1990, with permission.

Ca^{2+} (Rupert et al., 1986). Thus, at 25°C (i.e., below T_c of DDP) and in the presence of 1 mM Ca^{2+} (which is actually below the threshold concentration for symmetric DDP vesicle fusion), asymmetric fusion occurs between DDP vesicles and vesicles prepared from PS. Obviously, the PS vesicles can engage in symmetric fusion at these conditions. However, for reasons yet to be determined, preferential fusion occurs with DDP bilayers. The occurrence of asymmetric fusion also implies that, most likely, a 'trans' $PS-Ca^{2+}-DDP$ complex must be formed. The molecular nature of this complex is unknown but it is interesting to note that the bilayer structure of the fusion was stabilized in this case, in contrast to a massive transformation of bilayer to non-bilayer (H_{II}-phase) structures in case of Ca^{2+}-induced symmetric DDP vesicle fusion (see above; Rupert et al., 1987; Fonteijn et al., 1991). Similarly, we also observed (unpublished observations, Fonteijn et al., in preparation) that DDP vesicles readily fuse with Sendai virus without collapse of the bilayer structure.

Apart from its fundamental, molecular interest, asymmetric fusion systems may also be of interest for application as carrier system for hydrophobic and hydrophilic molecules of cell biological interest. For example, the massive fusion of fusion-susceptible phospholipid vesicles with each other, prior to their fusion with the cellular plasma membrane (see e.g. Struck et al., 1981), has greatly hampered a 'fusogenic' application of such systems. Although optimal conditions need to be worked out further, DDP vesicles have been shown to fuse readily with erythrocyte membranes (Fonteijn et al., 1990). Furthermore, cationic amphiphiles also attract attention as potential carrier system. In one of these studies (Felgner et al., 1987, 1989), effective transfection could be accomplished by using vesicles composed of, among others, N-[2,3-(dioleyloxy)propyl]-N,N,N-trimethylammonium (DOTMA). Other quarternary ammonium derivatives have been used for similar purposes (Pinnaduwage et al., 1989). The recombinant DNA vectors were added exogenously to the vesicles, rather than that they were encapsulated during preparation. Therefore, the extent to which fusion was relevant in achieving transfection remains to be determined. We (Rupert et al., see above) and others (Düzgünes et al., 1989) have shown that vesicles composed of, and/or containing, cationic synthetic amphiphiles are quite fusogenic. The role of the relatively low extent of hydration of the ammonium headgroup, apart from the electrostatic interactions with the nucleic acid, in bringing about efficient transfection needs further investigation,

In summary, work carried out thusfar on membrane mimetic properties of synthetic amphiphiles justifies the conclusion that these derivatives can be seen as very interesting and valuable tools for unraveling processes, involved in membrane fusion. Also, mechanistic studies of polymorphism, aimed at understanding and defining molecular parameters in bilayer-to- nonbilayer transitions may benefit from the ability to systematically modify the structure of amphiphiles undergoing such transitions. It also becomes apparent that synthetic amphiphiles may serve as attractive target membranes for studying fundamental aspects of the fusion of biological membranes, such as the fusion properties of enveloped viruses. Finally, although obviously requiring further work, synthetic amphiphiles, when compared to phospholipids, may also provide an attractive alternative as carrier system. Specific 'targeting' of fusion of vesicles composed of synthetic amphiphiles ('asymmetric' fusion), a condition frustrating application of phospholipid vesicles, may be of particular interest in this respect.

ACKNOWLEDGMENTS

This work was supported by the Netherlands Foundation for Chemical Research (SON), with financial aid from the Netherlands Organization of Pure Research (NWO) and by grant AI 255534 from the National Institutes of

Health (to DH). We are very grateful to Rinske Kuperus for excellent
support in preparing this manuscript.

REFERENCES

Arnold, K., Herrmann, A., Pratsch, L., and Gawrisch, K., 1985, The
    dielectric properties of aqueous solutions of poly(ethylene glycol)
    and their influence on membrane structure, Biochim. Biophys. Acta,
    815:515.
Ataman, M., 1987, Properties of aqueous salt solutions of poly(ethylene
    oxide). Cloud points, Θ temperatures, Coll. & Polym. Sci., 265:19.
Blumenthal, R., 1988, Membrane fusion, Current Top. in Membr. and Transp.
    29:203.
Boni, L.T., Stewart, T.P., Alderfer, J.L., and Hui, S.W., 1981,
    Lipid-polyethylene glycol interactions: II. Formation of defects in
    bilayers, J. Membr. Biol., 62:71.
Boni, L.T., Stewart, T.P., and Hui, S.W., 1984, Alterations in phospho-
    lipid polymorphism by polythylene glycol, J. Membr. Biol., 80:91.
Boucher, E.A., and Hines, P.M., 1978, Properties of aqueous salt solutions
    of poly(ethylene oxide): Thermodynamic quantities based on viscosity
    and other measurements, J. Polym. Sci. Polym. Phys. Ed., 16:501.
Carmona-Ribeiro, A.M., Yoshida, L.S., and Chaimovich, H., 1985, Salt
    effects on the stability of dioctadecyldimethylammoniumchloride and
    sodium dihexadecylphosphate vesicles, J. Phys. Chem., 89:2928.
Cevs, G., Zeks, B., and Podgornik, R., 1981, The undulations of hydrated
    phospholipid multibilayers may be due to water-mediated
    bilayer-bilayer interactions, Chem. Phys. Letters, 84:209.
Düzgünes, N., Goldstein, J.A., Friend, D.S., and Felgner, P.L., 1989,
    Fusion of liposomes containing a novel cationic lipid, N-
    2,3-(dioleoyloxy)propyl N,N,N,trimethylammonium: Induction by
    multivalent anions and asymmetric fusion with acidic phospholipid
    vesicles, Biochemistry, 28:9179.
Felgner, P.L., Gadek, T.R., Holm, M., Roman, R., Chan, H.W., Wenz, M.,
    Northrop, J.P., Ringold, G.M., and Danielsen, M., 1987, Lipofection-A
    highly efficient lipid mediated DNA-transfection procedure. Proc.
    Natl. Acad. Sci. USA, 84:7413.
Felgner, P.L., and Ringold, G.M., 1989, Cationic liposome-mediated
    transfection, Nature, 337:387.
Fendler, J.H., 1982, Membrane mimetic chemistry, John Wiley, USA.
Florin, E., Kjellander, R., and Eriksson, J.C., 1984, Salt effect on the
    cloud point of the poly(ethylene oxide) + water system, J. Chem.
    Soc., Faraday Trans. 1, 80:2889.
Fonteijn, T.A.A., Hoekstra, D., and Engberts, J.B.F.N., 1990, Specific
    asymmetric fusion between artificial and biological model membranes,
    J. Am. Chem. Soc., 112:8870.
Fonteijn, T.A.A., Engberts, J.B.F.N., and Hoekstra, D., 1991, Asymmetric
    fusion between phospholipid vesicles and vesicles formed from
    synthetic di-n-alkylphosphates, Biochemistry, in press.
Gruen, D.W.R., and Marcelja, S., 1983, Spatially varying polarization in
    water, J. Chem. Soc., Faraday Trans. 2, 79:225.
Helfrich, W., 1975, Out-of-plane fluctuations of lipid bilayers,
    Z. Naturforsch. C, 30:841.
Hoekstra, D., 1982, Role of lipid phase separations and membrane hydration
    in phospholipid vesicle fusion, Biochemistry, 21:2833.
Hoekstra, D., and Kok, J.W., 1989, Entry mechanisms of enveloped viruses.
    Implications for fusion of intracellular membranes, Bioscience
    Reports, 9:273.
Hoekstra, D., and Wilschut, J., 1989, Membrane fusion of artificial and
    biological membranes. Role of local membrane dehydration, in: "Water
    Transport in Biological Membranes", G. Benga, ed., CRC Press, Boca
    Raton, pp. 143.

Hoekstra, D., Klappe, K., Hoff, H., and Nir, S., 1989, Mechanism of fusion
of Sendai virus: Role of hydrophobic interactions and mobility
constraints of viral membrane proteins. Effects of poly(ethylene)-
glycol, J. Biol. Chem., 264:6786.

Hoekstra, D., 1990, Membrane fusion of enveloped viruses: Especially a
matter of proteins, J. Bioenerg. and Biomembr., 22:121.

Israelachvili, J.N., Marcelja, S., and Horn, R.G., 1980, Physical
principles of membrane organization, Q. Rev. Biophys., 13:121.

Israelachvili, J.N., and Sornette, D., 1985, The interdependence of
intra-aggregate and inter-aggregate forces, J. de Physique, 46:2125.

Israelachvili, J.N., and Wennerström, H., 1990, Hydration or steric
forces between amphiphilic surfaces?, Langmuir, 6:873.

Jönsson, B., and Wennerström, H., 1983, Image-charge forces in
phospholipid bilayer systems, J. Chem. Soc., Faraday Trans. 2, 79:19.

Klibanov, A.L., Maruyama, K., Torchilin, V.P., and Huang, L., 1990,
Amphipathic polyethylene glycols effectively prolong the circulation
time of liposomes, FEBS Lett., 268:235.

Kunitake, T., and Okahata, Y., 1977, A totally synthetic bilayer membrane,
J. Am. Chem. Soc., 99:3860.

Kurland, R.J., Hammoudah, M., Nir, S., and Papahadjopoulos, D., 1979,
Binding of Ca^{2+} and Mg^{2+} to phosphatidylserine vesicles: Different
effects on P-31NMR shifts and relaxation times, Biochem. Biophys.
Res. Commun., 88:927.

MacDonald, R.I., 1986, Membrane fusion due to dehydration by polyethylene
glycol, dextran, or sucrose, Biochemistry, 24:4058.

Marcelja, S., and Radic, N., 1976, Repulsion of interfaces due to boundary
water, Chem. Phys. Letters, 42:129.

Marra, J., and Israelachvili, J., 1985, Direct measurement of forces
between phosphatidylcholine and phosphatidylethanolamine bilayers in
aqueous electrolyte solutions, Biochemistry, 24:4608.

Novick, S.L., and Hoekstra, D., 1990, Significance of hydrophobic
interactions in membrane fusion of enveloped viruses. A comparison
with model membranes, Springer Series in Biophysics, 5:237.

Parente, R.A., and Lentz, B.R., 1986, Rate and extent of poly(ethylene-
glycol)-induced large vesicle fusion monitored by bilayer and
internal contents mixing, Biochemistry 25:6678.

Pinnaduwage, P., Schmitt, L., and Huang, L., 1989, Use of a quartenary
ammonium detergent in liposome mediated DNA transfection of mouse
L-cells, Biochim. Biophys. Acta, 985:33.

Pullman, B., Pullman, A., Berthod, H., and Gresh, N., 1975,
Quantum-mechanical studies of environmental effects on biomolecules.
VI Abinitio studies on the hydration scheme of the phosphate group,
Theoret. Chim. Acta, 40:93.

Rand, R.P., Fuller, N., Parsegian, V.A., and Ran, D.C., 1988, Variation in
hydration forces between neutral phospholipid bilayers: Evidence for
hydration attraction, Biochemistry, 27:7711.

Rand, R.P., and Parsegian, V.A., 1989, Hydration forces between
phospholipid bilayers, Biochim. Biophys. Acta, 988:351.

Rupert, L.A.M., Hoekstra, D., and Engberts, J.B.F.N., 1985, Fusogenic
behaviour of didodecyldimethylammonium bromide bilayer vesicles,
J. Am. Chem. Soc., 107:2628.

Rupert, L.A.M., Engberts, J.B.F.N., and Hoekstra, 1986, Role of membrane
hydration and membrane fluidity in the mechanism of anion-induced
fusion of didodecyldimethylammonium bromide vesicles, J. Am. Chem.
Soc., 1081:3920.

Rupert, L.A.M., Hoekstra, D., and Engberts, J.B.F.N., 1987, Ca^{2+}-mediated
fusion of didodecylphosphate vesicles, J. Colloid Interface Sci.,
120:125.

Rupert, L.A.M., Van Breemen, J.F.L., Van Bruggen, E.F.J., Engberts,
J.B.F.N., and Hoekstra, D., 1987a, Calcium-induced fusion of

    didodecylphosphate vesicles: The lamellar to hexagonal II (H_{II}) phase transition, J. Membr. Biol., 95:255.

Rupert, L.A.M., Engberts, J.B.F.N., and Hoekstra, D., 1988, Effect of poly(ethylene glycol) on the Ca^{2+}-induced fusion of didodecylphosphate vesicles, Biochemistry, 27:8232.

Rupert, L.A.M., Van Breemen, J.F.L., Hoekstra, D., and Engberts, J.B.F.N., 1988a, pH-Dependent fusion of didodecylphosphate vesicles. Role of hydrogen-bond formation and membrane fluidity, J. Phys. Chem., 92:4416.

Saeki, S., Kuwahara, N., Nakata, M., and Kaneko, M., 1976, Upper and lower critical solution temperatures in poly(ethylene glycol) solutions, Polymer, 17:685.

Siegel, D.P., 1986, Inverted micellar intermediates between lamellar, cubic, and inverted hexagonal lipid phases. I. Mechanism of the L_{α}-H_{II} phase transitions, Biophys. J., 49:1155.

Siegel, D.P., 1986a, Inverted micellar intermediates and the transition between lamellar, cubic, and inverted hexagonal lipid phases. II. Implications for membrane-membrane interactions and membrane fusion, Biophys. J., 49:1171.

Sornette, D., and Ostrowsky, N., 1986, Importance of membrane fluidity on bilayer interactions, J. Chem. Phys., 84:4062.

Struck, D.K., Hoekstra, D., and Pagano, R.E., 1981, Use of resonance energy transfer to monitor membrane fusion, Biochemistry, 20:4093.

Tanford, D., 1980, The hydrophobic effect, Wiley-Interscience, New York.

Wagenaar, A., Rupert, L.A.M., Engberts, J.B.F.N., and Hoekstra, D., 1989, Synthesis and vesicle formation of identical- and mixed-chain di-n-alkyl phosphate amphiphiles, J. Org. Chem., 54:2638.

KINETICS OF INTERMEMBRANE INTERACTIONS LEADING TO FUSION

Dimiter S. Dimitrov and Robert Blumenthal

Section on Membrane Structure and Function
LMB, NCI, NIH, Bethesda, MD 20892

INTRODUCTION

Fusion of membranes requires close apposition,
destabilization and appropriate structural reorganizations
leading to membrane mixing (Blumenthal, 1987). The time
period between the application of the fusogen and the first
indication of membrane mixing, i.e. the delay in fusion, may
vary widely for different systems. In viral fusion it is in
the range of hundreds of milliseconds to seconds and even
minutes (Morris et al., 1989; Sarkar et al., 1989; Hoekstra
et al., 1989; Clague et al., 1990). Delays in electrofusion
can be as short as the time resolution of the currently used
videomicroscopy method, which is tens of milliseconds, and as
long as minutes (Dimitrov and Sowers, 1990a).

There are several major factors which may account for
the relatively wide range of variations of the characteristic
times of adhesion, destabilization and fusion: i) the
geometry of the systems, ii) the viscous dissipation of
energy in the medium surrounding the membranes or/and in the
membranes and iii) the forces which drive these processes.
The different geometry of approaching membranes may lead to
different time periods of binding. The viscous resistance of
the thin liquid layer between the approaching membranes is
high and may play a role not only in binding but also in
fusion when the membranes need to further approach to make
molecular contact. The forces which drive adhesion,
destabilization and fusion may vary widely (Rand and
Parsegian, 1989; Ohki and Arnold, 1990), but in general, an
increase in the driving force leads to a decrease in the
characteristic time periods of these processes. For large
driving forces the processes can go through kinetically
favorable pathways, which involve small membrane areas and
therefore can be fast. Such pathways are well known for

colloid systems, e.g., rupture of draining thin liquid films
(Ivanov and Dimitrov, 1974).

A major goal of this chapter is to present a theoretical
framework for description of the kinetics of membrane
approach, destabilization and formation of structures leading
to fusion. The emphasis will be on two relatively well
characterized fusogens - viral fusion proteins and high
voltage electric pulses. They represent the limiting cases of
specific (viral) fusion and non-specific (electro-) fusion.
In spite of the different nature and behavior of these
systems they show similarities in the time coarse of fusion.
This may reflect non-specific factors operating in any kind
of fusion. The differences can be attributed to intermediate
structures which are specific for the type and conditions of
fusion. By finding similarities and comparing differences one
can learn more for the fundamental mechanisms of fusion.

STAGES OF CELL-CELL AND VIRUS-CELL INTERACTIONS LEADING TO
FUSION

It is convenient to divide the process leading to fusion
of cells with cells or viruses into six stages (Blumenthal,
1987)(Fig. 1):

a) Approach of weakly interacting non-deformed
membranes. Membranes are at relatively large separations.
They approach each other either by Brownian motion or long-
range attractive intermembrane forces, including those
induced by external electric fields.

b) Deformation of membranes. Membrane deformation can be
induced by strong intermembrane forces at close approach. The
membrane deformation significantly decreases the rate of
membrane approach due to the high viscous resistance of the
thin film between the membranes.

c) Stable apposition of membranes. At this stage the
membrane system is in equilibrium with respect to mutual
approach. There are energy barriers which prevent further
approach (Fig. 2). In many cases the membranes are in stable
apposition at the so-called "primary" minimum in the
intermembrane interaction energy. The primary minimum is
separated by an "adhesion" energy barrier from the
"secondary" minimum, where the membranes can easily be
separated by Brownian motion.

d) Destabilization of interacting membranes. In order to
overcome energy barriers ("fusion" barriers (Fig. 2)) the
free energy of the system should be increased by applying
external agents, as high voltage electric pulses to induce
electrofusion, pH or temperature changes to promote viral
fusion. This can lead to decrease of the intermembrane
interaction energy, which results in further approach of the
membranes to reach molecular contact needed for fusion. The
destabilizion of the membranes themselves can occur before
or after reaching molecular contact.

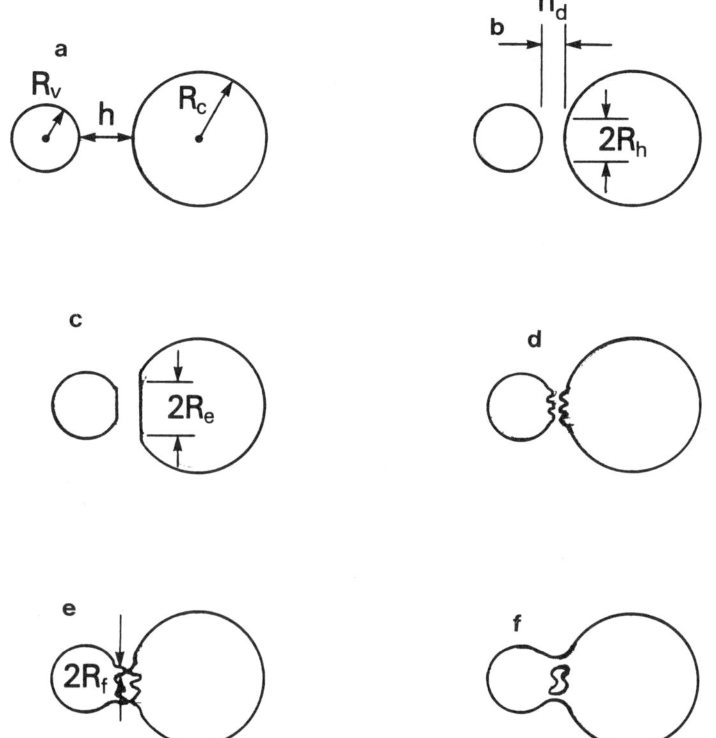

Fig. 1. Stages of membrane interactions leading to fusion.
(a) Approach; (b) deformation;
(c) apposition; (d) destabilization;
(e) fusion intermediates; (f) fusion

e) Formation of fusion intermediate structures. The interacting membranes change their structure in an appropriate way to accommodate for the next stage of fusion. Fusion intermediates can form during or after the application of the destabilizing fusion trigger. Their life-time may be determined by the time needed to overcome the decreased barrier to fusion if any.

f) Membrane fusion. The two membranes fuse, which allow mixing of membrane components and water-soluble substances in the compartments enclosed by the membranes. Fluorescence markers eventually will diffuse at this stage from the labeled to the unlabeled membranes. The time period between the application of the fusion trigger and the diffusion of fluorescence probes reflects the lag time (delay) of fusion and can be measured by spectrofluorimetry or fluorescence video microscopy.

The next sections describe the kinetics of each of these stages in more detail. The emphasis will be on a theoretical approach in comparison with experimental data on delays in viral fusion and electrofusion. We show the major role of the liquid layer between the membranes and suggest a formula for description of delays in fusion.

APPROACH, DEFORMATION AND ADHESION OF MEMBRANES

Approach of Spherical and Deformable Membranes

Rates of approach, V, of non-deformed spherical membranes toward each other can be described by an interpolation formula (Dimitrov, 1983)

$$V = -dh/dt = F/6\pi\mu R(1+R/h), \quad R = R_v R_c/(R_c+R_v), \quad (1)$$

where h is the smallest separation between the two membranes (see Fig. 1), t - time, F - driving force, commonly F = -dE/dh, E - intermembrane energy of interaction (see Fig. 2), μ - viscosity, R_c and R_v are the radii of the interacting cells or the cells and the viruses. The time of approach can be obtained by integrating (1) from the initial to the final separation. It should be noted that in the general case the driving force F depends on separation which may not allow integration of Eq. (1) in analytical form. In spite of this, several qualitative conclusions can be made directly from Eq. (1): i) the time of approach is proportional to the medium viscosity, ii) it increases with increasing the size of the particle and decreasing the separation h and iii) it decreases with increasing the driving force. The second conclusion is valid if the driving force does not increase significantly with decreasing the separation and decreasing the particle size. When F < 0 Eq. (1) describes separation of membranes.

Equation (1) was used to calculate the intermembrane forces F between erythrocyte ghosts induced by AC electric fields from the measured intermembrane separations h, membrane radii R and rates of membrane approach V (Dimitrov et al., 1990). The observations showed that at separations on the order or larger than the ghost radii the interaction

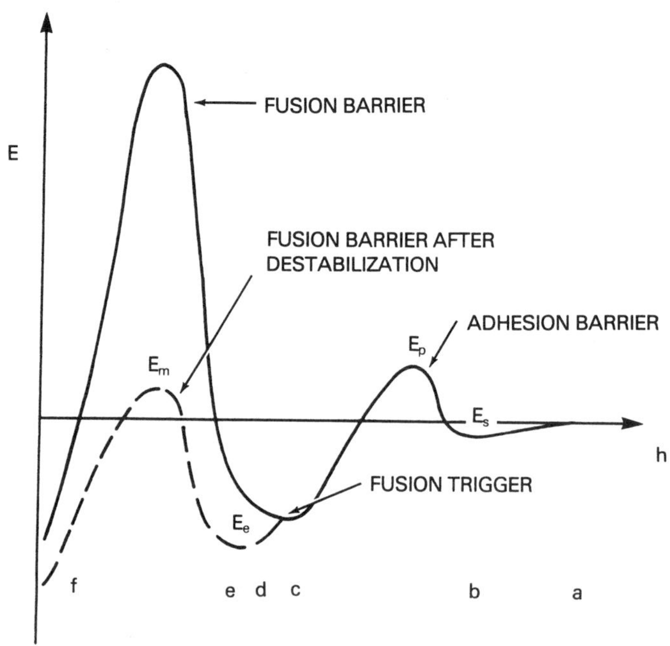

Fig. 2. Energy of intermembrane interaction, E, as function
of intermembrane interaction, h. The letter bellow the
curves denote the stages shown in Fig. 1. The
continuous line represent the system pathway without
fusion trigger. The broken line reflects the system
behavior after application of a fusion trigger. $E_{m,e,p,s}$
are the values of E at the respective maxima and
minima.

force can be described as due to dipole-dipole interaction. This is reasonable from theoretical point of view and indicates that Eq. (1) can be used to calculate the rate of spherical cell-cell approach.

In the same set of experiments (Dimitrov et al., 1990) it was found that when the intermembrane separation was of the order or less than 1 μm, the membranes deformed (see Fig. 1b). Other observations (Wu and Weinbaum, 1982) indicate that a nearly plane-parallel film can form between membranes, which thins at much slower rate than that given by Eq. (1). The rate of approach of plane-parallel membranes can be estimated by the Reynolds' equation (see, e.g. in (Dimitrov and Ivanov, 1978; Dimitrov, 1983))

$$V = 2Fh^3/3\pi\mu R_h^4, \qquad\qquad (2)$$

where R_h is the radius of the plane-parallel membranes. This radius depends primarily on the interaction energy between the membranes and the membrane tension T and can be approximated (Dimitrov and Ivanov, 1978) by the equilibrium radius of contact R_e, given by Eq. (4). The separation, h_d, at which the deformation begins can be estimated by (Dimitrov, 1983)

$$h_d = F/4\pi T, \qquad T = T_v T_c/(T_c+T_v), \qquad\qquad (3)$$

where T_c and T_v are the tensions of the two interacting membranes. Equation (3) is in very good agreement with the measured separation of deformation and force of interaction induced by AC fields (Dimitrov et al., 1990). It shows that membrane deformation increases with increasing the interaction force and decreasing the membrane tension. This indicates one of the interesting features of approach of deformable membranes. The increase in the driving force may not lead to proportional increase in the rate of approach because of the membrane deformation.

The analysis of Eq. (2) shows that the time of approach of deformable membranes is proportional to the liquid viscosity as in the case for approach of spherical membranes. However, the membrane deformation leads to much longer times of approach and more complicated dependence on the driving force than for spherical membranes.

<u>Adhesion</u>

Adhesion occurs when the driving force $F = F_a - F_r = 0$ and the intermembrane interaction energy, E, is much larger than the thermal energy kT, k being the Boltzman's constant and T - absolute temperature. The intermembrane separation can be estimated by equalizing the attractive, F_a, and repulsive, F_b, forces. The radius of contact can be evaluated from an integral balance of forces which leads to (Dimitrov, 1983)

$$R_e = (F_a R/2\pi T)^{1/2}. \qquad\qquad (4)$$

Here, the attractive force, F_e, can be calculated as a derivative of the intermembrane energy of attraction at equilibrium separation and contact angles were assumed small.

The time of establishing a state of adhesion can be estimated
by integrating Eqs. (1) and (2) or an equations which
combines them (Dimitrov, 1983).

When two equilibrium states of adhesion are separated by
a potential energy barrier (see Fig. 2) a stochastic approach
should be used to calculate the time of transition between
these states. A typical example is the interplay between van
der Waals attraction forces, electrostatic and hydration
repulsion forces which may lead to two minima in the
intermembrane energy function. The absolute value of the
interaction energy at the "secondary" minimum is comparable
or smaller than kT, and the adhered cells can be easily
separated by the thermal motion. The adhesion of cells at the
"primary" minimum is much more stable. To reach the primary
minimum, and respectively stable adhesion, the cells should
overcome the potential energy barrier (Fig. 2). They can not
do this transition by a directional motion caused by driving
force F = -dE/dh, because F is negative when climbing the
barrier. However, cells are able to overcome the potential
barrier between the secondary and the primary minima by a
stochastic process similar to Brownian motion (Ruckenstein et
al., 1976).

In this case the cellular flux over the potential
barrier, j, is given by

$$j = - dc/dt = pc*. \tag{5}$$

where c is the cell concentration at any given moment, c* is
the number of cells at close proximity to the secondary
minimum and p is the probability per unit time that a cell
will escape over the potential barrier. If one assume that
cells arrives at the secondary minimum faster than the rate
of escape over the potential barrier, c = c*. Then the
concentration of adhered cells $C = c_o - c$, where c_o is the
total concentration of individual cells which equals the
initial cell concentration. The solution of Eq. (5) with
these assumptions is

$$C = c_o(1-\exp-pt). \tag{6}$$

The probability p, which reflects the characteristic
life-time of the intermediate state between the secondary and
the primary minima, can be calculated by an equation derived
from a solution of a modified Fokker-Plank equation by
approximating the energy barrier by a parabolic function
(Ruckenstein et al., 1976)

$$P = [(dE^2/dh^2)_s (dE^2/dh^2)_p]^{1/2} M_p \exp(-E_{aa}/kT)/2\pi, \tag{7}$$

where E is the free energy of intermembrane interaction, h -
local separation of the membranes, the subscripts s and p
denote that the respective quantities are taken at the
secondary minimum and at the maximum free energy between the
secondary and the primary minima, E_{aa} is the activation
energy which is equal to the difference between the maximum
free energy and the free energy at the secondary minimum. The
mobility M of the cells is defined as

$$M = V/F, \tag{8}$$

where the rate of approach, V, is given by Eqs. (1) or (2).

Commonly, mobilities are calculated on the basis of the Stokes' formula for motion of non-interacting spherical particles in an infinite viscous liquid (Bentz et al., 1985). At close membrane approach the Stokes equation is no longer valid and Eqs. (1) or (2) must be used. This is especially important in the case of existence of energy barriers, when membranes are at very close apposition for long periods of time. The increase in the time period for adhesion of deformable membranes can be of the order of $(R/h)^3$, compared to approach of non-interacting membranes. For h = 10 nm and R in the range of 100 nm to 0.01 mm this gives 1000 to 10^9 times longer periods of time than those obtained by the Stokes' formula.

DESTABILIZATION OF INTERACTING MEMBRANES AND FORMATION OF FUSION INTERMEDIATES

Application of a "fusion trigger", e.g., electric pulse, change in pH or temperature, to adhered membranes leads to increase of the free energy of the membrane system. This can decrease or eliminate the energy barrier to fusion (Fig. 2) and result in destabilization of the membranes and/or the liquid film between them.

Destabilization of Single Membranes

Experimental data have shown that destabilization of single membranes by high voltage electric pulses can be very fast. It commonly occurs during the pulse (Dimitrov and Sowers, 1990b) which usually lasts from few microseconds to milliseconds. The rate of conformational changes of viral fusion proteins leading to membrane destabilization have not been measured, but they can be also fast - not longer than the delays in fusion which are of the order of or less than hundreds of milliseconds or seconds (Clague et al., 1990). The destabilization of the target membrane, if at molecular contact with the viral protein, can occur during or very shortly after the fusion protein conformational change. Viral fusion occurs at optimal conditions within 1 s (Clague et al., 1990).

The time of destabilization should be proportional to the membrane viscosity (Dimitrov, 1984; Dimitrov and Jain, 1984). The membrane viscosity is about two orders of magnitude higher than the medium viscosity. In spite of this the rate of destabilization is much higher than the rate of adhesion. Therefore, the high rate of destabilization can be due to other factors as: i) large driving force because of the high increase in the free energy and ii) involvement of small membrane areas. In the case of electric fields commonly pores are formed during or after destabilization. Viral fusion proteins also can induce pores or locally destabilized membranes. This demonstrates an important feature of destabilization of membrane systems. At large driving forces and high viscous resistance the system can use a pathway, which involves small membrane area. This feature is well documented for thin liquid films (Ivanov and Dimitrov, 1974).

An equation for the kinetics of membrane destabilization in electric fields was suggested for the case of absence of energy barriers (Dimitrov, 1984). It predicts that the time of destabilization is proportional to the membrane viscosity and decreases with increasing the driving force (the pulse voltage) and the membrane tension. The predictions of this formula are in qualitative and semi-quantitative agreement with experimental data for cell (Benz and Zimmermann, 1980) and model membranes (Abidor et al., 1979). Another model was based on the concept of pre-existing small pores and related the time of destabilization (electric breakdown) to the energy barrier needed to overcome to form large pores (Chizmadzhev et al., 1979). It also can describe data on bilayer membranes breakdown (Abidor et al., 1979). Both approaches can be used to construct a model which can be valid for cases with or without pre-existing pores and with or without energy barriers.

Destabilization of membranes by viral fusion proteins can be considered in a similar theoretical framework. One can imagine that the fusion protein induces high local increase in the free energy, either by increasing the enthalpy (commonly due to electrostatic interactions between the membrane and the protein), or decreasing the entropy (possibly by hydrophobic effects). For example, positive charges on the surface of the fusion proteins can induce high local transmembrane voltage which may be comparable in magnitude to the transmembrane voltage induced by external electric fields leading to formation of pores. Hence, in the case of protein induced destabilization the role of the strength of the external electric fields may be played by the specific conformation and concentration of the fusion protein.

Instability of Liquid Layers between Membranes

Destabilization of the liquid layer between membranes differs from that of single membranes mainly by the magnitude of the applied forces. While the low dielectric constant and conductivity of the membranes can lead to accumulation of charges and huge transmembrane voltage, and therefore large driving force, the high dielectric constant and conductivity of the liquid layer between the membranes lead commonly to small driving forces. This can increase tremendously the time of reaching molecular contact in spite of the relatively low viscosity of the water solutions. The small interaction force may also lead to possibility of uniform draining of the layer between membranes which is much slower than the localized one. (Remember that the rate of approach according to Eqs. (1) and (2) strongly increases with decreasing the size of the area involved in the motion.) Hence, both kinetic pathways, thinning of plane-parallel films and wavy, localized approach, can occur. Which will dominate depends primarily on the driving force and the membrane properties.

For example, in the absence of an energy barrier the time of localized approach, t_l can be estimated by (Dimitrov, 1982; Dimitrov, 1983)

$$t_{l_T} = 24mT/h^3(dP_d/dh)^2 \quad \text{when} \quad B(dP_d/dh)/T^2 \ll 1 \; ; \qquad (9)$$

$$t_{lB} = 9m(6B)^{1/2}/h^3(dP_d/dh)^{3/2} \quad \text{when} \quad B(dP_d/dh)/T^2 \gg 1, \quad (10)$$

where μ is the viscosity of the liquid layer between the membranes, T is the membrane tension, B - bending elasticity, P_d - driving pressure, which is the driving force per unit area, and h - intermembrane separation. The rate of uniform approach can be evaluated by Eq. (2). By using $F = \pi R_e^2 P_d$, $dP_d/dh = P_d/h$ and $t_{lT} = h/v$, v - being the rate of localized approach, we obtain from (2) and (9) an estimate for the ratio of both rates

$$v/V = P_d R_e^2/hT \qquad\qquad (11)$$

Equation (7) shows that for large driving forces, large area of contact and small separation and membrane tensions, the localized approach pathway can dominate over uniform approach. An example with $P_d = 1$ dyn/cm^2, $R = 10$ μm, $h = 10$ nm, $T = 1$ mdyn/cm and $\mu = 0.01$ P gives $v/V = 1000$ and $t_{lT} = 10$ s. Similar conclusions and estimates hold when the time of localized approach is determined by the bending elasticity (Eq. (10)).

<u>Formation of Fusion Intermediates</u>

Energy barriers slow down the processes of destabilization and establishment of equilibrium states. They can exist when the fusion trigger does not provide sufficient free energy to overcome the barriers to fusion or/and when fusion requires appropriate structural reorganization. In the latter case the energy barrier can be the energy of formation of the intermediate structures needed for fusion. For a localized type of fusion, which can be expected both on theoretical grounds and experimental evidence, the number of intermediates n leading to fusion can be estimated by a theoretical framework similar to that developed to describe slow aggregation of particles (see also the subsection "Adhesion" in the preceding section). One can assume that application of the trigger creates N_o potential fusion sites. In electrofusion N_o can be the number of fusogenic pores or other fusogenic structures. In viral fusion N_o represents the number of activated fusion proteins which may be proportional to the concentration of the viral protein in the membrane. The flux of fusogenic membrane sites over the potential barrier, J, leading to fusion is given by

$$J = -\,dN/dt = PN* \qquad\qquad (12)$$

where N is the number of fusogenic sites at any given moment, $N*$ is the number of fusogenic sites at close proximity to the energy barrier and P is the probability per unit time that a fusogenic site will escape over the potential barrier. If one assume that fusogenic sites formation is faster than the rate of escape over the potential barrier, $N = N*$. This assumption implies that we neglect energy barriers, e.g., due to slow lateral diffusion and aggregation possibly needed to form fusogenic sites. Then the number of fusion intermediates converted to fusion junction $n = N_o - N$. The solution of Eq. (12) with these assumptions is

$$n = N_o(1-\exp-Pt). \qquad\qquad (13)$$

The probability P, which reflects the characteristic time of fusion junctions formation, can be calculated by an equation (Dimitrov and Blumenthal, to be published) derived from a solution of a modified Fokker-Plank equation (Ruckenstein et al., 1976) by approximating the energy barrier by a parabolic function

$$P = [(d^2E/dh^2)_e (d^2E/dh^2)_m]^{1/2} m_m \exp(-E_a/kT)/2\pi, \qquad (14)$$

where E is the energy of intermembrane interaction leading to fusion intermediate formation, h - local separation of the membranes, the subscripts e and m denote that the respective values are taken at the bottom of the energy well and at the maximal energy, E_a is the activation energy which is equal to the difference between the maximal and minimal energy (see Fig. 2), and T - absolute temperature. The mobility m of the fusogenic site is defined as

$$m = v/F, \qquad (15)$$

where the rate of motion of the fusogenic site toward the target membrane or each other, v, can be estimated by geometrical models for the membrane configuration during approach. There are two major limiting cases. One of them is based on approach of spherical membranes. Then Eq. (1) at h<<R can be used to yield

$$m = h/6\pi\mu R^2. \qquad (16)$$

Here R is the radius of curvature of the deformed membrane at the fusion site. The other limiting case is based on approach of flat membranes. Then Eq. (2) gives

$$m = 2h^3/3\pi\mu R_f^4 \qquad (17)$$

where R_f is the radius of the area involved in formation of the fusion intermediate. Similar formula can be derived based on the fluctuation wave mechanism of thin liquid film instability (see the preceding subsection and Eqs. (9,10)). They differ only by a numerical coefficient. This indicates that Eq. (17) may be safely used in a wide variety of conditions where the exact geometry of the approaching membranes at the fusion site is not known. Then the only parameters which reflect that geometry are the radius R_f and the separation h at the height of the energy barrier.

DELAYS AND LIFE-TIMES OF FUSION INTERMEDIATES

Fluorescent video microscopy of individual membranes (Morris et al., 1989; Sarkar et al., 1989; Dimitrov and Sowers, 1990a) and spectrofluorimetry of populations of membranes (Hoekstra et al., 1989; Clague et al., 1990) gave fusion kinetics dependencies consisting of three major parts: i) lag times (delays), ii) fluorescence changes and iii) plateaux which give the maximum fusion yields. Figure 3 shows an example of a spectrofluorimetric record of kinetics of fusion of Sendai virus with erythrocyte ghosts at two different viscosities of the medium. Similar dependencies were observed by fluorescence video microscopy in

electrofusion of erythrocyte ghosts (see, e.g., Fig. 2 in
(Dimitrov and Sowers, 1990a)) and for individual cell fusion
events induced by viral fusion proteins (see, e.g. Fig. 5 in
(Morris et al., 1989) for fusion induced by the influenza
hemagglutinin).

<u>Delays in Fusion</u>

It was suggested that cell fusion involves six possible
stages (Rand and Parsegian, 1986): (i) stable membrane
apposition, (ii) triggering of fusion, (iii) contact, (iv)
focused destabilization, (v) membrane coalescence, and (vi)
restabilization. The delay is the time period between the
triggering of fusion and the membrane coalescence. Therefore,
it includes the stages of making contact and focused
destabilization in the scheme proposed by Rand & Parsegian
(1986). For the scheme shown in Fig. 1 the delay is the time
period between the destabilization of the system by the
fusion trigger (Fig. 1d) and the fusion itself (Fig. 1f).
Therefore, it consists of two major parts - destabilization
of the system and formation of intermediates. Depending on
the conditions either of them can be the determinant of the
rate of fusion.

Delays in fusion vary widely for different systems. They
can be less than 5 ms for the neurotransmitter release
(Heuser et al., 1979) or larger than 1 h for cell fusion
induced by the HIV envelope proteins (Dimitrov, Golding and
Blumenthal, to be published). For planar-planar bilayer
fusions (Chernomordik et al., 1987) the lifetimes of the
events leading to fusion ("waiting time" for fusion) are in
the range of seconds to minutes. Even for the same system,
but for different experimental conditions, the delays can
vary in orders of magnitude. For instance, delays in
electrofusion of erythrocyte ghosts are in the range of ms to
minutes (Dimitrov and Sowers, 1990a). Delays in viral fusion
are in the range of hundreds of seconds to minutes (Morris et
al., 1989; Clague et al., 1990; Hoekstra et al., 1989; Spruce
et al., 1989).

One important question in any study of fusion kinetics
is how to measure delays and how to define exactly when the
delay is over. The answer of this question varies in
dependence of the experimental system. One reasonable
approach is that the delay is the time interval between
trigger application and first indication of fusion. The first
indication of fusion occurs when the average signal increase
equals the level of noise. This definition evidently does not
apply to cases with flicker phenomena and large deviations of
the signal due to artifacts or other effects. Therefore any
comparison of delays obtained by different methods requires
careful examination of the experimental conditions of
measurements and definitions of delays.

In electrofusion the moment of trigger application can
be precisely located in time. For many other systems the
trigger is a chemical substance which requires time for
diffusion. This time can be long and not easily measured.
Another advantage of electrofusion is the use of a.c. fields
to achieve close membrane apposition. They can be precisely
regulated. This leads to rather reproducible and controllable

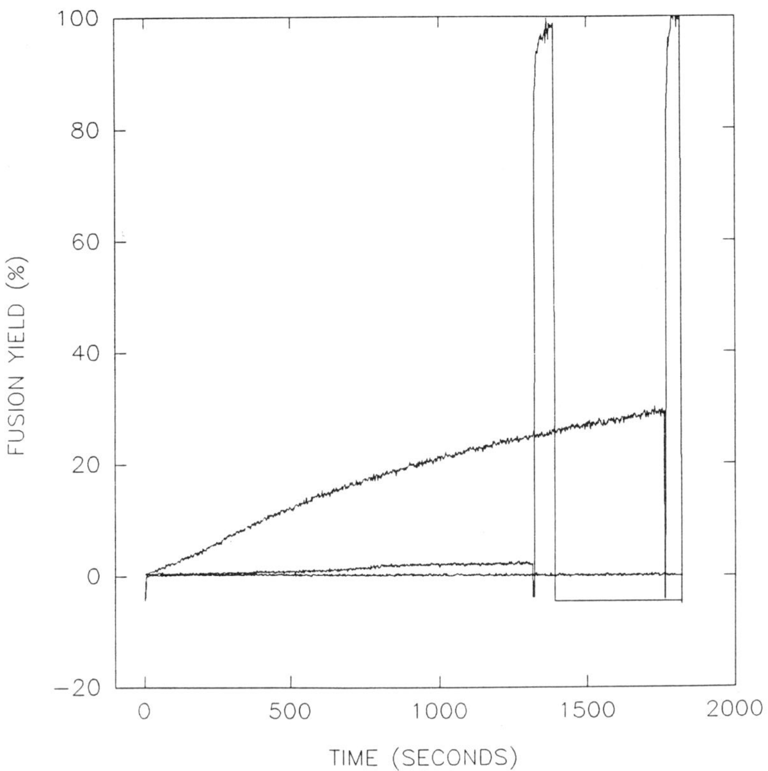#

Fig. 3. Fusion of Sendai virus with human erythrocyte ghosts as measured by a spectrofluorimetric method based on fluorescence dequenching (see, e.g., Hoekstra et al., 1989; Claque et al., 1990) in PBS (upper curve) and in 1.14 M sucrose solution (4 times the viscosity of PBS) (intermediate curve). The labeled Sendai virus itself does not show an increase in fluorescence (bottom).

initial starting positions of the membranes. In viral fusion
this is achieved by prebinding at low temperature. The
initial separation of the membranes, however, may not be
regulated.

A study of delays in electrofusion of erythrocyte ghosts
(Dimitrov and Sowers, 1990a) showed that they decreased over
the range 4 - 0.3 s with an increase in (i) the pulse
strength from 1 to 0.25 kV/mm, (ii) the pulse duration in the
range 0.073 - 1.8 ms, (iii) the dielectrophoretic force which
brings the membranes at close apposition before triggering
fusion. They increased proportionally to the increase in the
medium viscosity. The delays decreased 2 - 3 times with an
increase in temperature from 21 to 37°C. The Arrhenius plot
yielded straight lines. The calculated activation energy, 17
kcal/mol, does not depend on the pulse strength. (Dimitrov
and Sowers, 1990a).

The delays in fusion of fibroblasts, expressing
influenza hemagglutinin, with erythrocytes decreased with
decreasing the pH. They were of the order of seconds for pH
in the range 5.4 - 4.9 (Sarkar et al., 1989). The activation
energy is 18 kcal/mol for temperatures between 27 and 50°C
(Morris et al., 1989). The delays in fusion of intact viruses
with erythrocyte ghosts showed similar trends but were
shorter than for cell fusion (Clague et al., 1990). The delay
in fusion of Sendai virus with erythrocyte ghosts increased
with an increase in medium viscosity (Dimitrov and
Blumenthal, unpublished data, see also Fig. 3).

A Formula for Delays

Delays can be described by a formula derived by
combining Eqs. (13) and (14) and assuming that fluorescence
dye transfer can be detected after the formation of the first
fusion junction (Chen and Blumenthal, 1989). This can be the
case when fusion of single cells (Dimitrov and Sowers, 1990a)
or viruses (Lowy et al., 1990) is observed. Substituting n =
1 in Eq. (13) and assuming that the initial number of fusion
sites, $N_o \gg 1$, one obtains that the delay, t_d, is given by

$$t_d = [2\pi/N_o[(d^2E/dh^2)_e(d^2E/dh^2)_m]^{1/2}m_m]\exp(E_a/kT), \quad (18)$$

where m can be expressed either by Eq. (16) or (17).

For fusion in a population of cells (and/or viruses),
measured by spectrofluorimetry, sufficient number of cells
(viruses), N_{cf}, should fuse to reach the level of sensitivity
and signal-to-noise ratio of the system. After fusion these
cells produce signal $f_f = aN_{cf}$, where a is the fluorescence of
one cell (virus) after dequenching. For total dequenching of
all cells, e.g. after adding Triton, $f_t = aN_t$, where N_t is the
total number of labeled cells (viruses). The ratio f_f/f_t,
which is a measure for the fusion yield, must be equal to the
system sensitivity, f_c. Let assume that the delay ends when
the increase in signal, which can be detected, i.e. the
system sensitivity, f_c, equals the noise, f_n. Therefore,

$$f_n = f_c = f_f/f_t = N_{cf}/N_t. \quad (19)$$

Let assume that each cell (virus) requires one fusion
junction to allow dequenching of the fluorescent dye. Then

242

the number of fusion junctions, n, should be equal to the
critical number of fused cells, N_{cf}, necessary to reach the
sensitivity level of the system. Substituting N_{cf} for n in
Eq. (13), replacing N_o by the total number of fusion sites,
which is $N_o N_t$, and using Eq. (19) one gets the same equation
as (18), but the right side is multiplied by $f_{n,c}$. Because $f_{n,c}$
is always smaller than unity, typically it is of the order of
0.1 to 0.01, the delays measured for a population of cells
(viruses) could be shorter than those measured for single
viruses or cells, unless very large number of cells is
measured which is practically impossible.

It is difficult to make any rigorous quantitative
analysis of Eq. (18) and comparison with experimental data
unless the activation energy is known as function of the
intermembrane distance. However, some estimates and
qualitative predictions can be made. By using Eq. (17) for
the mobility and approximating the derivatives in (18) by
their finite differences one can reduce Eq. (18) to

$$t_d = (\mu R_f^4 / N_o E_a h) \exp(E_a / kT) \tag{20}$$

This formula is in reasonable agreement with published
data on delays in viral fusion and electrofusion. For
example, for fusion of fibroblasts, expressing hemagglutinin
(HA), with red blood cells (Morris et al., 1989) the
activation energy at 37°C is 27kT. Assuming $N_o = 10^4$, h = 1
nm, R_e = 10 nm one obtains from Eq. (20) values for the delay
of the order of 10 s, which is the order of magnitude of the
measured delays (Sarkar et al., 1989). Similar values were
also obtained for electrofusion (Dimitrov and Sowers, 1990a).

It is seen from Eq. (18) that the delay should be
proportional to the viscosity of the medium, which was found
experimentally for electrofusion of erythrocyte ghosts
(Dimitrov and Sowers, 1990a). One would expect that the
energy barrier decreases with decreasing the intermembrane
separation. Hence the delay should be decreasing with
decreasing the intermembrane separation. This was indeed
observed for electrofusion of erythrocyte ghosts starting
from different initial separations (Dimitrov and Sowers,
1990a). The delay should depend on the initial concentration
of fusogenic sites. This is in agreement with recent
observations of HA induced cell fusion (Clague et al., 1991).
The exact functional dependence of the delay on the HA
concentration may include dependence of the activation energy
and the mobility on the HA concentration. In addition, part
of the HA molecules may not be in fusogenic states for
various reasons.

<u>Life-times of Fusion Intermediates</u>

The life-time of fusion intermediates (fusogenic sites)
can be estimated if one assumes that the rate of their
disappearance, i.e. formation of fused membranes, is much
faster than the rate of their formation given by Eq. (13),
and that there is no another pathway of their destruction. In
this case most of their life-time will be spent in the energy
well before overcoming the potential barrier. Therefore, the

life-time, t_f, can be estimated as

$$t_f = 1/P, \tag{21}$$

where P is given by Eq. (14). It follows from (18) and (21) that

$$t_f = t_d N_o. \tag{22}$$

This means that when the number of fusogenic sites is larger than unity the life-times of fusion intermediates will be larger than the delays. This implies the possibilities for long-lived fusion intermediates. Such long-lived fusogenic states have been observed in electrofusion (Sowers, 1986; Teissie and Rols, 1986). However, Eq. (14) can not be used to describe them unless the membranes are at close separation. It should be modified, which is out of the scope of this chapter.

Equation (14) and respectively Eqs. (18,20) can be generalized to include several energy barriers, which may correspond to different intermediates. Unfortunately, at the present stage of development of the experimental techniques it is not possible to measure the life-times of short-lived intermediates and to compare with such generalized models. Hence, from theoretical point of view more fruitful direction seems to focus on the major fusion intermediate and to try to describe in more detail its structure and energy.

CONCLUSION

Membrane fusion is a multistage process. Dissecting steps of this process leads to understanding its mechanisms. Fusion kinetics is determined by the life-times of intermediates, which are characteristic for each of these steps. The delays in fusion are a measure of the overall life-time of the intermediates or the life-time of the major rate-determining intermediate. The theoretical framework for describing kinetics of membrane interactions leading to fusion, presented here, may help in understanding mechanisms of fusion in general and stimulate new experiments.

References

Abidor, I. G., Arakelyan, V. B., Chernomordik, L. V., Chizmadzhev, Yu. A., Pastushenko, V. F. and Tarnsevich, M. R., 1979, Electrical breakdown of bilayer lipid membranes: I. The main experimental facts and their qualitative discussion, <u>Bioelectrochem. Bioenerg.</u>, 6:37.

Bentz, J., Duzgunes, N. and Nir, S., 1985, Temperature dependence of divalent cation induced fusion of phosphatidylserine liposomes: Evaluation of the kinetic rate constants, <u>Biochemistry</u>, 24:1064.

Benz, R. and Zimmermann, U., 1980, Relaxation studies on cell membranes and lipid bilayers in the high electric field range, Bioelectrochem. Bioenerg., 7:723.

Blumenthal, R., 1987, Membrane Fusion, Curr. Top. Membr. Transp., 29:203.

Chen, Y. and Blumenthal, R., 1989, On the use of self-quenching fluorophores in the study of membrane fusion kinetics, Biophys. Chem., 34:283.

Chernomordik, L. V., Melikyan, G. B. and Chizmadzhev, Yu. A., 1987, Biomembrane fusion: a new concept derived from model studies using two interacting planar lipid bilayers, Biochim. Biophys. Acta, 906:309.

Chizmadzhev, Yu. A., Arakelyan, V. B. and Pastushenko, V. F., 1979, Electric breakdown of bilayer lipid membranes; III. Analysis of possible mechanisms of defect origination, Bioelectrochem. Bioenerg., 6:63.

Clague, M. J., Schoch, C., Zech, L. and Blumenthal, R., 1990, Gating kinetics of pH-activated membrane fusion of vesicular stomatitis virus with cells: stopped flow measurements by dequenching of octadecylrhodamine fluorescence, Biochemistry, 29:1303.

Clague, M. J., Schoch, C. and Blumenthal, R., 1991, The delay time for influenza hemagglutinin-induced membrane fusion depends on the haemagglutinin surface density, (UnPub).

Dimitrov, D. S and Jain, R. K., 1984, Membrane stability, Biochim. Biophys. Acta, 779:437.

Dimitrov, D. S., 1982, Instability of thin liquid films between membranes, Coll. Pol. Sci., 260:1137.

Dimitrov, D. S., 1983, Dynamic interactions between approaching surfaces of biological interest, Progr. Surface Sci., 14:295.

Dimitrov, D. S., 1984, Electrical breakdown of lipid bilayers and cell membranes - a thin viscoelastic film model, J. Membrane Biol., 78:53.

Dimitrov, D. S. and Ivanov, I. B., 1978, Hydrodynamics of thin liquid films. On the rate of thinning of microscopic films with deformable surfaces, J. Colloid Interface Sci., 64:97.

Dimitrov, D. S. and Sowers, A. E., 1990a, A delay in membrane fusion: Lag times observed by fluorescence microscopy of individual fusion events induced by an electric field pulse, Biochemistry, 29:8337.

Dimitrov, D. S. and Sowers, A. E., 1990b, Membrane electroporation - fast molecular exchange by electroosmosis, Biochim. Biophys. Acta, 1022:381.

Dimitrov, D. S., Apostolova, M. A. and Sowers, A. E., 1990, Attraction, deformation and contact of erythrocyte membranes induced by low frequency electric fields, Biochim. Biophys. Acta, 1023:389.

Heuser, J. E., Reese, T. S., Dennis, M. J., Jan, Y., Jan, L. and Evans, L., 1979, Synaptic vesicle exocytosis captured by quick freezing and correlated with quantal transmitter release, J. Cell Biol., 81:275.

Hoekstra, D., Klappe, K., Hoff, H. and Nir, S., 1989, Mechanism of fusion of sendai virus: role of hydrophobic interactions and mobility constraints of viral membrane proteins. Effects of PEG, J. Biol. Chem., 264:6786.

Ivanov, I. B. and Dimitrov, D. S., 1974, Hydrodynamics of thin liquid films. Effects of surface viscosity on thinning and rupture of foam films, Coll. Pol. Sci., 252:982.

Lowy, R. J., Sarkar, D. P., Chen, Y. and Blumenthal, R., 1990, Observation of single influenza virus-cell fusion and measurement by fluorescence video microscopy, Proc. Natl. Acad. Sci. USA, 87:1850.

Morris, S. J., Sarkar, D. P., White, J. M. and Blumenthal, R., 1989, Kinetics of pH-dependent fusion between 3T3 fibroblasts expressing influenza hemagglutinin and red blood cells, J. Biol. Chem., 264:3972.

Ohki, S. and Arnold, K., 1990, Surface dielectric contstants, surface hydrophobicity and membrane fusion, J. Membrane Biol., 114:195.

Rand, R. P. and Parsegian, V. A., 1986, Mimicry and mechanism in phospholipid models of membrane fusion, Annu. Rev. Physiol., 48:201.

Rand, R. P. and Parsegian, V. A., 1989, Hydration forces between phospholipid bilayers, Biochim. Biophys. Acta, 988:351.

Ruckenstein, E., Marmur, A. and Gill, W. N., 1976, Coverage dependent rate of cell deposition, J. theor. Biol., 58:439.

Sarkar, D. P., Morris, S. J., Eidelman, O., Zimmerberg, J. and Blumenthal, R., 1989, Initial stages of influenza hemagglutinin-induced cell fusion monitored simultaneously by two fluorescent events: cytoplasmic continuity and lipid mixing, J. Cell Biol., 109:113.

Sowers, A. E., 1986, A long-lived fusogenic state is induced in erythrocyte ghosts by electric pulses, J. Cell Biol., 102:1358.

Spruce, A. E., Iwata, A., White, J. M. and Almers, W., 1989, Patch clamp studies of single cell-fusion events mediated by a viral fusion protein, Nature, 342:555.

Teissie, I. and Rols, M. P., 1986, Fusion of mammalian cells in culture is obtained by creating the contact between cells after their electropermeabilization, <u>Biochem. Biophys. Res. Commun.</u>, 140:258.

Wu, R. and Weinbaum, S., 1982, , <u>J. Fluid Mech.</u>, 121:315.

SHORT-RANGE REPULSIVE INTERACTIONS BETWEEN THE SURFACES OF LIPID MEMBRANES

Thomas J. McIntosh[1], Alan D. Magid[1], and Sidney A. Simon[2]

Departments of Cell Biology[1], Neurobiology[2], and Anesthesiology[2]
Duke University Medical Center
Durham, North Carolina 27710

INTRODUCTION

Short-range repulsive interactions are critical to many properties of both cell and model membranes. For example, the close approach of apposing membrane surfaces is governed by short-range interactions. Therefore, for cell membranes, both the magnitude and range of these short-range forces are important in the numerous biological processes where membranes come together, such as cell-cell recognition, synaptic transmission, and protein secretion. Moreover, specificity in biochemical associations can be achieved through a balance of short-range attractive and repulsive interactions. For model membranes, interbilayer interactions determine to a large extent the hydration, aggregation, and fusogenic properties of lipid bilayers and lipid-protein assemblies. In this paper we discuss our recent work on the two principal short-range repulsive interactions acting between cell and model membranes, namely the hydration and steric pressures.

The hydration pressure (or more generally, the solvation pressure) forms a major barrier to the close approach of two polar or charged surfaces and is the dominant intermembrane repulsive pressure for membrane separations of about 5 to 20 Å (LeNeveu et al., 1977; Parsegian et al., 1979; McIntosh and Simon, 1986; Simon and McIntosh, 1989a). For a variety of lipid bilayer membranes it has been found experimentally that the hydration pressure, P_h, decays exponentially with increasing fluid separation between bilayers, d_f, such that $P_h = P_o \cdot \exp(-d_f/\lambda)$, where λ is the decay length (LeNeveu et al., 1977; Parsegian et al., 1979; McIntosh and Simon, 1986; McIntosh et al., 1989a &c; Rand and Parsegian, 1989; Simon and McIntosh, 1989a). Numerous theoretical treatments have been proposed to explain the magnitude and range of the hydration pressure (Marcelja and Radic, 1976; Gruen and Marcelja, 1983; Jonsson and Wennerstrom, 1983; Schiby and Ruckenstein, 1983; Cevc and Marsh, 1985; Graham et al., 1986; Henderson and Lozada-Cassou, 1986; Belaya et al., 1986; Kornyshev and Leikin, 1989). Most of these theories are general in that they do not consider details of the structure of the hydrated surface nor specific interactions between the surface and the water molecules. However, it is now generally accepted that the hydration pressure arises from the polarization or reorganization of water by the membrane surface. One goal of our work described here has been to determine those properties of the membrane surface and the solvent molecules that are responsible for the magnitude (P_o) and decay length (λ) of the hydration pressure.

Considerably less experimental information is available concerning the steric pressure between membrane surfaces, even though steric hindrance must play an important role in many cell-cell and receptor-ligand interactions. Although pressure-distance relationships have been obtained for the close approach of adjacent membranes, a major experimental problem has been the separation of the total pressure into its component pressures. In particular, at very small intermembrane distances, it has been extremely difficult to distinguish the hydration and steric pressures (Marra and Israelachvili, 1985). In this paper, we describe a method to separate hydration and steric pressures, and present experimental data showing the range and magnitude of the steric pressure between phosphatidylcholine and phosphatidylcholine:cholesterol bilayers.

MATERIALS AND METHODS

Phospholipids used in these studies were obtained from Avanti Polar Lipids, Inc., Alabaster, Ala. Cholesterol, poly(vinylpyrrolidone) (PVP), and formamide were purchased from Sigma Chemical Company, St. Louis, Mo., and 1,3-propanediol was obtained from Aldrich Chemical Co., Milwaukee, Wis. PVP solutions between 5 and 60% (w/w) were made in three solvents: triple-distilled water, formamide, and 1,3-propanediol.

Osmotic pressures in the range of 1.1×10^5 to 3.2×10^7 dyn/cm^2 were applied to unoriented multiwalled liposomes by the "osmotic stress" procedures of Parsegian, Rand, and colleagues (LeNeveu et al., 1977; Parsegian et al., 1979; Parsegian et al., 1986). In brief, an excess amount of the appropriate PVP solution was added to the dry lipid. The suspensions were covered with nitrogen and incubated for several hours with periodic vortexing above the lipid's main phase transition temperature. Because PVP molecules are too large to enter between the lipid multilayers, they compete for solvent and thereby compress the lipid lattice (LeNeveu et al., 1977; Parsegian et al., 1979; Parsegian et al., 1986). Osmotic pressures for the PVP-water, PVP-formamide, and PVP-1,3-propanediol solutions were measured with a custom built membrane osmometer as described previously (McIntosh et al., 1989b). For x-ray diffraction experiments, the lipid-polymer-solvent suspensions were sealed in quartz glass capillary tubes and mounted in a point-collimation X-ray diffraction camera.

Vapor pressures in the range of 2.8×10^7 to 2.2×10^9 dyn/cm^2 were applied to oriented lipid multilayers using established procedures (Parsegian et al., 1979; McIntosh et al., 1987). The oriented specimen was formed by placing a small drop of lipid/chloroform solution on a piece of aluminum foil and slowly evaporating the chloroform. The foil substrate was given a convex curvature and mounted in a controlled humidity chamber on a line-focussed single-mirror x-ray camera, where the x-ray beam was oriented at a grazing angle relative to the lipid multilayers. The humidity chamber consisted of a canister with two mylar windows for passage of the x-ray beam. The vapor pressure was controlled by means of a cup of saturated salt solution in the chamber. To speed equilibration, a gentle stream of nitrogen gas was passed through a flask of the saturated salt solution and through the chamber. For the salt solutions used in these experiments, the ratio of the vapor pressure (p) of the saturated salt solution to the vapor pressure of pure water (p_o) has been measured (O'Brien, 1948; Weast, 1984). The following saturated salt solutions were used to obtain the relative vapor pressures (p/p_o) indicated in parentheses: $CuSO_4$ (0.98), Na_2SO_4 (0.93), KCl (0.87), NH_4Cl (0.80), $NaNO_2$ (0.66), $CaCl_2$ (0.32), and $KC_2H_3O_2$ (0.20). The applied pressure is given by $P = -(RT/V_w) \cdot \ln(p/p_o)$ where R is the molar gas constant, T is the temperature (293 ^{o}K for the experiments reported here), and V_w is the molar volume of water.

For all specimens, oriented multilayers and unoriented lipid-polymer-solvent suspensions, X-ray diffraction patterns were recorded on Kodak DEF X-ray film. X-ray films were processed by standard techniques and densitometered with a Joyce-Loebl microdensitometer as described previously (McIntosh et al., 1989a&b). After background subtraction, integrated intensities, I(h), were obtained for each order h by measuring the area under each diffraction peak. For unoriented patterns, the structure amplitude F(h) was set equal to $(h^2I(h))^{1/2}$ (Blaurock and Worthington, 1966; Herbette et al., 1977). For the oriented line-focussed patterns the intensities were corrected by a single factor of h due to the cylindrical curvature of the multilayers (Blaurock and Worthington, 1966; Herbette et al., 1977), so that $F(h) = (hI(h))^{1/2}$.

Estimates for the widths of the bilayer and solvent layer between adjacent bilayers were obtained from the X-ray diffraction data by the use of electron density profiles. Electron density profiles, p(x), on a relative electron density scale were calculated from

$$\rho(x) = (2/d)\Sigma \exp\{\phi(h)\} \cdot F(h)\cos(2\pi xh/d) \qquad \text{Eqn 1.}$$

where x is the distance from the center of the bilayer, d is the lamellar repeat period, $\phi(h)$ is the phase angle for order h, and the sum is over h. Phase angles were determined by a sampling theorem analysis as described in detail previously (McIntosh and Holloway, 1987). All electron density profiles described in this paper are at a resolution of $d/2h_{max} \approx 5$ Å.

For measurements of the Volta potential (V), monolayers were formed by spreading 10 to 40 µL of a lipid/chloroform solution (25 mg/ml) onto the appropriate solvent containing 1 mM KCl. A Teflon trough with a surface area of about 30 cm^2 was used. Under these conditions, the surface monolayer is in equilibrium with liposomes in the subphase (MacDonald and Simon, 1987). To ensure that the surface was free of surface-active impurities, the KCl was roasted at 600 °C and the subphase surface was vacuum aspirated immediately before the monolayer was spread. The trough was thoroughly cleaned between runs. The Volta potential was measured between a Ag/AgCl electrode in the subphase and a polonium electrode in air which was connected to a Keithly electrometer, as previously described (MacDonald and Simon, 1987; McIntosh et al., 1989b). The reported values of Volta potential represent the differences in the potential of the subphase surface in the presence and absence of the monolayer.

RESULTS

Osmotic Stress Experiments

For all lipids analyzed in this study, the x-ray diffraction patterns consisted of several orders of a low-angle lamellar repeat period and one or two wide-angle bands. The low-angle data arise from the stacking of the lipid multilayers and are interpreted as described below. The wide-angle reflections provide information about the hydrocarbon chain packing within each bilayer. That is, a sharp wide-angle reflection centered at 4.1 to 4.2 Å indicates that the lipid hydrocarbon chains are ordered in a gel phase, whereas the presence of a broad wide-angle band centered at approximately 4.5 Å implies that the hydrocarbon chains are in a disordered or liquid-crystalline phase (Tardieu et al., 1973).

A typical example of an applied pressure versus lamellar repeat period relationship is shown in Figure 1. This figure shows data for egg phosphatidylcholine (EPC) multilayers in the liquid-crystalline phase. Note that for repeat periods from 60 Å to 52 Å the data points can be fit quite closely ($r^2 = 0.96$) by a single straight line. However, there is a sharp upward break in the plot at repeat period values of about 51 Å at ln P $\approx$ 17. The lamellar repeat period is the total width of the each unit cell, and so includes the bilayer and the fluid space between bilayers. To estimate both the bilayer thickness and the fluid layer thickness, we calculated electron density profiles for each applied pressure (McIntosh and Simon, 1986; McIntosh et al., 1987).

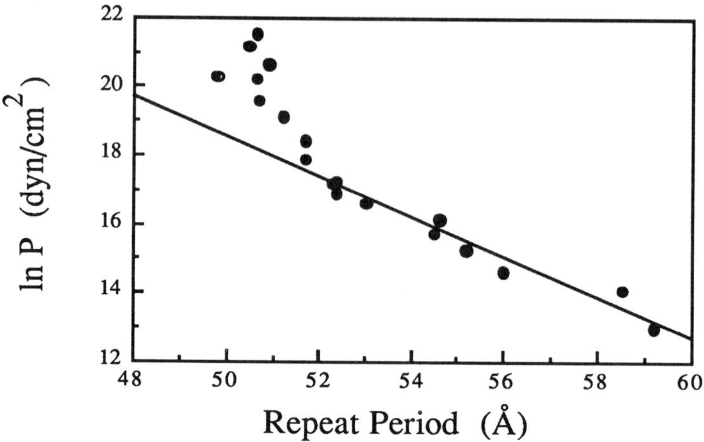

Figure 1. Natural logarithm of applied pressure plotted versus lamellar repeat period for EPC bilayers. The straight line is a least squares fit to the data for 12.5 < ln P < 17.5.

Examples of profiles for EPC in 0%, 20%, and 60% PVP are shown in Figure 2. For each profile, the geometric center of the bilayer is at the origin. The low electron density trough in the center of each bilayer corresponds to the localization of the low density terminal methyl groups of the lipid hydrocarbon chains. The medium density regions positioned about 10 Å from the bilayer center correspond to the lipid methyl groups and the high electron density peaks positioned about 18 Å from the bilayer center for each profile correspond to the lipid head groups. The high density peaks positioned at about 32 Å for 60% PVP, at about 38 Å for 20% PVP, and at about 44 Å for 0% PVP correspond to the head group peaks from the apposing bilayer. Two points should be noted from these profiles in Figure 2. First, the distance between head group peaks from adjacent bilayers decreases with increasing PVP concentration. This means that water is being removed from between adjacent bilayers with increasing osmotic pressure. Second, the three profiles superimpose quite closely from 0 Å to 20 Å, implying that the bilayer thickness does not change appreciably as water is removed from between adjacent bilayers (McIntosh and Simon, 1986).

Since at this resolution the high density head group peak in electron density profiles is known to be located between the phosphate moiety and the glycerol backbone of the lipid (Lesslauer et al., 1972; Hitchcock et al., 1974), these profiles can be used to estimate the location of the lipid/water interface. As noted previously (McIntosh and Simon, 1986; McIntosh et al., 1987; McIntosh et al., 1989a&b), the definition of the lipid/water interface is somewhat arbitrary, because the bilayer surface is not smooth, the lipid head groups are mobile (Hauser et al., 1981), and water penetrates into the head group region of the bilayer (Worcester and Franks, 1976; Simon et al., 1982). We operationally define the bilayer width as the total thickness of the bilayer assuming that the head group conformation is the same as it is in single crystals of dimyristoylphosphatidylcholine (Pearson and Pascher, 1979). That is, we assume that the phosphocholine (PC) group is, on average, oriented approximately parallel to the bilayer plane, so that the edge of the bilayer lies about 5 Å outward from the center of the high density peaks in the electron density profiles (McIntosh and Simon, 1986; McIntosh et al., 1987; McIntosh et al., 1989a&b).

Using electron density profiles and the above definition for the water/bilayer interface, we obtained the applied pressure-bilayer separation relationship for EPC bilayers shown in Figure 3. As with the pressure-repeat period data in Figure 1, there are two distinct regions in the pressure-bilayer separation relationship in Figure 3. That is, for

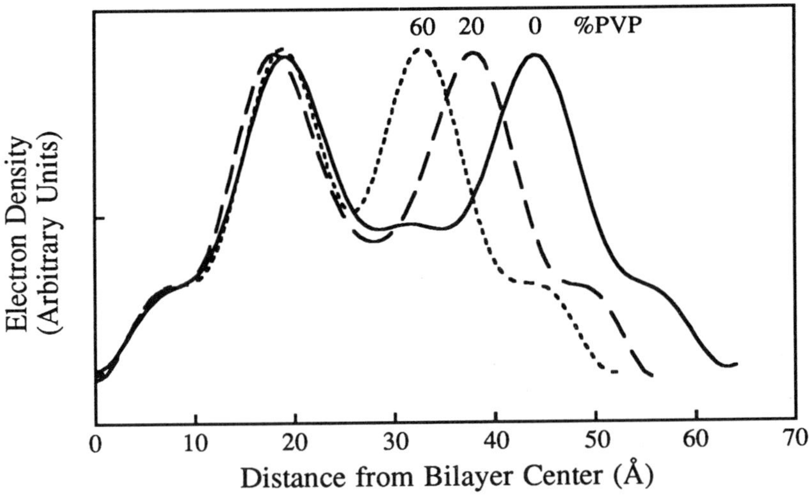

Figure 2. Electron density profiles of EPC bilayers in excess water (solid line), 20% PVP (dashed line), and 60% PVP (dotted line).

pressures such that ln P < 17.5 or bilayer separations > 4 Å the data points can be fit quite closely with a single straight line, such that $P = P_o \cdot \exp(-d_f/\lambda)$, where $P_o = 4.0 \times 10^8$ dyn/cm^2 and $\lambda = 1.7$ Å. This exponential decrease in pressure with increasing fluid spacing has been attributed to the hydration pressure (LeNeveu et al., 1976; LeNeveu et al., 1977; Parsegian et al., 1979; McIntosh and Simon, 1986; McIntosh et al., 1989a&b; Rand and Parsegian, 1989; Simon and McIntosh, 1989a). For ln P > 17.5 and bilayer separation < 4 Å, the pressure decays much more rapidly with increasing bilayer separation, so that the pressure decays exponentially with increasing d_f with a decay length of about 0.6 Å (McIntosh et al., 1987). This part of the pressure-separation relationship has been attributed to steric pressure between the head groups from apposing bilayers (McIntosh et al., 1987; McIntosh et al., 1989a). We will now consider the factors responsible for determining the magnitude and range of both the hydration and steric pressures which contribute to the total pressure-separation curve in Figure 3.

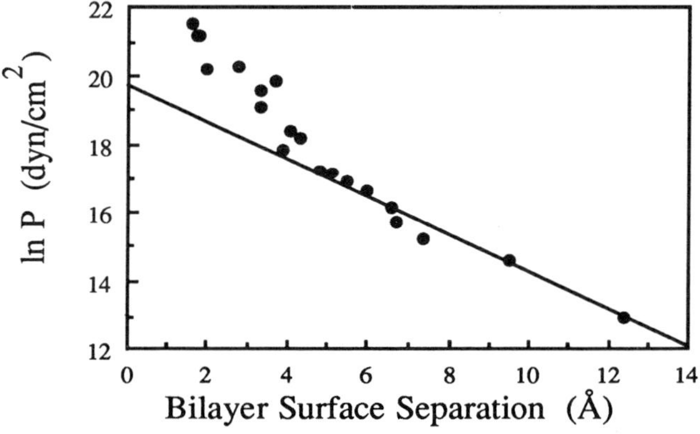

Figure 3. Natural logarithm of applied pressure plotted versus the separation between adjacent EPC bilayers. The solid line is a least-squares fit ($r^2 = 0.97$) to the data for bilayer separations greater than 4 Å.

<u>Hydration Pressure</u>

As seen in Figure 3, and found experimentally for a number of lipid bilayer systems, including phosphatidylcholines in the liquid-crystalline (LeNeveu et al., 1977; Lis et al., 1982; McIntosh and Simon, 1986), gel (Lis et al., 1982; McIntosh and Simon, 1986; Rand and Parsegian, 1989), and interdigitated phases (Simon et al., 1988), phosphatidylcholine:cholesterol mixtures (Lis et al., 1982; McIntosh et al., 1989a), monoglycerides (McIntosh et al., 1989c), and phosphatidylglycerol (McIntosh et al., 1990), the hydration force decays exponentially with increasing bilayer separation such that $P_h = P_o \cdot \exp(-d_f/\lambda)$, where P_o is the magnitude and λ is the decay length. We have been interested in determining the factors which govern P_o and λ.

First, consider the decay length, λ, of the hydration pressure. We found for both phosphatidylcholines (Simon et al., 1988) and monoglycerides (McIntosh et al., 1989c) that the decay length does not depend on the lipid's molecular area or acyl chain length. We also investigated how the decay length varies with different solvents, since theoretical treatments have suggested that λ should depend on specific properties of the solvent molecule, including its size (Schiby and Ruckenstein, 1983), its static or optical dielectric constant (Gruen and Marcelja, 1983), and the number and type of hydrogen-bonding defects in the solvent (Attard and Batchelor, 1988). We performed osmotic stress experiments on lipid multilayers formed in two nonaqueous solvents, formamide and 1,3-propanediol (1,3-PDO), whose dimensions, as well as several other physical properties, are quite different than water. Water has a molecular weight of 18, a surface tension of 72.8 dyn/cm, a dipole moment of 1.85 D, and a dielectric constant of 78; formamide has a molecular weight of 45, a surface tension of 57.4 dyn/cm, a dipole moment of 3.7 D, and a dielectric constant of 111; and 1,3-PDO has a molecular weight of 76, a surface tension of 45.6 dyn/cm, a dipole moment of 2.5 D, and a dielectric constant of 35. Thus, these three molecules provide a range of sizes, surface tensions, dipole moments, and dielectric constants.

Osmotic stress experiments were performed for both EPC and equimolar EPC:cholesterol bilayers. Figure 4 shows the natural logarithm of applied pressure (ln P) plotted versus the lamellar repeat period for EPC bilayers in water, formamide, and 1,3-PDO. Note that each of the three data sets can be closely fit to a straight line, implying that the solvation pressure decreases exponentially with increasing repeat period for each of these solvents. The slopes of the three lines are different for the three solvents. Electron density profiles were calculated for each of these bilayer systems. These profiles showed that for each solvent the bilayer thickness remained essentially constant for all osmotic pressures (see McIntosh et al., 1989b). By subtracting the bilayer thickness from the total repeat period, the separation between bilayers, d_f, was obtained. Plots of ln P versus d_f provided a value for λ and P_o for each solvent (McIntosh et al., 1989b). For both EPC and equimolar EPC:cholesterol bilayers, the decay length was found to increase with increasing size of the solvent molecule. For example, for equimolar EPC:cholesterol bilayers, λ was 2.1 Å, 2.9 Å, and 3.1 Å in water, formamide, and 1,3-PDO, respectively.

Since λ does not vary monotonically with either the static dielectric constant or bulk dipole moment of the solvent, the importance of these two parameters in determining the range of the solvation pressure must be small. However, the measured values of decay length do vary monotonically with both the molecular weight and surface tension of the solvent. We initially analyze λ in terms of the packing properties of the solvent. Although it is difficult to determine the precise packing arrangement of the solvent molecules in the interbilayer space, measurements of the partial specific volume of water in EPC bilayers (White et al., 1987) suggest that differences in packing between interbilayer water and bulk water are small. As a first approximation of the packing of the solvent molecules in the direction perpendicular to the bilayer surface, we use the cube root of the number of solvent molecules per volume ($n^{1/3}$) in bulk solution, where $n = N\rho/M_w$, where N is Avogadro's number, ρ is the solvent density, and M_w is the solvent molecular weight. This calculation makes no assumptions about the shapes of the solvent molecules. In Figure 5 we plot for

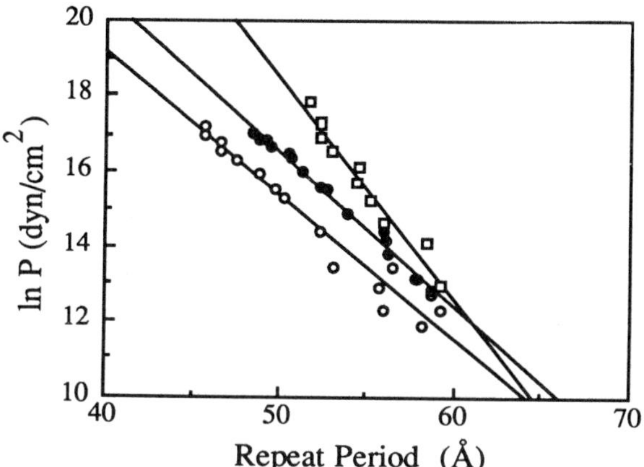

Figure 5. Decay length plotted versus the inverse cube root of the number of solvent molecules per unit volume.

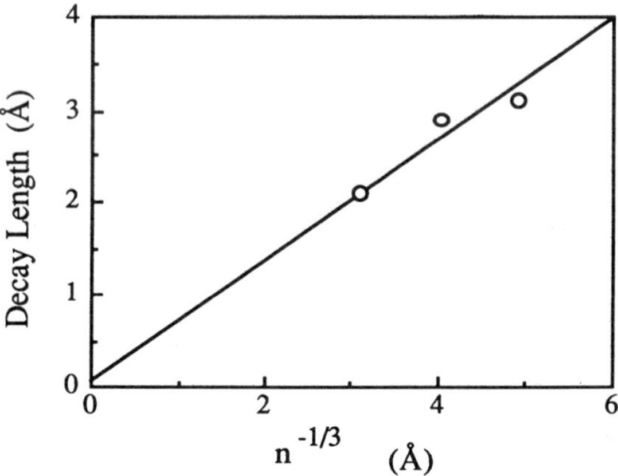

Figure 4. Natural logarithm of applied pressure versus lamellar repeat period for EPC multilayers in water (squares), formamide (solid circles), and 1,3-propanediol (open circles).

EPC:cholesterol multilayers in water, formamide, and 1,3-PDO the decay length versus $1/n^{1/3}$. The straight line in this graph, which has a slope of 0.66, is the least-squares fit ($r^2 = 0.99$) to the three data points plus a point at the origin. The point at the origin was included in the regression, since if we assume that the decay length does depend on the solvent's dimensions, then λ should approach zero in the limit of a vanishingly small solvent molecule. As can be seen, the three data points fall quite closely to this line, leading to the prediction that the relation between the decay length of the solvation pressure and the size of the solvent molecule is $\lambda = 0.66/n^{1/3}$. The factor of 0.66 might represent a measure of the packing distribution of the solvent molecules in the interbilayer space. For example, if the solvent molecules were spherical and hexagonally close packed, it can be shown (McIntosh et al., 1989b) that $R = 0.56/n^{1/3}$, where R is the "radius" of the solvent molecule. This implies that the decay of the solvation pressure, at least for the three solvents studied here, is approximately equal to the "radius" of the solvent molecule calculated assuming spherical symmetry and hexagonal close packing of the solvent molecules.

A recent model (Israelachvili and Wennerstrom, 1990) of the short-range interactions between bilayers predicts that there should be a correlation between λ and the interfacial tension at the bilayer/solvent interface. In this model the major short-range pressure arises from vertical protrusions of the lipid molecules from the bilayer plane; the greater the protrusion the greater the range of this pressure. The extent of the protrusion depends on the free energy required to transfer acyl chains from the bilayer to the solvent, which is proportional to the interfacial tension between the solvent and the acyl chains. Our measurements of λ for the three different solvents are consistent with the "protrusion force" model (Israelachvili and Wennerstrom, 1990) in that there seems to be a correlation between λ and the solvent surface tension. However, as detailed below, other experimental data are not consistent with the idea that molecular protrusions are the major source of short-range repulsive interactions between bilayers for the entire range of d_f from 0 to 15 Å.

Next, let us consider the magnitude of the hydration pressure. In their pioneering study to explain the hydration pressure between neutral bilayers, Cevc and Marsh (1985) extended the water polarization model developed by Gruen and Marcelja (1983) and derived the following expression for the magnitude of the hydration pressure:

$$P_o = 2\chi\,(\psi(0)/\lambda)^2 \qquad\qquad \text{Eqn. 2}$$

where $\psi(0)$ is the "hydration potential" at $d_f = 0$ and χ is orientational susceptibility of the solvent and is equal to $\varepsilon_o(\varepsilon - 1)/\varepsilon$ where ε is the solvent's bulk dielectric constant and ε_o is the permittivity of free space. This expression is similar in form to that developed for the magnitude of the electrostatic pressure between surfaces containing fixed charges (Israelachvili, 1985), where the magnitude of the electrostatic pressure is proportional to the square of the surface potential. That is, in equation 2 the magnitude of the hydration pressure is proportional to the square of the potential that arises from the perpendicular components of the "multipole surface charge densities" in the polar head groups of the lipids (Cevc and Marsh, 1985). Cevc and Marsh (1987) calculated $\psi(0)$ for a variety of lipids by summing the different dipole moments of the polar components of the lipid head group. Recently the work of Marcelja and coworkers (Marcelja and Radic, 1976; Gruen and Marcelja, 1983) and Cevc and Marsh (1985) has been extended by Belaya et al. (1986) and Dzhavakhidze et al. (1986, 1988) by the use of a nonlocal electrostatic approach to calculate P_o. They noted that the hydration pressure could arise "as a result of nonlocal polarization of water by permanent dipole moments of the surface--either due to oriented polar groups of the surface itself or due to chemisorbed water molecules" (Dzhavakhidze et al., 1986).

We have performed experiments to test these ideas concerning the relationship of the magnitude of the hydration pressure and the electric field caused by oriented dipoles. That is, we have considered Equation 2, but have equated the hydration potential, $\psi(0)$, to the Volta potential, V, measured for monolayers in equilibrium with liposomes (MacDonald and Simon, 1987). The measured Volta includes contributions from all fixed charges and all oriented dipoles and multipoles, of both the lipid and solvent molecules. This identification predicts that

$$P_o = 2\chi \ (V/\lambda)^2 \qquad\qquad \text{Eqn. 3}$$

For zwitterionic lipids, such as those discussed in this paper, V can arise only from the vector sum of perpendicular components of the dipole (and multipole) moments of the lipid *and* solvent molecules in the bilayer.

To test this hypothesis, we measured P_o by x-ray diffraction and V by Volta potential measurements for the same lipids with the same solvents (Simon and McIntosh, 1989a). To vary V, we studied zwitterionic phospholipids with different polar head groups (phosphatidylcholine and dimethylphosphatidylethanolamine), different structure (gel and liquid crystalline phases), with different amounts of cholesterol, and with different solvents (water, formamide, and 1,3-propanediol).

We found that the Volta potential depended on *both* the solvent and the lipid. For example, the Volta potential for EPC was 415 mV in water, 266 mV in formamide, and 223 mV in 1,3-PDO (McIntosh et al., 1989b). V was considerably higher in the gel (575 mV) than in the liquid crystalline phase (415 mV) (Simon and McIntosh, 1989a), consistent with the previous observation that V is inversely proportional to the area per lipid molecule (MacDonald and Simon, 1987). Thus, V increased with increasing number of polar head groups per unit area at the lipid/solvent interface. The incorporation of cholesterol slightly increased V, as V was 415 mV for EPC and 493 mV for equimolar EPC:cholesterol monolayers over water (Simon and McIntosh, 1989a). Taken together, these data imply that the Volta potential depends on *both* the lipid head group and the solvent molecules in the head group region.

Osmotic stress experiments showed that P_o increased with increasing Volta potential. Figure 6 shows a plot of P_o as obtained by x-ray diffraction measurements plotted versus $2\chi \ (V/\lambda)^2$. As can be seen, there is an excellent correlation, indicating that the magnitude of the hydration pressure can be accurately predicted by measurements of the Volta potential. The close agreement between the measured and predicted values of P_o also support our choice of the edge of the bilayer as the plane of origin of the hydration pressure. The results shown in Figure 6 imply that the hydration pressure arises primarily from the polarization of solvent molecules in the interbilayer space by fields from oriented dipoles and multipoles in the lipid head group region. These results may explain why structurally diverse molecules, such as polyhydric alcohols, fatty acids, and detergents, act as fusogens between lipid bilayers and/or biological membranes (Howell and Lucy, 1969; Maggio et al., 1976; Papahadjopoulos et al., 1976; Lucy, 1978). All of these fusogenic molecules reduce the Volta potential (Maggio et al., 1976; Maggio and Lucy, 1976) and would therefore be expected to reduce the magnitude of the repulsive hydration force (Simon et al., 1988).

Steric Pressure

We now consider the pressure-bilayer separation relation shown in Figure 3 for high applied pressures (ln P > 17.5) and small separations ($d_f < 4$ Å). Two causes could be postulated for the observed upward break in the pressure-separation curve at ln P $\approx$ 17.5: (1) a discontinuity in the hydration pressure, that is an abrupt increase in the energy required to remove water from between bilayers at low water contents, and (2) the onset of steric repulsion between lipid head groups from apposing bilayers. Possibility (1) can be can be excluded since the pressure required to remove water is an exponential function of

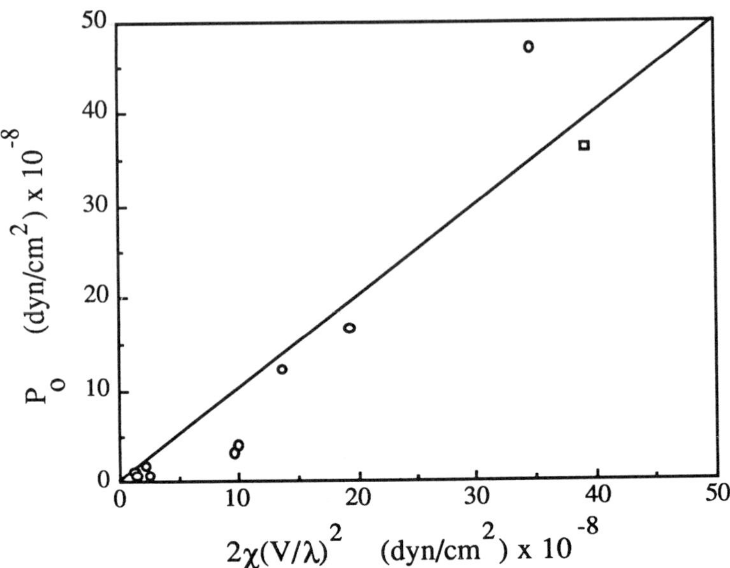

Figure 6. A plot of the magnitude of the hydration pressure (P_o) versus the quantity $2\chi\,(V/\lambda)^2$. P_o was obtained from x-ray diffraction experiments and V was obtained from Volta potential measurements. The circles represent data from phosphatidylcholine and phosphatidylcholine:cholesterol bilayers and the square represents results from dimethylphosphatidylethanolamine bilayers. The straight line is a plot of the theoretical prediction $P_o = 2\chi\,(V/\lambda)^2$. This figure is reproduced from the *Biophysical Journal.*, Simon et al., (1991), by copyright permission of the Biophysical Society.

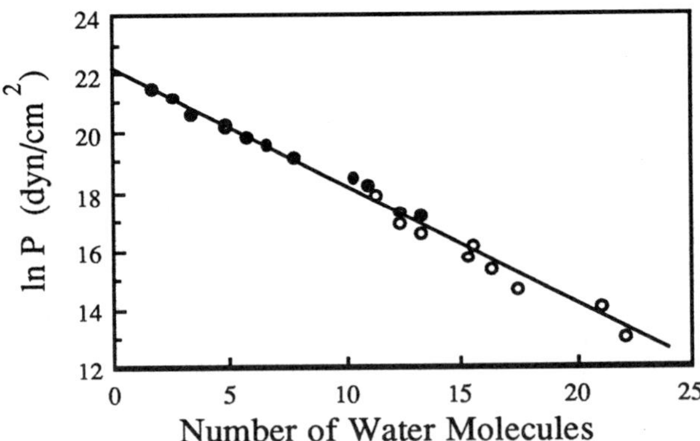

Figure 7. Natural logarithm of applied pressure plotted versus the number of water molecules per EPC molecule. The solid circles are taken from the adsorption isotherms of Jendrasiak and Hasty (1974) and the open circles are taken from our x-ray phase diagram of EPC in water (McIntosh, et al., 1989b) by the method of LeNeveu et al. (1977). This figure is taken from (McIntosh et al., 1987) and reprinted with permission of the American Chemical Society.

water content and has no apparent discontinuities from 23 to 2 waters per lipid molecule (Figure 7). On the other hand, as shown below, possibility (2) fits all of the available experimental data.

The position of the observed break in the pressure versus separation relationship shown in Figure 3 can be explained in terms of steric interactions. That is, the onset of steric repulsion at a fluid separation of about 4 to 6 Å is consistent with the NMR data of Hauser (1981) which indicate that the phosphocholine group rotates from its preferred position, approximately parallel to the bilayer, to a position where it extends 2 to 3 Å farther into the fluid space (see Figure 8). Thus, on the basis of Hauser's model for head-group motion, trimethylammonium groups from apposing bilayers would be expected to come into steric hindrance at $d_f \approx 4$ to 6 Å. Local variations in bilayer thickness, or "breathing" modes, would tend to "soften" the edge of the steric barrier barrier and could extend the range of steric interactions. Such local variations in bilayer thickness have been incorporated into the recently proposed "protrusion force" model for steric interactions (Israelachvili and Wennerstrom, 1990).

Thus, we argue that the upward break in the plot of ln P versus d_f (Figure 3) occurs because of steric hindrance between the bulky EPC head groups from apposing bilayers. This analysis implies that the extent of this upward break should depend on the volume fraction of PC head groups at the hydrocarbon/water interface. To quantitatively test this hypothesis, we systematically varied the volume fraction of interfacial PC head groups by incorporating various concentrations of cholesterol into EPC bilayers. Cholesterol is an amphipathic molecule with a relatively bulky hydrophobic steroid ring region which is embedded into the bilayer, but a rather small hydrophilic moiety (an OH group) which is located at the hydrocarbon water interface (Worcester and Franks, 1976; McIntosh, 1978). In a bilayer, the surface area occupied by a cholesterol molecule is about one-half the area occupied by an EPC molecule (Lecuyer and Dervichian, 1969). Therefore, the incorporation of cholesterol into EPC bilayers separates adjacent PC head groups and decreases their volume fraction at the bilayer/water interface (McIntosh et al., 1989a; Simon and McIntosh, 1989b).

Figure 9 shows the natural logarithm of applied pressure (ln P) plotted versus bilayer separation for EPC and equimolar EPC:cholesterol bilayers. The solid line corresponds to a least-squares fit to the the equimolar EPC:cholesterol data. For ln P < 17.5, the data points for both EPC and equimolar EPC:cholesterol fall quite closely to this line, indicating that the hydration pressure is similar for both systems (McIntosh et al., 1989a). However, the pressure-separation data are quite different for ln P > 18. As noted above, for EPC bilayers, the data points deviate upward from the straight line due to steric hindrance between the bulky PC head groups from apposing bilayers. However, for equimolar EPC:cholesterol bilayers the data points fall closely to the same straight line up to the highest applied pressures (ln P =21.5). That is there is a larger decrease in d_f at high applied for EPC:cholesterol bilayers than for EPC bilayers. For example, at the highest applied pressure (ln P = 21.5) the value of d_f is about 6 Å smaller for equimolar EPC:cholesterol bilayer than for EPC bilayers. In fact, the values of d_f are negative for equimolar EPC:cholesterol bilayers at the highest applied pressures, meaning that the PC head groups from apposing bilayers have interpenetrated (McIntosh et al., 1989a).

These observations can be explained in terms of a simple model in which cholesterol spreads the EPC molecules apart in the plane of each bilayer, reducing the volume fraction of PC head groups at the interface, thereby decreasing steric repulsion between adjacent bilayers (McIntosh et al., 1989a). For equimolar EPC:cholesterol bilayers the EPC molecules are spread far enough apart that head groups from apposing bilayers can interpenetrate. This interpenetration of apposing head groups can be observed in the electron density profiles of Figure 10. That is, at applied pressures ln P =13.9 and 17.2, the electron density profiles show a fluid space between high density head-group peaks from apposing bilayers. However, at the high applied pressure of ln P = 21.2, no fluid

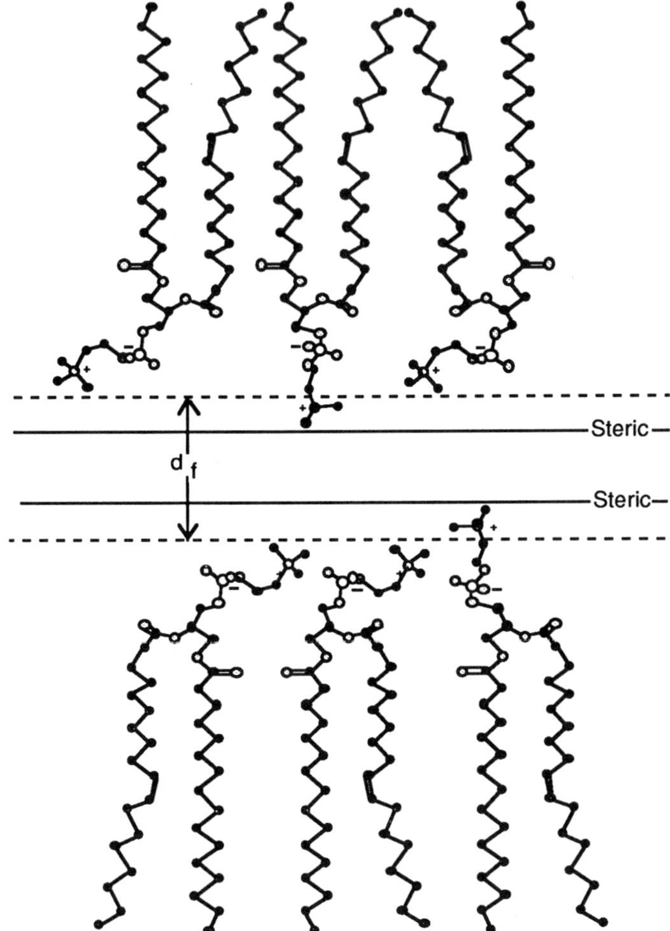

Figure 8. Schematic diagram showing the approximate location of the plane where steric interactions are first observed for EPC (solid line). The dotted lines delimit the fluid space, d_f, as determined by electron density profiles. assuming the parallel orientation of the lipid head groups. The head groups rotate so that the trimethylammonium moieties extend 2 to 3 Å into the fluid space. Depicted are the two extremes of head-group orientation, with the phosphocholine moiety parallel and perpendicular to the plane of the bilayer (Hauser, 1981). This figure is taken from (McIntosh et al., 1987) and reprinted with permission of the American Chemical Society.

space is observed and the head group peaks from apposing bilayers have merged. Calculations show that equimolar concentrations of cholesterol can indeed increase the area per PC head group enough to allow interpenetration of apposing head groups. Thus, with equimolar EPC:cholesterol bilayers apposing head groups have enough room to slide past each other and cause a reduction in d_f of about 6 Å beyond what is observed with EPC bilayers (Figure 9). That is, it appears that high concentrations of cholesterol can effectively reduce the steric hindrance between apposing bilayers by reducing the density of PC head groups in the plane of the bilayers. In plots of ln P versus bilayer separation, data points for EPC bilayers containing 0.2 and 0.33 mole fraction cholesterol fall in between the extreme cases of 0 and 0.5 mole fraction cholesterol (McIntosh et al., 1989a). This indicates that the magnitude of steric repulsion depends on the amount of cholesterol in the bilayer, or on the volume fraction of PC head groups at the interface.

Although the data in Figure 9 can be explained by a combination of steric and hydration pressures as described above, these data present several difficulties for the "protrusion force" model of Israelachvili and Wennerstrom (1990). First of all, for EPC bilayers, the data exhibit two exponentially decaying force regimes, whereas the "protrusion force" model, as it stands, predicts only one. Second, the incorporation of equimolar cholesterol into EPC bilayers would be expected to reduce vertical fluctuations by its condensing effect on acyl chain packing. Yet, for ln P < 17.5, the pressure-distance relationship is similar for EPC and equimolar EPC:cholesterol (Figure 9). In particular, the total repulsive pressure extends to 15 Å both in the presence and absence of cholesterol. Third, when the upward break in the pressure-distance curve at ln P $\approx$ 17.5 is eliminated by the incorporation of equimolar cholesterol, there is still an underlying pressure which decays exponentially with increasing fluid separation with a decay constant of about 2 Å. Moreover, as the head groups from apposing bilayers interpenetrate, this pressure extends all of the way from $d_f = 15$ Å to $d_f = -5$ Å. This implies, that if the protrusion force were the only force present, the protruding lipid molecules would have to extend at least 10 Å from each bilayer surface, a distance which corresponds to at least 8 CH_2 groups per acyl chain or 16 CH_2 groups per lipid molecule. Exposure of that much hydrocarbon to water would be quite energetically unfavorable (Tanford, 1980). Therefore, although protrusions of lipid molecules from the bilayer surface may contribute to the short-range steric interaction, we argue that these protrusions can not be the *sole* cause of the total repulsive pressure-distance relationships shown in Figures 3 and 9.

From the data shown in Figure 7, it can be calculated that the energy required to dehydrate EPC bilayers from 23 to 2 waters per lipid molecule is 51 erg/cm^2, or about 4 times the thermal energy, RT (McIntosh et al., 1987). In order for two membranes to fuse, this large energy barrier must be overcome. One possible source for a sufficiently large attractive energy is the "hydrophobic" interaction (Helm et al., 1989), which results when hydrophobic groups are exposed to water. As argued by Helm et al. (1989), such exposed hydrophobic areas could result from inhomogeneous ionic or osmotic stresses or local packing strains induced by integral membrane proteins. The attractive hydrophobic energy is potentially large, as the creation of two hydrocarbon-water interfaces generates an energy of about -100 erg/cm^2. Along these lines, Ohki (1985, 1988) emphasizes the relationship between hydrophobicity of the membrane surface and membrane fusion. He notes that an increase in interfacial tension and a decrease in dielectric constant of the membrane polar region are correlated with the extent of membrane fusion, and that both the increase in interfacial tension and decrease in dielectric constant are related to an increase in hydrophobicity of the membrane surface. Thus, the ability of lipid vesicles to fuse depends on a delicate balance between attractive interactions (such as the hydrophobic and van der Waals pressures, hydrogen bonds, and salt bridge formation) and short-range repulsive interactions (hydration and steric pressures). Factors which tend to either decrease the exposure of hydrophobic surfaces to water or increase the steric and hydration pressures would tend to decrease the probability of fusion.

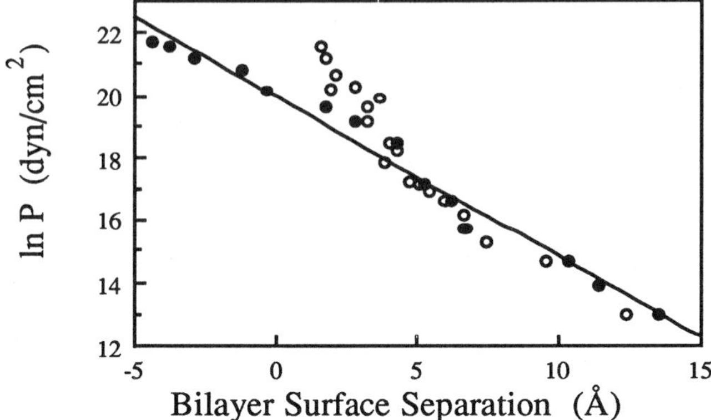

Figure 9. Natural logarithm of applied pressure plotted versus the separation between adjacent bilayer surface for EPC (open circles) and equimolar EPC:cholesterol bilayers (solid circles). The solid line is a linear least-squares fit to the equimolar EPC:cholesterol data.

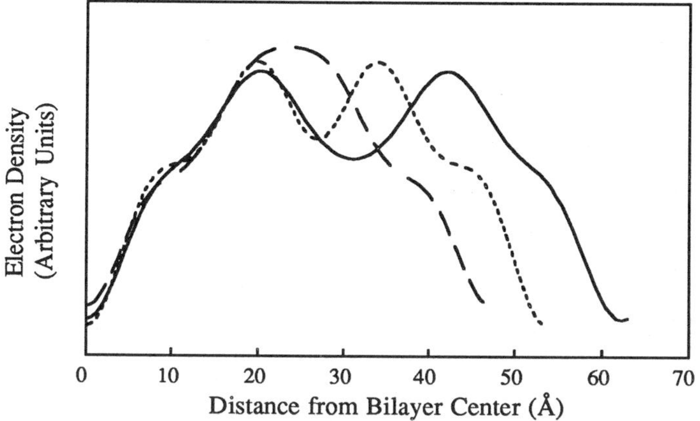

Figure 10. Electron density profiles of equimolar EPC:cholesterol subjected to the following applied pressures (in dyn/cm^2): $\ln P = 13.9$ (solid line), $\ln P = 17.2$ (dotted line), and $\ln P = 21.2$ (dashed line). The geometric center of each bilayer is at the origin.

CONCLUSIONS

The data presented here indicate that there are two short-range pressures that govern the fluid spaces between electrically neutral lipid bilayers--steric pressure and hydration pressure. For phosphatidylcholine bilayers, both the steric pressure and the hydration pressure decay exponentially with increasing fluid spacing. The range of the steric repulsion is significantly shorter than that of the hydration repulsion, since the measured decay length is about 0.6 Å for the steric pressure and about 2 Å for the hydration pressure. Although neither of these pressures is well understood from first principles for rough, fluid surfaces, we have experimentally found parameters that determine the range and magnitude of these pressures. That is, the magnitude of the hydration pressure depends on the square of the Volta potential and its range depends on the packing density of the solvent molecules, whereas the magnitude of the steric pressure depends on the volume fraction of polar head groups at the membrane/water interface and its range is a function of the extension of the head groups into the fluid space. These observations provide a basis for the manipulation of these repulsive pressures, which could potentially be used to aid or inhibit events which involve membrane adhesion.

ACKNOWLEDGEMENTS

This work was supported by Grant GM-27278 from the National Institutes of Health.

REFERENCES

Attard, P. and Batchelor, M. T., 1988, A mechanism for the hydration force demonstrated in a model system, Chem. Phys. Letts. 149: 206-211.

Belaya, M. L., Feigel'man, M. V. and Levadny, V. G., 1986, Hydration forces as a result of non-local water polarizability, Chem. Phys. Letts. 126: 361-364.

Blaurock, A. E. and Worthington, C. R., 1966, Treatment of low angle X-ray data from planar and concentric multilayered structures, Biophys. J. 9: 305-312.

Cevc, G. and Marsh, D., 1985, Hydration of noncharged lipid bilayer membranes. Theory and experiments with phosphatidylethanolamine, Biophys. J. 47: 21-32.

Cevc, G. and Marsh, D., 1987, "Phospholipid Bilayers. Physical Principles and Models", John Wiley & Sons, New York.

Dzhavakhidze, P. G., Kornyshev, A. A. and Levadny, V. G., 1986, The role of the interface in the nonlocal electrostatic theory of hydration force, Phys. Letts. A 118: 203-208.

Dzhavakhidze, P. G., Kornyshev, A. A. and Levadny, V. G., 1988, The structure of the interface in the solvent-mediated interaction of dipolar surfaces, Il Nuovo Cimento 10D: 627-654.

Graham, I. S., Georgallas, A. and Zuckermann, M. J., 1986, Forces between charged lipid bilayers: a theoretical model, J. Chem. Phys. 85: 6010-6021.

Gruen, D. W. R. and Marcelja, S., 1983, Spatially varying polarization in water, J. Chem. Soc. Faraday Trans. 2 79: 225-242.

Hauser, H., 1981, The polar group conformation of 1,2-dialkyl phosphatidylcholines, an NMR study, Biochim. Biophys. Acta 646: 203-210.

Hauser, H., Pascher, I., Pearson, R. H. and Sundell, S., 1981, Preferred conformation and molecular packing of phosphatidylethanolamine and phosphatidylcholine, Biochim. Biophys. Acta 650: 21-51.

Helm, C. A., Israelachvili, J. N., and McGuiggan, P. M., 1989, Molecular mechanisms and forces involved in the adhesion and fusion of amphiphilic bilayers, Science 246: 919-922.

Henderson, D. and Lozada-Cassou, M., 1986, A simple theory for the force between spheres immersed in a fluid, J. Colloid Interface Sci. 114: 180-183.

Herbette, L., Marquardt, J., Scarpa, A. and Blasie, J. K. 1977, A direct analysis of lamellar x-ray diffraction from hydrated oriented multilayers of fully functional sarcoplasmic reticulum, Biophys. J. 20: 245-272.

Hitchcock, P. B., Mason, R., Thomas, K. M. and Shipley, G. G., 1974, Structural chemistry of 1,2 dilauroyl-DL-phosphatidylethanolamine: molecular conformation and intermolecular packing of phospholipids, Proc. Nat. Acad. USA 71: 3036-3040.

Howell, J. I. and Lucy, J. A., 1969, Cell fusion induced by lysolecithin, FEBS Lett. 4: 147-150.

Israelachvili, J. N., 1985, "Intermolecular and Surface Forces", Academic Press, Inc., London.

Israelachvili, J. N. and Wennerstrom, H., 1990, Hydration or steric forces between amphiphilic surfaces?, Langmuir 6: 873-876.

Jendrasiak, G. L. and Hasty, J. H., 1974, The hydration of phospholipids, Biochim. Biophys. Acta 337: 79-91.

Jonsson, B. and Wennerstrom, H., 1983, Image-force forces in phospholipid bilayer systems, J. Chem. Soc. Faraday Trans. 2 79: 19-35.

Kornyshev, A. A. and Leikin, S., 1989, Fluctuation theory of hydration forces: The dramatic effects of inhomogeneous boundary conditions, Phys. Rev. A 40: 6431-6437.

Lecuyer, H. and Dervichian, D. G., 1969, Structure of aqueous mixtures of lecithin and cholesterol, J. Mol. Biol. 45: 39-57.

LeNeveu, D. M., Rand, R. P. and Parsegian, V. A., 1976, Measurement of forces between lecithin bilayers, Nature 259: 601-603.

LeNeveu, D. M., Rand, R. P., Parsegian, V. A. and Gingell, D., 1977, Measurement and modification of forces between lecithin bilayers, Biophys. J. 18: 209-230.

Lesslauer, W., Cain, J. E. and Blasie, J. K., 1972, X-ray diffraction studies of lecithin bimolecular leaflets with Incorporated fluorescent probes, Proc. Nat. Acad. Sci. USA 69: 1499-1503.

Lis, L. J., McAlister, M., Fuller, N., Rand, R. P. and Parsegian, V. A., 1982, Interactions between neutral phospholipid bilayer membranes, Biophys. J. 37: 657-666.

Lucy, J. A., 1978, in "Membrane Fusion," Poste, G and Nicholson, G. L., ed., pp 267-304, Elsevier/North Holland, The Netherlands.

MacDonald, R. C. and Simon, S. A., 1987, Lipid monolayer states and their relationship to bilayers, Proc. Nat. Acad. Sci. USA 84: 4089-4094.

Maggio, B., Ahkong, O. F. and Lucy, J. A., 1976, Poly(ethylene glycol), surface potential and cell fusion, Biochem. J. 158: 647.

Maggio, B. and Lucy, J. A., 1976, Polar-group behavior in mixed monolayers of phospholipids and fusogenic lipids, Biochem. J. 155: 353-364.

Marcelja, S. and Radic, N., 1976, Repulsion of interfaces due to boundary water, Chem. Phys. Lett. 42: 129-130.

Marra, J. and Israelachvili, J., 1985, Direct measurements of forces between phosphatidylcholine and phosphatidylethanolamine bilayers in aqueous electrolyte solutions, Biochemistry 24: 4608-4618.

McIntosh, T. J., 1978, The effect of cholesterol on the structure of phosphatidylcholine bilayers, Biochim. Biophys. Acta 513: 43-58.

McIntosh, T. J. and Holloway, P. W., 1987, Determination of the depth of bromine atoms in bilayers formed from bromolipid probes, Biochemistry 26: 1783-1788.

McIntosh, T. J., Magid, A. D. and Simon, S. A., 1987, Steric repulsion between phosphatidylcholine bilayers, Biochemistry 26: 7325-7332.

McIntosh, T. J., Magid, A. D. and Simon, S. A., 1989a, Cholesterol modifies the short-range repulsive interactions between phosphatidylcholine membranes, Biochemistry 28: 17-25.

McIntosh, T. J., Magid, A. D. and Simon, S. A., 1989b, Range of the solvation pressure between lipid membranes: dependence on the packing density of solvent molecules, Biochemistry 28: 7904-7912.

McIntosh, T. J., Magid, A. D. and Simon, S. A., 1989c, Repulsive interactions between uncharged bilayers. Hydration and fluctuation pressures for monoglycerides, Biophys. J. 55: 897-904.

McIntosh, T. J., Magid, A. D. and Simon, S. A., 1990, Interactions between charged, uncharged, and zwitterionic bilayers containing phosphatidylglycerol, Biophys. J. 57: 1187-1197.

McIntosh, T. J. and Simon, S. A., 1986, The hydration force and bilayer deformation: a reevaluation, Biochemistry 25: 4058-4066.

O'Brien, F. E. M. (1948) J. Sci. Instrum. 25: 73-76.

Ohki, S., 1985, Membrane fusion: theory and experiment, Studia Biophysica 110: 95-104.

Ohki, S., 1988, Membrane fusion, hydration energy and hydrophobicity, Studia Biophysica 127: 89-97.

Papahadjopoulos, D., Hui, S., Vail, W. J. and Poste, G., 1976, Studies on membrane fusion. I. interactions of pure phospholipid membranes and the effect of myristic acid, lysolecithin, protein and dimethylsulfoxide, Biochim. Biophys. Acta 448: 245-264.

Parsegian, V. A., Fuller, N. and Rand, R. P., 1979, Measured work of deformation and repulsion of lecithin bilayers, Proc. Nat. Acad. Sci. USA 76: 2750-2754.

Parsegian, V. A., Rand, R. P., Fuller, N. L. and Rau, R. C., 1986, Osmotic Stress for the direct measurement of intermolecular forces, Methods in Enzymology 127: 400-416.

Pearson, R. H. and Pascher, I., 1979, The molecular structure of lecithin dihydrate, Nature 281: 499-501.

Rand, R. P. and Parsegian, V. A., 1989, Hydration forces between phospholipid bilayers, Biochim. Biophys. Acta 988: 351-376.

Schiby, D. and Ruckenstein, E., 1983, The role of the polarization layers in hydration forces, Chem. Phys. Lett. 95: 435-438.

Simon, S. A., Fink, C. A., Kenworthy, A. K. and McIntosh, T. J., 1991, The hydration pressure between lipid bilayers: a comparison of measurements using x-ray diffraction and calorimetry, Biophys. J. in press.

Simon, S. A. and McIntosh, T. J., 1989a, Magnitude of the solvation pressure depends on dipole potential, Proc. Nat. Acad. USA 86: 9263-9267.

Simon, S. A. and McIntosh, T. J., 1989b, Steric repulsion between lipid membranes, Comments Mol. Cell. Biophys. 6: 175-195.

Simon, S. A., McIntosh, T. J. and Latorre, R., 1982, Influence of cholesterol on water penetration into bilayers, Science 216: 65-67.

Simon, S. A., McIntosh, T. J. and Magid, A. D., 1988, Magnitude and range of the hydration pressure between lecithin bilayers as a function of head group density, J. Colloid Interface Sci. 126: 74-83.

Tanford, C., 1980, "The Hydrophobic Effect: Formation of Micelles & Biological Membranes," John Wiley & Sons, New York.

Tardieu, A., Luzzati, V., and Reman, F. C., 1973, Structure and polymorphism of the hydrocarbon chains of lipids: a study of lecithin-water phases, J. Mol. Biol. 75: 711-733.

Weast, R. C. (1984) "Handbook of Chemistry and Physics, 65th Edition," E-42, CRC Press, Boca Raton, Florida..

White, S. H., Jacobs, R. F. and King, G. I., 1987, Partial specific volumes of lipid and water in mixtures of egg lecithin and water, Biophys. J. 52: 663-666.

Worcester, D. L. and Franks, N. P., 1976, Sturctural analysis of hydrated egg lecithin and cholesterol bilayers. II. Neutron diffraction, J. Mol. Biol. 100: 359-378.

PHYSICO-CHEMICAL FACTORS UNDERLYING MEMBRANE

ADHESION AND FUSION

Shinpei Ohki

Department of Biophysical Sciences
State University of New York at Buffalo
Buffalo, New York U.S.A.

INTRODUCTION

Membrane fusion is an essential event in many cellular processes, such as exocytosis, endocytosis, cell membrane assembly, fertilization, and virus infection[1-6]. In spite of extensive studies of membrane fusion, its basic molecular mechanism is not well understood in biological systems, probably because there may be several molecular pathways possible for fusion of biological membranes[7,8]. However, two distinctly different pathways have been described by which two cell membranes fuse[3,4,9,10]. These involve mechanisms active through lipid membrane fusion, and those active through a protein which interacts directly with two membranes.

Since lipid model membranes are physically and chemically well defined, many investigators have attempted to elucidate the phenomena of membrane fusion using model membrane systems for the last one and half decades[9-11].

Since Ca^{2+} seems to be involved in many exocytotic membrane fusion reactions, the effects of divalent cations have been studied extensively. It has been shown that acidic phospholipid vesicles fuse at certain concentrations of divalent cations[12-19]. These fusion reactions are directly related to the strong binding of divalent metal ions to negatively charged sites on the membrane[18-21].

It has been shown also by ^2H NMR studies that Ca^{2+} has a strong tendency to dehydrate bound water from acidic phospholipid membrane surfaces[22-25], and that divalent cations cause an increase in surface tension of acidic phospholipid membranes[18,26,27]. In all acidic lipid membrane systems examined so far, the increase in interfacial tension of the membrane upon the application of fusogenic cations has been observed[18,19,28], and there was a good correlation between the extent of vesicle fusion and the increased interfacial tension of the lipid membrane. (Table 1) It is noted, however, that the values of the increase in surface tension of the membrane at the threshold concentration of fusogenic ion to induce vesicle fusion were slightly different for the membranes which were made of different lipids (see Tables 1 and 2). Recently, we have found the surface dielectric constant to be another physical property relating to membrane fusion, where the decrease in the surface dielectric constant of the membrane caused by fusogenic

TABLE 1

PHOSPHATIDYLSERINE (PS) MEMBRANE IN 0.1M NaCl (pH 7.0)

(Refs.18 & 28)

ADDED IONS	FUSION THRESHOLD CONC. OF IONS	AGGREGATION	FUSION	INCREASED SURFACE TENSION (dynes/cm)	SURFACE DIELECTRIC CONSTANT
0	none	no	no	reference value	32
Na^+	none (up to 1M)	yes	no	−0.2 (max)	32
K^+	none (up to 1M)	yes	no	0	32
Li^+	none (up to 1M)	yes	no	a few (max)	32
$Spermine^{4+}$	none (up to 1M)	yes	no	a few (max)	30
$Spearmidine^{3+}$	none (up to 1M)	yes	no	1–2 (max)	30
$Putresceine^{2+}$	none (up to 1M)	yes	no	0	30
Mg^{2+}	7mM	yes	yes	7.2	13–14
Ca^{2+}	1mM	yes	yes	7.6	10
Mn^{2+}	0.6mM	yes	yes	7.6	12
La^{2+}	3–5mM	yes	yes	7.6	10–12
Tb^{3+}	3–6mM	yes	yes	7.6	10–12
H^+	2mM	yes	yes	8.0	10

TABLE 2

PHOSPHATIDIC ACID (PA) MEMBRANE IN 0.1 M NaCl (pH 6.0)

(Ref. 29)

ADDED IONS	FUSION THRESHOLD CONC. OF IONS	AGGREGATION	FUSION	INCREASED SURFACE TENSION (dynes/cm)	SURFACE DIELECTRIC CONSTANT
0	none	none	none	reference value	32
Na^+	none (up to 1M) (pH 6.0)	yes	none	0.0	32
Ca^{2+}	0.45mM	yes	yes	6.0	15
Mg^{2+}	0.44mM	yes	yes	6.0	14
Sr^{2+}	0.45mM	yes	yes	6.0	15

substances correlated well with the extent of membrane fusion [30,31] (Tables 1 & 2). The same observation has recently been made by use of a poly-ethylene glycol-induced phospholipid vesicle fusion system[30,31].(Table 3).

TABLE 3

PHOSPHATIDYLSERINE (PS) MEMBRANE IN 0.1M NaCl (pH 7.0)

(Ref. 31)

ADDED IONS + MOLECULES	FUSION THRESHOLD CONC. OF FUSOGENS	AGGREGATION	FUSION	SURFACE DIELECTRIC CONSTANT
0	none	no	no	32
PEG CA^{2+}	0 1.1mM	yes	yes	11
PEG CA^{2+}	5 wt% 0.8mM	yes	yes	10
PEG CA^{2+}	10wt% 0.45mM	yes	yes	12
PEG CA^{2+}	15wt% 0.22mM	yes	yes	10
PEG CA^{2+}	30wt% 0	yes	yes	12

All these observations indicate that the membrane surface becomes more hydrophobic in nature as a result of fusogenic agent interaction with membranes either directly (non-induced fusion) or indirectly (for PEG-induced fusion).

Uncharged macromolecules, such as poly-ethylene glycol (PEG), can induce fusion of both charged and neutral lipid vesicles[32-39]. The action of PEG in facilitating vesicle fusion results from the ability to force vesicles together by removing water from the intermembraneous space. This is due to the high osmotic pressure of PEG[40]. The mechanism of PEG-induced membrane fusion seems to be different from the ion-induced membrane fusion. However, the removal of water which in turn causes the membrane surface to become more hydrophobic seems to be a common physical condition[30,39,41] for both processes.

We have developed a theory[41,42] accounting for a close adhesion of two interacting membranes which is related to membrane fusion. Our theory[30,41,42] states that when a membrane surface attains a certain hydrophobicity, the energy due to hydration repulsive forces exerting to the membrane becomes small and therefore other attractive forces become comparable to the repulsive forces. The hydration force is one of the strongest repulsive forces[43-45] when the membrane is sufficiently hydrophilic. As a membrane surface attains a certain hydrophobicity which may be slightly different for different species of lipids, two interacting membranes can become closely apposed. Such a close apposition of two membranes, however, may not be sufficient to induce membrane fusion. When two membranes are adhered closely, the membrane molecules at the boundary between the close contact and non-contact regions may receive greater physical and chemical stresses, and may form a higher energy region or create a region of greater hydrophobicity which may well be the most susceptible site for instability of these membranes [30].

In this paper, the results of some fusion experiments using lipid vesicles containing molecules to create higher surface energy regions in the membrane are given which may support the above mentioned fusion mechanism. Also a theoretical analysis to demonstrate a possibility of such a fusion mechanism is given.

MATERIALS AND METHODS

Materials

Egg phosphatidylcholine (PC) and egg phosphatidylethanolamine (PE) (Avanti Polar Lipids, USA) were chromatographically pure as shown by TLC. Cholesterol (Fisher, USA) was recrystallized from ethanol. 1,2-Dimyristoyl-rac-Glycerol(DAG)(Sigma) was used without further purification. Fluorophore-labelled phospholipids(dansylphosphatidylethanolamine (DPE), 1-4-nitrobenzo-2-oxa-1,3-diazole-PE (NBD-PE) and lissamine rhodamine B sulfonyl-PE (Rh-PE)) were obtained from Avanti Polar Lipids. Hepes (N-2-hydroxyethylpiperazine-N'2-ethanesulfonic acid, Ultrol grade, Calbiochem) was used as buffer to all solutions. Poly-ethylene glycol (MW 6000) from Fluka Chemical Co., Switzerland, was used without further purification. All other chemicals used were of reagent grade and obtained from the Baker Chemical Company. The water used was distilled three times, including an alkaline permanganate process. Small unilamellar vesicles were prepared by hydrating lipids in 3 mM Hepes/0.1 M NaCl/0.05 mM EDTA/pH = 7.3, vortexing for 10 min and then sonicating for 40 min.[18]. The lipid concentration of these solutions were about 3 m mol/l.

Measurements of the Dielectric Constant

DPE was solved together with the phospholipids, cholesterol or diacylglycerol in chloroform at a molar ratio of phospholipid/DPE of 200-300. The sample was evaporated, then suspended in the above mentioned buffer solution. After vortexing for 10 min the sample was sonicated for 40 min. An aliquot of this vesicle stock solution was suspended in an appropriate buffer solution. The fluorescence signal of DPE was detected by a spectrofluorimeter (Perkin-Elmer, LS-5). The excitation wave length was 340 nm and the emission was measured in the range of 400 - 600 nm. From the shifts of the emission spectra maxima the dielectric constants of the DPE environment in the lipid membrane were calculated using the Stokes shift equation which relates the wave length at the maximum value of the emission spectrum and its dielectric properties (Kimura & Ikegami[46], Ohki & Arnold[31]).

Vesicle Fusion Assay

The fusion of the vesicles was followed by using the fluorescence energy transfer method, using NBD-PE and Rh-PE (Struck et al.[47]). The vesicles were composed of PC/Cholesterol, PC/DAG or PC/PE and 1% (mole/mole PC) of both NBD-PE and Rh-PE. They were prepared by sonication in the above mentioned buffer solution. One part (0.05 μ mole lipid) of the fluorophore-incorporated vesicles and two parts (0.1 μ mole lipid) of the unlabelled vesicles were suspended in at least 0.25 ml of the appropriate buffer solution without PEG 6000. PEG 6000 (40 wt%) containing buffer solution was then added to achieve the PEG concentration required. After intensive shaking, the fluorescence measurements were carried out by exciting at 470 nm and recording the fluorescence from 500 to 620 nm. The extent of fusion F was evaluated from the intensity of NBD at 525 nm:

$$F = \frac{I_{525} - I^0_{525}}{I_{525}}$$

where I_{525} was the fluorescence amplitude from the solution containing PEG 6000, and I^0_{525} without PEG. The experiments were done at room temperature. The 100% value of fusion was defined as the value obtained by solubilization of the vesicles in 0.2% Triton X-100. This value was corrected by the factor 1.5 because of the quenching effect of Triton X-100.

EXPERIMENTAL RESULTS

Cholesterol-containing PC Liposomes

It is known that the incorporation of cholesterol into egg PC liposomes is possible up to 50 mol % cholesterol (Ladbrooke et al.[48]). In this range we prepared egg PC liposomes containing cholesterol. In Fig. 1, the surface dielectric constants of the egg PC/cholesterol liposome in dependence of the PEG 6000 concentration is given. The surface dielectric constants were calculated according to Ohki and Arnold[31]. At 0 wt% PEG 6000, there were only slight differences in the surface dielectric constant for pure egg PC liposomes and in 42 mol% cholesterol containing egg PC liposomes. The higher cholesterol content in the liposomes, does not lead to lower surface dielectric constants of the membranes. By increasing the PEG 6000 concentration in the surrounding medium, the difference between the decreases in the surface dielectric constants for the membrane containing cholesterol and cholesterol free membrane was observed. Cholesterol containing membranes exhibited a greater decrease in surface dielectric constant than the cholesterol free membrane when exposed to high PEG 6000 concentrations; pure PC liposomes exhibited a surface dielectric constant of about 15 in the presence of 25 wt% PEG 6000, while PC liposomes containing 17 mol% cholesterol showed a dielectric constant of about 13 at the same PEG concentration, and PC liposomes containing 42 mol% chloesterol showed dielectric constant values

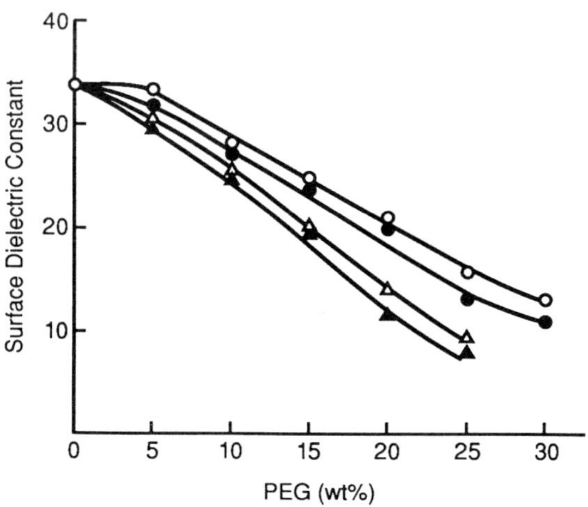

Fig. 1.   Surface dielectric constant of PC/cholesterol SUV as measured by the anti-Stokes Shifts of DPE in the presence of various amounts of PEG 6000 in 0.1 M NaCl, pH 7.0 (Ref. 54).
O : 0 mol% cholesterol   ● : 17 mol% cholesterol
△ : 42 mol% cholesterol   ▲ : 50 mole% cholesterol

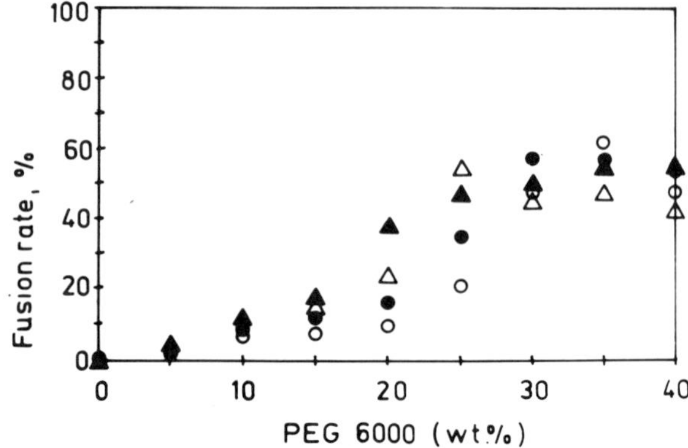

Fig. 2.   Fusion of PC/cholesterol SUV induced by PEG 6000
          as measured by the NBD/Rb fluorescence assay (Ref.54)
          O : 0 mol% cholesterol   ● : 17 mol% cholesterol
          △ : 29 mol% cholesterol   ▲ : 38 mol% cholesterol

below 10.   Fig. 2 shows the results of the NBD/Rh fluorescence fusion
assay measured using egg PC liposomes containing 0-48 mol% cholesterol
with respect to various amounts of PEG 6000 added.   Pure egg PC liposomes
started with fusion at about 25 wt% PEG 6000.   Cholesterol containing
liposomes (38 mol%) started at about 19 wt% PEG 6000 with fusion. However,
in both cases, the surface dielectric constants were about 15.   The extent
of fusion induced by PEG depended directly on the cholesterol content in
the membranes;   the higher the cholesterol, the lower the PEG 6000
concentration necessary to induce fusion.   However, at the threshold of
fusion, the surface dielectric constant was about 15 at each case (Table
4).

<u>Liposomes containing phosphatidylethanolamine</u>

Among the PC vesicles containing different PE contents up to 50%, no

TABLE 4

PEG-INDUCED FUSION AND ITS SURFACE DIELECTRIC CONSTANT OF PHOSPHATIDYLCHOLINE
(PC) + CHOLESTEROL SUV (SMALL UNILAMALLAN VESICLE) IN 0.1M NaCl (pH 7.0)

LIPID COMPONENTS	THRESHOLD CONC. OF PEG	AGGREGATION	FUSION	SURFACE DIELECTRIC CONSTANT
PC	none	none	none	33
PC	PEG 27(wt%)	yes	yes	15
PC/Chol(5/1)	PEG 24	yes	yes	15
PC/Chol(5/?)	PEG 20	yes	yes	14
PC/Chol(5/3)	PEG 17	yes	yes	15

differences in the surface dielectric constant was detected. Additionally, we checked the influence of different PE concentrations in PC SUV on
the surface dielectric constant at different pH and no change in surface
dielectric constant was detected (data not shown).   PEG 6000 caused a
reduction of the surface dielectric constant of the PC/PE SUV.   The
changes in dielectric constant for the different PC/PE liposomes after
addition of PEG 6000 were about the same as the pure PC.   On the other
hand, the threshold concentration of PEG necessary to induce vesicle
fusion was slightly decreased with the increase of PE content in PC
vesicles (Table 5). Compared to pure PC vesicles which fused at about 18
wt% PEG 6000, PC/PE SUV (46 mol% PE) fuse at about 14 wt% PEG.   The fusion
characteristics of the PC/PE SUV are shown in Fig. 3.   The extent of
fusion of the PE containing liposomes was increased compared to pure PC
SUV. The higher the PE content,the greater extent of fusion was observed.

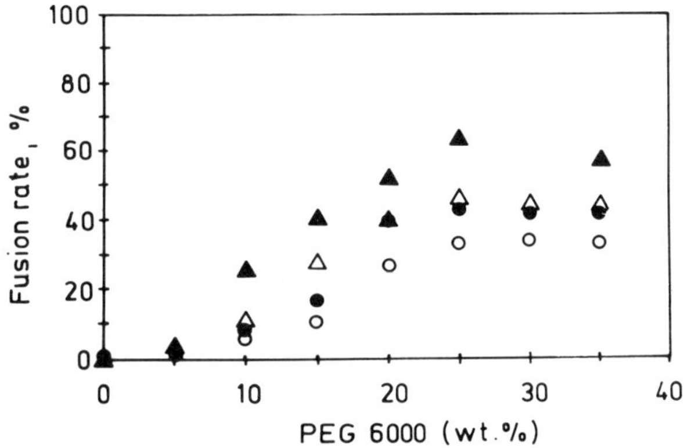

Fig. 3.   Fusion of PC/PE SUV induced by PEG 6000 as
measured by the NBD/Rh fluorescence assay (Ref.54).
○:0 mol% PE   ●: 30 mol% PE
△:39 mol% PE   ▲: 46 mol% PE

TABLE 5

PEG-INDUCED FUSION AND ITS SURFACE DIELECTRIC CONSTANT OF PHOSPHATIDYLCHOLINE
(PC) + PHOSPHATIDYLETHANOLAMINE (PE) IN 0.1 M NaCl (pH 7.0)

LIPID COMPONENTS	THRESHOLD CONC. OF PEG	AGGREGATION	FUSION	SURFACE DIELECTRIC CONSTANT
PC	none	none	none	33
PC	PEG 18 (wt%)	yes	yes	15
PC/PE(2/1)	PEG 18	yes	yes	16
PC/PE(3/2)	PEG 17	yes	yes	16
PC/PE(7/6)	PEG 17	yes	yes	16

<u>Liposomes containing diacylglycerol</u>

We observed no appreciable change in the surface dielectric constant of the vesicle of egg PC containing DAG. In the presence of PEG 6000 there were no great differences in the decrease of surface dielectric constant of egg PC liposomes with different contents of DAG. However, the presence of PEG 6000 in the solution lowers, in all cases, the surface dielectric constant of the PC/DAG SUV investigated. The fusion experiments were done with the NBD/Rh fluorescence assay. The results of these measurements are given in Fig. 4. In the range of DAG content in

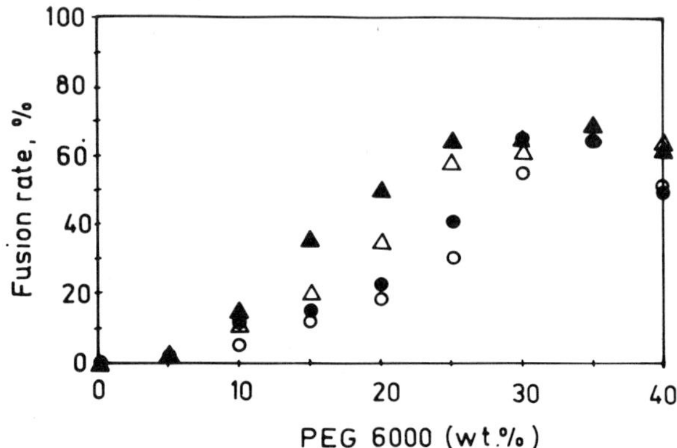

Fig. 4.   Fusion of PC/DAG SUV induced by PEG 6000 as
           measured by the NBD/Rh fluorescence assay (Ref.54).
           O : 0 mol% DAG   ● : 7 mol% DAG
           △ : 13 mol% DAG   ▲ : 19 mol% DAG

the egg PC liposomes, the differences in the fusion characteristics were observed. Compared to the pure PC liposomes, PC SUV containing DAG started to fuse at lower PEG concentrations and the decrease in the surface dielectric constant at the fusion threshold (Table 6) was less than that of the pure PC liposome. Also, the extent of fusion was higher. The higher the DAG concentration in the membrane, the lower was the PEG 6000 concentration necessary to induce fusion.

TABLE 6

PEG-INDUCED FUSION AND ITS SURFACE DIELECTRIC CONSTANT OF PHOSPHATIDYLCHOLINE
(PC) AND DIACYLGLYCEROL (DAG) SUV IN 0.1 M NaCl (pH 7.0)

LIPID COMPOSITION (M%)	THRESHOLD CONC. OF PEG	AGGREGATION	FUSION	SURFACE DIELECTRIC CONSTANT
PC	none	none	none	33
PC	PEG 25 (wt%)	yes	yes	15
DAG/PC+DAG (7%)	PEG 22	yes	yes	16
DAG/PC+DAG (13.3%)	PEG 20	yes	yes	18
DAG/PC+DAG (18.7%)	PEG 16	yes	yes	20

THEORY FOR MEMBRANE INTERACTION

When two spherical shells, like two lipid vesicles, are in interaction at a certain separation distance (Fig. 5a), the total interaction energy, W, for the two spherical shells having polar surface layers may be expressed as a sum of van der Waals attractive interaction energy U_A, electrostatic and hydration repulsive interaction energies V^{el} and V^{hd}, respectively.

$$W^{tot} = U_A + V^{el} + V^{hd} \qquad (1)$$

These quantities are a function of the interseparation distance R between the two spheres. When the separation distance is small enough (e.g. less than 20 Å) compared to the size of the sphere (e.g. the radius of the vesicle may be greater than 200 Å, the interaction energy of these spherical shells may be replaced with that of the two parallel flat plates having a polar surface layer. In addition, when the thickness of the

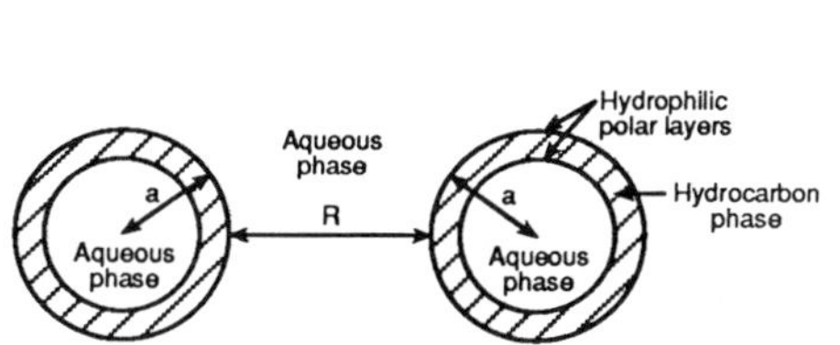

Fig. 5a. A schematic diagram of two interacting spherical shells of the radium a at the separation distance R.

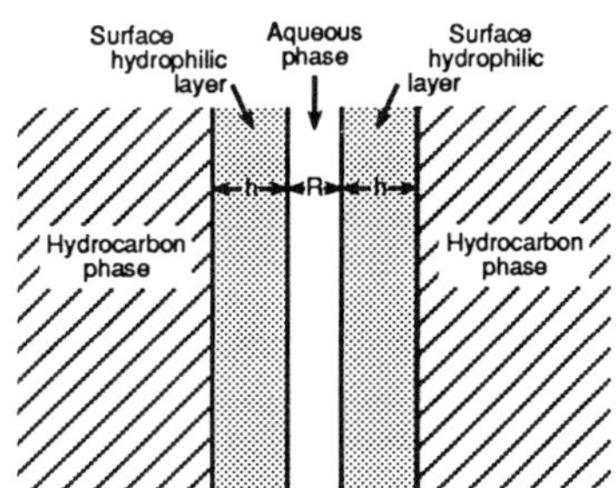

Fig. 5b. Schematic diagram of two interacting membranes at a small separation distance R, where h is the thickness of the surface polar layers.

spherical shell (that of lipid bilayers is about 50 Å) is greater than the separation distance R, the thickness of the plate can be considered as infinite (see Fig. 5b). Then, the van der Waals energy for such a system is expressed by

$$U_A(R) = \frac{1}{12\pi} (\sqrt{A_w} - \sqrt{A_m})^2 \left\{ \frac{P^2}{R^2} + \frac{p(1-p)}{(R+h)^2} + \frac{(1-p)^2}{(R+2h)^2} \right\} \qquad (2)$$

where A is the Hamaker constant (e.g. $A_{\frac{water}{water}} = 2.4 \times 10^{-13}$ erg and $A_{\frac{hydrocarbon}{hydrocarbon}} = 9.49 \times 10^{-14}$ erg)[49] R the interlayer separation distance, h the thickness of the surface polar layer and p is the degree of hydrophobicity of the polar layer ($0 \leq p \leq 1$); the case of $p = 0$ corresponds to the hydrophobicity of the water-like phase and that of $p = 1$ the hydrocarbon phase.

Electrostatic energy of the system is expressed by

$$V^{el}(R) = \frac{64\, n\, kT\, \gamma^2}{\kappa} \exp(-2\kappa R)$$

$$\text{where} \quad \gamma = \frac{\exp(z e \Psi_0/2kT) - 1}{\exp(z e \Psi_0/2kT) + 1} \tag{3}$$

In the equation, n is the number density of the electrolyte in the solution, Ψ_0 is the surface potential of the membrane and κ is the Debye constant. The value of Ψ_0 is obtainable with use of various experimental methods[29].

For the estimation of the hydration repulsive interaction energy with respect to the interseparation distance, we consider the following: The hydration repulsive interaction forces exerting two interacting lipid bilayers have been studied extensively[43-45,50]. It was found that it decreases exponentially with respect to the separation distance at a decay constant ℓ (ℓ = 1.7 - 2.5 Å according to the literature[43,44,50,51]):

$$P = P_0\, e^{-R/\ell} \tag{4}$$

where P is the force exerting on the interacting bilayer surface per unit area. Therefore, by integrating the forces P in eq. (4) from ∞ to R, we would obtain the energy of squeezing out water by bringing the two interacting plates from the infinite separation distance to a finite distance R:

$$V^{hd}(R) = -\int_{\infty}^{R} P dx = -P_0 \exp(-R/\ell). \tag{5}$$

According to the published work, P_0 is estimated to be 7×10^{-9} erg/cm^2 for phospholipid membranes (e.g. PC) in 0.1 M NaCl solution[44]. Therefore, the energy to remove the interlayer water (e.g. to make the two membranes close contact)per unit area would be

$$V^{hd}(0) = -\int_{\infty}^{0} P dR = P_0\, \ell \tag{6}$$

On the other hand, the energy to remove water from the interlayer space is considered to be equivalent to those expressed in terms of interfacial tension of the interacting phases[41,42]: First, the energy W^{sep} to separate the water phase from the membrane phase would be

$$W^{sep} = \gamma_{m/a} + \gamma_{w/a} - \gamma_{m/w} \tag{7}$$

where $\gamma_{m/a}$ and $\gamma_{w/a}$ are the surface tension of the membrane and water phases, respectively and $\gamma_{m/w}$ is the interfacial tension of the membrane/ water interface. Then, the energy to remove water surface from its system would be

$$W^{hd} = \gamma_{m/a} - \gamma_{m/w}. \tag{8}$$

Since these are two such interfaces for the two interacting membranes, the
energy E^{hd} to remove water from the intermembrane space would be

$$E^{hd} = 2W^{hd} = 2 (\gamma_{m/a} - \gamma_{m/w}) . \tag{9}$$

This expression corresponds to the disjoining pressure proposed by
Derjaguin. This energy would be equal to $V^{hd}(0)$:

$$E^{hd} = V^{hd}(0) . \tag{10}$$

The surface tension of the membrane $\gamma_{m/a}$ and the interfacial tension of the
membrane $\gamma_{m/w}$ are not easily obtainable quantities, although they can be
measured to a limited degree. Thus, for simplicity, we assume that the
surface of the membrane consists of the mixtures of water-like phase and
hydrocarbon-like phase at a certain ratio, which would vary depending on
the environmental conditions:

$$\gamma_m = p \ \gamma_H + (1-p) \ \gamma_w , \tag{11}$$

where p is the fractional value($0 \leq p \leq 1$) which we may call the hydrophobic index and γ_H and γ_w are the surface tensions of hydrocarbon and
water-like phases, respectively. Then, the dehydration energy E^{hd} is
rewritten in terms of γ_H and γ_w with eqs. (10) and (11).

$$E^{hd} = 2(p \ \gamma_{H/a} + (1-p) \ \gamma_{w/a} - p \ \gamma_{H/w}) , \tag{12}$$

where $\gamma_{H/a} \sim 20$ dynes/cm, $\gamma_{w/a} \sim 72$ dynes/cm, and $\gamma_{H/w} \sim 50$ dynes/cm.
Therefore, the hydration repulsive energy $V^{hd}(R)$ can be expressed by the
interlayer separation distance R and the above known surface tension
values. With eqs. (5), (10), and (12) we have

$$V^{hd}(R) = V^{hd}(0)\exp(-R/\ell) = E^{hd} \exp(-R/\ell) . \tag{13}$$

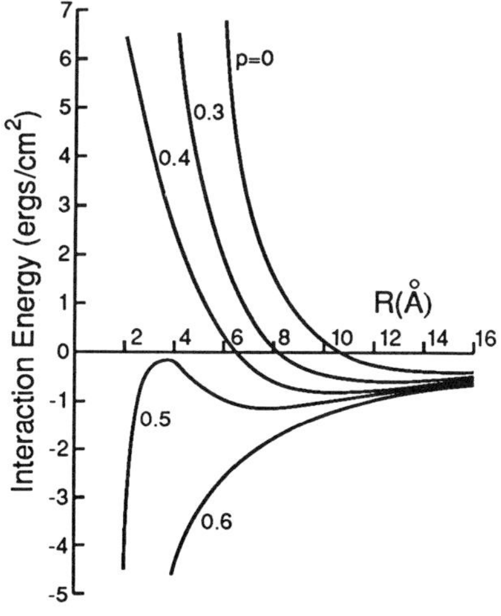

Fig. 6. The total interaction
energy of the two closely
apposed phosphatidylserine
membranes in 0.1 M NaCl and
appropriate divalent cation
concentration (e.g. 1 mM Ca^{2+}
for p = 0.4) as depicted in
Fig. 5b as a function of the
separation distance R (in Å)
and various hydrophobicity
index (p=0, 0.3, 0.4, 0.5 and
0.6) of the surface polar
layer. The thickness of the
surface polar layer was
assumed to be 5 Å.

Fig. 6 shows the calculated total interaction energy, $W^{tot} = U_A(R) + V^{el}(R) + V^{hd}(R)$, for the two interacting spherical membranes as a function of the separation distance R(Å) in the cases of various degrees of the surface hydrophobicity (p = 0, 0.3, 0.4 and 0.6), respectively. It is seen from Fig. 6, that in the case of p = 0, the hydration repulsive energy is too large, and the two vesicles will not make a close contact. Even when the hydrophobic index p is 0.4 which corresponds to the interfacial tension of the membrane/water interface to be about 20 dynes/cm, the membranes would have a stable adhesion at about 6 - 8 Å but they do not come to a close molecular contact (e.g. a few Å distance).

The interfacial tension of lipid bilayer of 20 dynes/cm may correspond to those of the small unilamellar phosphatidylserine vesicle having 1 - 2 mM Ca^{2+} in the 0.1 M NaCl solution of pH 7.0. At the hydrophobic index in between 0.4 and 0.5, there would be a large change in the total interaction energy; a critical point where the total energy becomes negative and the two membranes would adhere at a close molecular distance.

DISCUSSION

According to the theoretical analysis described above, when the membrane surface is sufficiently polar and hydrophilic (e.g. similar to that of the water phase, p = 0), the interfacial tension of the membrane/ water interface is nearly zero and, in such a case, a large hydration repulsive force would exert to the two interacting membranes which prevents close approach of the two membranes. In such a case, even if there is a secondary minimum energy point existing in between the two interacting membranes, the separation distance of this minimum point is greater than 15 Å (see Fig. 6). Such a separation distance would be too large for allowing the direct molecular exchange interaction between the two membranes. However, when the membrane surface attains a certain hydrophobicity (e.g. p = 0.4), although there still exists a large hydration repulsive force (as seen in Fig. 6), the interlayer separation distance at the secondary minimum point becomes small (i.e. approximately 6 - 8 Å). At this condition (p = 0.4), there would still be large amounts of water existing in between the two interacting membranes, and, therefore, the direct molecular exchange interaction between the membranes would not occur easily. The case of p = 0.4 corresponds to the inter-facial tension of the membrane to be about 20 dynes/cm. The equivalent state for the interfacial tension of 20 dynes/cm can be created for the small unilamellar phosphatidylserine vesicle suspended in 0.1 M NaCl solution at pH 7.0 in the presence of 1 mM $CaCl_2$; the interfacial tension of the small unilamellar vesicle (~ 300 Å in diameter) is about 10 dynes/cm because the outer monolayer is in the expanded state 80 $Å^2$/lipid molecule[52] without considering the effect of the Ca^{2+} binding to the membrane, and in addition, the effect of divalent cation binding to the membrane (1 mM Ca^{2+}) induces about 8 dynes/cm of the interfacial tension, provided that the surface area (area per molecule) of the membrane does not change, and the effect of the curvature on the interfacial tension (a few dynes), would contribute about 10 dynes/cm to the interfacial tension. The increase in interfacial tension of a phosphatidylserine monolayer was measured at the condition of a constant area per molecule.(18, 19)

The calculated results of the interlayer interaction energy shows that at some point around p = 0.5 of the membrane hydrophobicity, the hydration energy becomes considerably small and the attractive energies exceed the hydration repulsive energy in quantity. The values of

interfacial tension corresponding to such hydrophobicity value would be
greater than 25 dynes/cm.

However, all vesicle fusions observed so far seem to occur at a
lower interfacial tension of about 20 dynes/cm. From the theoretical
analysis and experimental results, it is likely that the relatively close
adhesion (6 - 8 Å of the interlayer distance) is attained at the fusion
threshold concentration of divalent cation (e.g. 1 mM Ca^{2+} in 0.1 M NaCl
at pH 7.0), but the adhesion area which is separated by a considerable
amount of water, may not be the fusion site of the two interacting
membranes. Instead, the region between the close contact and non-close
contact membranes (e.g. the edge of the close contact region) (see Fig. 7)
may become a higher energy area (more than 25 dynes/cm) than the rest of
the membrane; the curvature of this membrane region is greatest. Also
this region is exerted by an additional asymmetrical force (line
tension)[53] The latter tension becomes larger as the size of vesicles
becomes smaller, provided that the separation distance of the closely
adhered membranes remains the same [18]. This higher hydrophobic region is
then possible for undergoing the molecular exchange between the two
membranes inspite of the separation distance of about 6 - 8 Å. This
interpretation of vesicle fusion mechanism is also supported by the above
described vesicle fusion experiments of different types of membranes.

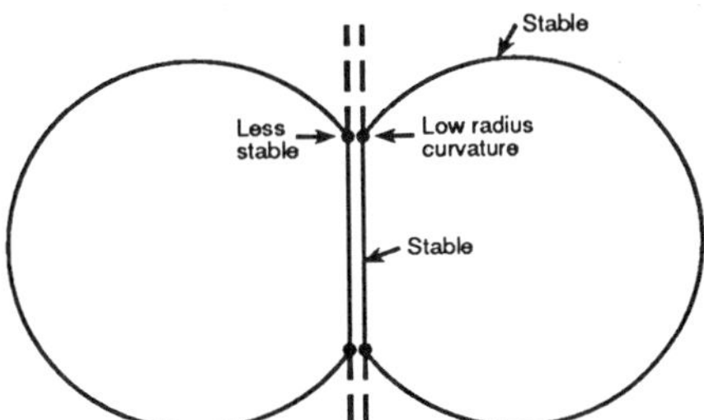

Fig. 7. A possible fusion site for lipid membrane
         vesicle.

The threshold concentrations for vesicle fusion should be slightly
different for different membranes, with their sizes and molecular
composition. The threshold concentrations of divalent cations for the
phosphatidic acid vesicles were lower than that obtained for the
phosphatidylserine vesicles. One of the reasons for the low threshold
concentration value for the PA vesicles may be the smaller size of the
vesicle than the PS vesicles and therefore, the dielectric constant at the
fusion threshold was greater than that (12) for the PS vesicles. Another
interesting fact for the PA vesicle fusion, the threshold concentration
was the same for all three divalent cations (Ca^{2+}, Mg^{2+} and Sr^{2+}). The
fusion experiments using lipid vesicles containing molecules contributing
a high surface energy. e.g. DAG in the membrane also supports the above
fusion mechanisms. The incorporation of these molecules into the membrane
reduced the threshold concentration of PEG to induce vesicle fusion and
also the surface dielectric constant corresponding to the fusion threshold
was increased. This means that these surface high energy molecules
residing at the fusion site effectively would be raised to a higher energy
state or high hydrophobicity and it is possible to induce fusion through

this fusion site inspite of the lower interfacial energy state of the rest of the membranes. On the other hand, the PC/chlesterol vesicles showed the tendency similar to the pure PC vesicles with respect to fusion and surface dielectric constant changes induced by PEG. However, the fusion threshold concentration of PEG for the PC/cholesterol vesicles was lower than those for the pure PC vesicles, and the greater the cholesterol content, the lower was the threshold concentration of PEG to induce vesicle fusion. Also, the extent of vesicle fusion was greater as the cholesterol content was increased. An interesting observation was that the surface dielectric constant of the PC/cholesterol vesicle was not appreciably affected by the amounts of cholesterol incorporated (up to 50 mol%) into the PC membrane in the absence of PEG. These experimental results suggest that cholesterol per sei may not alter the surface dielectric nature but the cholesterol molecule may easily release the adsorbed water molecules from its surface than the phosphatidylcholine molecule when an external driving force to remove the adsorbed water is applied. The phosphatidylethanalamine molecule seems to have an intermediate property with respect to fusion capability and surface dielectric constant changes of the other two molecules, DAG and cholesterol.

In summary, the following conclusion may be mentioned from our experimental results and theoretical analysis; when the lipid membrane is made of a single component, the correlation between the threshold of vesicle fusion and increased surface tension or decreased surface dielectric constant is held well. The increased interfacial tension or decreased surface dielectric constant of the membrane corresponds to the increased hydrophobicity of the membrane surface.

The overall increased surface hydrophobicity coresponding to the observed interfacial tension increase or lowered surface dielectric constant of the membrane at the fusion threshold may induce a close approach of the two interacting membranes due to the reduction of repulsive hydration forces caused by the increased surface hydrophobicity, but may not be sufficient enough to induce the intermixing of molecules of the two interacting membranes, except for through the high energy molecules in the boundary region of the apposed membranes ("the fusion site").

However, when a molecule (DAG) contributing a high surface energy to the membrane, is mixed with the lipid composing vesicle membranes, the fusion threshold is reduced to a lower value and the surface dielectric constant value is also altered to a slightly higher value than those with the lipid vesicles of a single component. This indicates that the vesicle fusion starts at an overall lower surface hydrophobicity because the high surface energy molecule like DAG located at the fusion site would contribute to make this region more hydrophobic inspite of overall lower hydrophobicity of the membrane which is necessary to induce vesicle fusion through the fusion site. The above argument for the fusion mechanism is based on the experiments done with use of lipid vesicles made of a few variations of types of lipids and "high surface" energy molecules. Therefore, further studies would be necessary in order to confirm the above hypothesis of the mechanism of lipid vesicle fusion.

ACKNOWLEDGEMENTS

This work was partly supported by a grant from the U. S. National Institutes of Health (GM-24840).

REFERENCES

1. G. Poste and A.C. Allison, "Membrane Fusion". Biophys.Acta 300:300 (1973).
2. H. Rasmussen, "Cell Communication, Calcium Ion, and Cyclic Adenosine Monophosphate", Science 170-:404 (1970).
3. H. Plattner, "Regulation of Membrane Fusion During Exocytosis". Intern.Rev.Cytol. 119:197 (1989).
4. R. Blumenthal, "Membrane Fusion", in: "Current Topics in Membranes and Transport". vol. 29, R.D. Klausner and C. Kemp, eds., Academic Press, NY, p. 203, (1987).
5. S. Ohnishi, "Fusion of Viral Envelopes with Cellular Membranes", in: "Current Topics in Membranes and Transport", vol. 32, Academic Press, NY, p. 257, (1988).
6. J. Meldolesi, N. Borgese, P. DeCamilli and B. Ceccarelli, "Cytoplasmic Membranes and the Secretory Process". in: "Membrane Fusion", G. Nicholson, ed., North Holland, Amsterdam (1978).
7. N. Duzgunes and F. Bronner, eds. in: "Current Topics in Membranes and Transport". vol. 32, Academic Press (1988).
8. M.D. Housley and K.K. Stanley, "Dynamics of Biological Membranes". John Wiley & Sons, NY (1984).
9. D. Papahadjopoulos, "Calcium-Induced Phase Changes and Fusion in Natural and Model Membranes", in: "Membrane Fusion", G Poste and G Nicolson eds, Elsevier/North Holland, Amsterdam, p. 265 (1978).
10. S. Ohki, D. Doyle, T.D. Flanagan, S.W. Hui and E. Mayhew, eds. "Molecular Mechanisms of Membrane Fusion", Plenum Publ.Co., NY (1988).
11. A.E. Sower, ed, "Cell Fusion", Plenum Publ. Co., NY (1987)
12. D. Papahadjopoulos, G. Poste, B.E. Schaeffer and W.J. Vail, "Membrane Fusion and Molecular Segregation in Phospholipid Vesicles.", Biochim.Biophys.Acta 352:10 (1974).
13. D.Papahadjopoulos, W.J. Vail, C. Newton, S. Nir, K. Jacobson, G. Poste, and R. Lazio, "Studies on Membrane Fusion: III. The Role of Calcium-Induced Phase Changes.", Biochim.Biophys.Acta 465:579 (1977).
14. F.J. Martin and R.C. MacDonald, "Phospholipid Exchange Between Bilayer Membrane Vesicles.", Bioche. 15:321 (1975).
15. T.D. Ingolia and D.E. Koshland, "The Role of Calcium in Fusion of Antificial Vesicles.", J. Biol.Chem. 253:3821 (1978).
16. M.J. Liao and N.H. Prestegard, "Fusion of Phosphatidic Acid-Phosphatidylcholine Mixed Lipid Vesicles.", Biochim.Biophys.Acta 550:157 (1979).
17. J. Wilschut and D. Papahadjopoulos, "Ca^{2+}-Induced Fusion of Phospholipid Vesicles Monitored by Mixing of Aqueous Contents." Nature 281:690 (1979).
18. S. Ohki, "A Mechanism of Divalent Ion-Induced Phosphatidylserine Membrane Fusion". Biochim.Biophys.Acta 689:1 (1982).
19. S. Ohki, "Effects of Divalent Cations Temperature, Osmotic Pressure Gradient and Vesicle Curvature on Phosphatidylserine Vesicle Fusion." J. Membrane Biol. 77:265 (1984).
20. N. Duzgunes, S. Nir, J. Wilschut, J. Bentz, C. Newton, A. Portis and D. Papahadjopoulos, "Calcium- and Magnesium-Induced Fusion of Mixed Phosphatidylserine/Phosphatidylcholine Vesicles: Effect of Ion Binding." J. Membrane Biol. 59:115 (1981).
21. S. Nir, J. Bentz, J. Wilschut and N. Duzgunes, "Aggregation and Fusion of Phospholipid Vesicles", in "Progress in Surface Science", vol. 13, Pergamon Press, NY, p. 54 (1983).
22. E.G. Finer and A. Darke, "Deuteron Magnetic Resonance Studies of Water Associated with phospholipids." Chem. Phys. Lipids 12:1 (1974).

23. H. Hauser, M.C. Phillips and M.D. Barratt, "Differences in the Interaction of Inorganic and Organic (Hydrophobic) Cations with Phosphatidylserine Membranes." Biochem. Biophys. Res.Comm. 506:281 (1975).

24. H. Hauser, E.G. Finer and A. Darke, "Crystalline Anhydrous Ca-Phosphatidylserine Bilayer." Biochim.Biophys.Acta 506:281 (1977).

25. S. Ohki and K. Arnold, in preparation, (1990).

26. D. Papahadjopoulos, "Surface Properties of Acidic Phospholipids: Interaction of Monolayers and Hydrated Liquid Crystals with Uni-and Bi-Valent Metal Ions." Biochim.Biophys.Acta 163:240 (1968).

27. R.C. MacDonald, "Mechanisms of Membrane Fusion in Acidic Lipid-Cation Systems." in "Molecular Mechanisms of Membrane Fusion" S. Ohki et al. eds., Plenum Publ.Co., NY (1988).

28. S. Ohki and J. Duax, "Interaction of Polyamines with Phosphatidyl serine Membranes." Biochim.Biophys.Acta 861:177 (1986).

29. S. Ohki and H. Ohshima,"Divalent Cation-Induced Phosphatidic Acid Membrane Fusion", Biochim.Biophys.Acta 812:142 (1985).

30. S. Ohki and K. Arnold, "Phospholipid Vesicle Fusion Induced by Cations and Poly-(Ethylene Glycol)", in "Biophysics of Cell Surface", R. Glaser and D. Gingell, eds., Springer-Verlag, London, (1989).

31. S. Ohki and K. Arnold, "Surface Dielectric Constant, Surface Hydrophobicity and Membrane Fusion." J. Membrane Biol. 114:195 (1990).

32. A.M.J. Blow, G.M. Botham, D. Fisher, A.H. Goodall, C.P.A. Tilcock and J.A. Lucy, "Water and Calcium Ions in Cell Fusion Induced by Poly(Ethylene Glycol)." FEBS Lett 94:305 (1978).

33. C.P.A. Tilcock and D. Fisher, "Interaction of Phospholipid Membranes with Poly(Ethylene Glycols)s." Biochim.Biophys. Acta 577:53 (1979).

34. L.T. Boni, T.S. Stewart, J.L. Aldefer and S.W. Hui, "Lipid-Polyethylene Glycol Interactions: I. Induction of Fusion Between Lipsomes." J.Memb.Biol. 62:65 (1981).

35. R.I. MacDonald, "Membrane Fusion Due to Dehydration by Polyethylene Glycol, Dextran or Sucrose." Biochemistry 24:4058 (1985).

36. R.A. Parente and B.R. Lentz, "Rate and Extent of Poly(Ethylene Glycol)-Induced Large Vesicle Fusion Monitored by Bilayer and Internal Content Mixing." Biochemistry 25:6678 (1986).

37. D. Hoekstra, "Role of Lipid Phase Separation and Membrane Hydration in Phospholipid Vesicle Fusion." Biochemistry 21:2833 (1982).

38. K. Arnold, A. Herrmann, K. Gawrisch and L.Pratsch, "Water-Mediated Effects of PEG on Membrane Properties and Fusion", in "Molecular Mechanisms of Membrane Fusion", S. Ohki et al., eds., Plenum Publ.Co., NY, p. 255 (1988).

39. K. Arnold, A. Herrmann, L. Pratsch and K. Gawrisch, "The Dielectric Properties of Aqueous Solution of Poly-(Ethylene Glycol) and Their Influence on Membrane Structure." Biochim.Biophys. Acta 815:515 (1985).

40. C.J. van Oss, K. Arnold, R.J. Good, K. Gawrisch, and S. Ohki, "Interfacial Tension and the Osmotic Pressure of Solutions of Polar Polymers." J. Macromol.Sci.A27:563 (1990).

41. S. Ohki, "Surface Tension, Hydration Energy and Membrane Fusion", in "Molecular Mechanisms of Membrane Fusion", S. Ohki et al., eds., Plenum Publ.Co., NY, p. 123 (1988).

42. S. Ohki, "Membrane Fusion: Theory and Experiments", Studia Biophysica 10:95 (1985).

43. D.M. LeNeveu, P. Rand and V.A. Parsegian, "Measurement of Forces Between Lecithin Bilayers." Nature (Lond) 259:601 (1976).

44. P. Rand, "Interacting Phospholipid Bilayers: Measured Forces and Induced Structural Changes." Rev.Biophys.Bioeng 10:277 (1981).

45. S. Marcelja and N. Radic, "Repulsion of Interfaces Due to Boundary Water.", Chem.Phys.Letter 42:129 (1976).

46. Y. Kimura and A. Ikegam, "Local Dielectric Properties Around Polar Regions of Lipid Bilayers." J. Membrane Biol. 85:225 (1985).

47. D.E. Struck, D. Hoekstra and R.E. Pagano, "Use of Resonance Energy Transfer to Monitor Membrane Fusion." Biochemistry 20:4093 (1981).

48. B.D. Ladbrooke, R.M. Williams and D. Chapman, "Studies of Lecithin-Cholesterol-Water Interactions by Differential Scanning Calorimetry and X-Ray Diffraction", Biochim.Biophys.Acta 150:333 (1968).

49. S. Nir, "On the Analysis of Aggregation of Acidic Phospholipid Vesicles with Regard to Cation Environment, Temperature and Their Free Energy of Interaction", Chem. Phys.Lipids 16:195 (1976).

50. T.J. McIntosh and S.A. Simon, "The Hydration Force and Bilayer Deformation: A Reevaluation", Biochemistry 25:4058 (1986).

51. T.J. McIntosh, A.D. Magid and S.A. Simon, "Short-Range Repulsive Interactions Between the Surfaces of Lipid Membranes", in: Biological and Model Membranes Interaction.", S. Ohki, ed., Plenum Publishing Co., NY, (1991).

52. C.G. Brouillette, J.P. Sogrest, T.C. Ng and J.L. Jones, "Minimal Size phosphatidylcholine Vesicles: Effects of Radius of Curvature on Head Group Packing and Conformation". Biochemistry 21:4569 (1982).

53. I. Ivanov and P.A. Kralehevsky, "Mechanics and Thermodynamics of Curved Thin Films.", in: Thin Liquid Films", I.B. Ivanov, ed., Marcel Dekker, N.Y., p. 49 (1985).

54. O. Zschornig and S. Ohki, "Effect of Cholesterol, Diocylglycerol and Phosphatidylethanolamine on PEG Induced Fusion and Surface Dielectric Constant of Phospholipid Membranes."